Foundations of
Quantum Field Theory

World Scientific Lecture Notes in Physics

ISSN: 1793-1436

*For the complete list of published titles, please visit
http://www.worldscientific.com/series/wslnp

World Scientific Lecture Notes in Physics – Vol. 84

Foundations of Quantum Field Theory

Klaus D Rothe

University of Heidelberg, Germany

 **World Scientific**

NEW JERSEY · LONDON · SINGAPORE · BEIJING · SHANGHAI · HONG KONG · TAIPEI · CHENNAI · TOKYO

Published by

World Scientific Publishing Co. Pte. Ltd.

5 Toh Tuck Link, Singapore 596224

USA office: 27 Warren Street, Suite 401-402, Hackensack, NJ 07601

UK office: 57 Shelton Street, Covent Garden, London WC2H 9HE

Library of Congress Cataloging-in-Publication Data
Names: Rothe, Klaus D. (Klaus Dieter) author.
Title: Foundations of quantum field theory / Klaus D. Rothem, University of Heidelberg, Germany.
Description: Hackensack : World Scientific, [2021] | Series: World scientific lecture notes in physics,
    1793-1436 ; vol. 84 | Includes bibliographical references and index.
Identifiers: LCCN 2020037035 | ISBN 9789811221927 (hardcover) | ISBN 9789811223006 (paperback) |
    ISBN 9789811221934 (ebook)
Subjects: LCSH: Relativistic quantum theory.
Classification: LCC QC174.24.R4 R68 2020 | DDC 530.14/3--dc23
LC record available at https://lccn.loc.gov/2020037035

British Library Cataloguing-in-Publication Data
A catalogue record for this book is available from the British Library.

For any available supplementary material, please visit
https://www.worldscientific.com/worldscibooks/10.1142/11873#t=suppl

Printed in Singapore

To my wife, Neusa Maria,
and my son, Thomas

PREFACE

Quantum Field Theory (QFT) emerged in the 1930's as a natural extension of Quantum Mechanics to include Special Relativity and particle creation in its second quantized formulation.

Over the years QFT has gone through an extensive evolution with regard to the role of the quantum fields involved, the range of applicability (elementary particles, phase transitions in solid state physics), the treatment of infinities (renormalization) resulting from its local structure (microcausality), the asymptotic behaviour of Green functions (Callan–Symanzik equation, asymptotic freedom), and the analyticity property of transition amplitudes (S-matrix theory).

In particular one has learned why Quantum Electrodynamics (QED) is so successful at low energies, whereas perturbative Quantum Chromodynamics (QCD), the theory of the strong interactions, is successful at high energies (asymptotic freedom, deep inelastic scattering). One has further learned that phenomena such as spontaneous symmetry breaking observed in solid state physics (ferromagnet) also plays a role in the theory of weak interactions in particle physics, where it is referred to as "Higgs mechanism".

The present lectures essentially represent the content of a two-semester course held by the author at the University of Heidelberg, and thus provides an adequate time-frame for the lecturer and student. As such it was intended to be a compact book providing a bird's eye view of the very basic foundations of QFT, including the traditional operator, as well as the more modern path integral approach, and should serve as a good basis for post-graduate students, and as orientation for lecturers. Very extensive treatises of the subject can be found in the still excellent book of Bjorken and Drell, as well as in more up-to-date books, such as by C. Itzykson and J.-B. Zuber, E. Peskin and D.V. Schroeder, Lewis H. Ryder and S. Weinberg, which have also served as a basis for these lectures.[1] Aside from the author's point of view in presenting, choosing and arranging the material, most of it can be found in some or other way in the existing literature. We have tried to present the material in reasonable detail, with emphasis on transparency and repeated cross references, at the expense of being sometimes pedantic. We therefore believe that the reader will be able to follow the material without engaging in detailed calculations, which are cumbersome at times. Though we have exemplified various regularization procedures (Pauli–Villars, Dimensional, Taylor-subtraction), we have dominantly used the traditional Pauli–Villars regularization as being the most intuitive one.

We paid much attention in Chapter 2 to the Lorentz group and its representations in Hilbert-space, since they play a fundamental role in Chapters 3 and 5, where some knowledge of Group Theory on the part of the student is assumed. Much of these particular chapters is based on a series of remarkable articles by Steven Weinberg in *Physical Review* 1964, which underline the fundamental ideas behind the construction of a Quantum Field Theory from the operator point of view.

---

[1] J.D. Bjorken and S.D. Drell, *Relativistic Quantum Fields* (McGraw-Hill, 1965); C. Itzikson, and J.-B. Zuber *Quantum Field Theory* (McGraw-Hill, 1980); E. Peskin and D.V. Schroeder, *Frontiers in Physics*, 1995; Lewis H. Ryder, *Quantum Field Theory* (Cambridge University Press, 1985 and 1996); S. Weinberg, *The Quantum Theory of Fields* (Cambridge University Press, 1996).

Many interesting topics could not be covered in these lectures. Thus subjects such as the conformal group and analyticity of scattering amplitudes have only been marginally touched. In particular, Quantum Chromodynamics and the Weak Interactions were not included. They would have exceeded the intended size of this book, and have been left to other more extensive treatises. In turn, we left much room for the Callan–Symanzik equation, which has received considerable attention in the seventies, and will also be revisited in Chapter 18 with regard to the renormalization group. Although the renormalization group actually preceded chronologically the work of Callan and Symanzik, we have preferred to first present the latter, since it connects directly with the chapter on renormalization in Chapter 16.

As for the figures, they were drawn with the aid of the program "METAFONT" developed by Thorsten Ohl and others.[2] Although the diagrams are perhaps not as professional as those of publishing companies, the procedure to generate them with "METAFONT" is nevertheless of remarkable simplicity in limited cases. This is the reason for having restricted ourselves to presenting only examples of diagrams directly related to the text. Furthermore, only references related directly to the text were quoted. An extensive list of references can be found in the above cited books.

I would like to thank Dr. Elmar Bittner for being always ready to help me with his expertise to solve computer related problems, and to Thorsten Ohl for helping me with some more complicated diagrams.

---

[2]Thorsten Ohl, CERN Computer Newsletter 220 April 1995, 221 October 1995 and December 1996.

Contents

Chapter 1

The Principles of Quantum Physics

Quantum Field Theory is a natural outgrowth of non-relativistic Quantum Mechanics, combining it with the Principles of Special Relativity and particle production at sufficiently high energies. We therefore devote this introductory chapter to recalling some of the basic principles of Quantum Mechanics which are either shared or not shared with Quantum Field Theory.

1.1 Principles shared by QM and QFT

We briefly review first the principles which non-relativistic Quantum Mechanics (NRQM), relativistic Quantum Mechanics (RQM) and Quantum Field Theory (QFT) have in common.

(1) *Physical states*

Physical states live in a Hilbert space $\mathcal{H}_{phys}$ and are denoted by $|\Psi\rangle$.

(2) *Time development*

In the Schrödinger picture, operators O_S are independent of time and physical states $|\Psi(t)\rangle_S$ obey the equation,

$$i\hbar\frac{\partial}{\partial t}|\Psi(t)\rangle_S = H|\Psi(t)\rangle_S, \quad \partial_t O_S = 0$$

with H the Hamiltonian.

In the Heisenberg picture physical states $|\Psi\rangle_H$ are independent of time and operators $O(t)_H$ obey the Heisenberg equation

$$i\hbar\partial_t O_H(t) = [O_H(t), H], \quad \partial_t|\Psi>_H = 0.$$

1

The states in the two pictures are related by the unitary transformation

$$|\Psi(t) >_S = e^{-iHt}|\Psi >_H .$$

(3) *Completeness*

Eigenstates $|\Psi_n >$ of H,

$$H|\Psi_n >= E_n|\Psi_n >$$

are assumed to satisfy the completeness relation

$$\sum_n |\Psi_n >< \Psi_n| = 1$$

with n standing for a discrete or continuous label.

(4) *Observables*

To every observable corresponds a hermitian operator; however, not every hermitian operator corresponds to an observable.

(5) *Symmetries*

Symmetry transformations are represented in the Hilbert space $\mathcal{H}$ by *unitary* (or *anti-unitary*) operators.

(6) *Vector space*

The complete system of normalizable states $|\Psi\rangle \in \mathcal{H}$ defines a linear vector space.

(7) *Covariance of equations of motion:*

If S and S' denote two inertial reference frames, then covariance means that the equation

$$i\hbar \frac{\partial}{\partial t}|\Psi(t)\rangle = H|\Psi(t)\rangle$$

implies

$$i\hbar \frac{\partial}{\partial t'}|\Psi'(t')\rangle = H|\Psi'(t')\rangle.$$

Furthermore, there exists a unitary operator U which realizes the transformation $S \rightarrow S'$:

$$U|\Psi(t)\rangle = |\Psi'(t)\rangle$$

(8) *Physical states*

All physical states can be gauged to have positive energy[1]

$$E = \langle\Psi_p|H|\Psi_p\rangle, \quad |\Psi_p\rangle \in \mathcal{H}_{phys}$$

---

[1] In a relativistic theory, $\mathcal{H}$ can also contain negative energy states, which then require a particular interpretation or must decouple altogether from the "physical sector" of the theory.

(9) *Space and time translations*

Space-time translations are realized on $|\Psi\rangle$ respectively by[2]

$$U_S(\vec{a}) = e^{+\frac{i}{\hbar}\vec{a}\cdot\vec{P}} \quad \text{resp.} \quad U_T(t) = e^{-\frac{i}{\hbar}tH}$$

where $\vec{P}$ is the momentum operator, and rotations are realized on $|\Psi\rangle$ by

$$U_{rot}(\vec{\alpha}) = e^{+\frac{i}{\hbar}\vec{\alpha}\cdot\vec{L}} \quad [L_i, L_j] = i\hbar\epsilon_{ijk}L_k, \quad \epsilon_{123} = 1 \ ,$$

with $\vec{L}$ the generator of rotations

$$\vec{L} = \vec{r} \times \vec{p} \ .$$

1.2   Principles of NRQM not shared by QFT

The following principles of non-relativistic quantum mechanics must be abandoned in the case of QFT:

(1) *Probability amplitude*

In NRQM we associate with the state $|\Psi(t)\rangle$ a wave function

$$\psi(\vec{r}, t) = \langle\vec{r}|\Psi(t)\rangle \ .$$

$|\Psi(\vec{r}, t)|^2 d^3r$ then represent the probability of finding a particle in the interval $[\vec{r}, \vec{r}+d\vec{r}]$ at time t. (Notice the treatment of space and time on unequal footing.) In QFT we can have particle production, that is, we are dealing with "many-particle" physics. Hence notions linked to a one particle picture must be abandoned in the relativistic case.

(2) *Galilei transformations*

For a scalar function $\psi(\vec{r}, t)$ and a Galilei transformation $\vec{r}\,' = \vec{r} - \vec{v}t, t' = t$ we must have

$$\psi'(\vec{r}\,', t) = \text{phase} \times \psi(\vec{r}, t) = \text{phase} \cdot \psi(\vec{r}\,' + \vec{v}t, t).$$

---

[2]In Quantum Mechanics

$$\langle\vec{r}|e^{+\frac{i}{\hbar}\vec{a}\cdot\vec{P}}|\vec{p}\rangle = \langle\vec{r}+\vec{a}|\vec{p}\rangle \ , \quad \langle\vec{r}|\vec{P}|\vec{r}\,'\rangle = \frac{\hbar}{i}\vec{\nabla}_r\delta(\vec{r}-\vec{r}\,')$$

$$e^{+\frac{i}{\hbar}\vec{a}\cdot\vec{P}}|\vec{r}\rangle = |\vec{r}-\vec{a}\rangle$$

$$\langle\vec{r}|e^{\frac{i}{\hbar}\vec{\alpha}\cdot\vec{L}}|\Psi\rangle = \int d^3r'\,\langle\vec{r}|e^{+\frac{i}{\hbar}\vec{\alpha}\cdot\vec{L}}|\vec{r}\,'\rangle\langle\vec{r}\,'|\Psi\rangle$$

$$\langle\vec{r}|\vec{L}|\vec{r}\,'\rangle = (\vec{r}\times\frac{\hbar}{i}\vec{\nabla}_r)\delta(\vec{r}-\vec{r}\,')$$

$$e^{+\frac{i}{\hbar}\vec{\alpha}\cdot\vec{L}}|\vec{r}\rangle = |R^{-1}(\vec{\alpha})\vec{r}\rangle, \quad (R^{-1}\vec{r})_i = (R^{-1})_{ij}r_j.$$

In this chapter we use everywhere lower indices, repeated indices being summed over.

Consider in particular a plane wave $\psi_{\vec{p}}(\vec{r}, t)$ in S as seen by an observer in S' moving with a velocity $\vec{v}$ with respect to S:

$$\psi_{\vec{p}'}(\vec{r}', t) = e^{\frac{i}{\hbar}(\vec{p}' \cdot \vec{r}' - \frac{\vec{p}'^2}{2m}t)},$$

where

$$\vec{r}' = \vec{r} - \vec{v}t, \quad \vec{p}' = \vec{p} - m\vec{v}.$$

We have

$$\psi'(\vec{r}', t) = \psi_{\vec{p}-m\vec{v}}(\vec{r}', t) = \eta(t)e^{-\frac{i}{\hbar}m\vec{v}\cdot\vec{r}}\psi_{\vec{p}}(\vec{r}, t) \tag{1.1}$$

with

$$\eta(t) = e^{\frac{i}{\hbar}m\frac{\vec{v}^2}{2}t}. \tag{1.2}$$

We seek an operator $U_{\vec{v}}(t)$ with the property

$$U_{\vec{v}}(t)|\Psi_{\vec{p}}(t)\rangle = |\Psi_{\vec{p}-m\vec{v}}(t)\rangle.$$

From

$$\psi_{\vec{p}-m\vec{v}}(\vec{r}', t) = \langle \vec{r}' | U_{\vec{v}}(t)|\Psi_{\vec{p}}(t)\rangle$$

and

$$\langle \vec{r}' | e^{it\vec{v}\cdot\vec{P}} = \langle \vec{r}' + \vec{v}t | = \langle \vec{r} |$$

we conclude, by comparing with (1.1),

$$U_{\vec{v}}(t) = \eta(t)e^{\frac{i}{\hbar}t\vec{v}\cdot\vec{P}}e^{-\frac{i}{\hbar}m\vec{v}\cdot\vec{R}} \tag{1.3}$$

where $\vec{R}$ is the position operator. Now

$$[R_i, P_j] = i\hbar\delta_{ij}.$$

Hence we may write $U_{\vec{v}}(t)$ in the form

$$U_{\vec{v}}(t) = e^{\frac{i}{\hbar}[\vec{v}\cdot(t\vec{P} - m\vec{R})]} \tag{1.4}$$

where we have used

$$e^A e^B = e^{\frac{1}{2}[A,B]}e^{A+B}.$$

Denoting by $|\vec{p}\rangle$ the eigenstates of the momentum operator

$$\vec{P}|\vec{p}\rangle = \vec{p}|\vec{p}\rangle$$

we have

$$e^{-\frac{i}{\hbar}m\vec{v}\cdot\vec{R}}|\vec{p}\rangle = |\vec{p} - m\vec{v}\rangle. \tag{1.5}$$

For the solution

$$|\Psi_{\vec{p}}(t)\rangle = |\vec{p}\rangle e^{-\frac{i}{\hbar}\frac{\vec{p}^2}{2m}t}$$

of the "free" Schrödinger equation $i\hbar\partial_t|\psi_{\vec{p}}(t)\rangle = H_0|\psi_{\vec{p}}(t)\rangle$ we obtain from (1.3),

$$U_{\vec{v}}(t)|\Psi_{\vec{p}}(t)\rangle = \eta(t)e^{\frac{i}{\hbar}t\vec{v}\cdot\vec{P}}|\vec{p} - m\vec{v}\rangle e^{-\frac{i}{\hbar}\frac{\vec{p}^2}{2m}t}$$

$$= |\vec{p} - m\vec{v}\rangle e^{-\frac{i}{\hbar}\frac{(\vec{p}-m\vec{v})^2}{2m}t},$$

or

$$U_{\vec{v}}(t)|\Psi_{\vec{p}}(t)\rangle = |\Psi_{\vec{p}-m\vec{v}}(t)\rangle \ . \tag{1.6}$$

Furthermore, making use of

$$e^{-B}Ae^{B} = A + [A, B], \quad \text{for} \quad [A, B] = c - \text{number},$$

we have

$$e^{+\frac{i}{\hbar}m\vec{v}\cdot\vec{R}}\frac{\vec{P}^{2}}{2m}e^{-\frac{i}{\hbar}m\vec{v}\cdot\vec{R}} = \frac{(\vec{P} - m\vec{v})^{2}}{2m}$$

or

$$U_{\vec{v}}^{-1}(t)H_{0}U_{\vec{v}}(t) = \frac{(\vec{P} - m\vec{v})^{2}}{2m} \tag{1.7}$$

with

$$H_{0} = \frac{\vec{P}^{2}}{2m} \ .$$

Correspondingly we have from (1.7)

$$E_{0}' = \langle\Psi_{\vec{p}'}(t)|H_{0}|\Psi_{\vec{p}'}(t)\rangle = \langle\Psi_{\vec{p}}(t)|U_{\vec{v}}^{-1}(t)H_{0}U_{\vec{v}}(t)|\Psi_{p}(t)\rangle$$

$$= \langle\Psi_{\vec{p}}(t)|\frac{(\vec{P} - m\vec{v})^{2}}{2m}|\Psi_{\vec{p}}(t)\rangle = \frac{(\vec{p} - m\vec{v})^{2}}{2m} \ ,$$

in accordance with expectations.

(3) *Covariance of equations of motion*

From

$$H_{0}|\Psi_{\vec{p}}(t)\rangle = i\hbar\partial_{t}|\Psi_{\vec{p}}(t)\rangle$$

follows

$$U_{\vec{v}}(t)H_{0}|\Psi_{\vec{p}}(t)\rangle = i\hbar U_{\vec{v}}(t)\partial_{t}|\Psi_{p}(t)\rangle \ ,$$

which we rewrite as

$$(U_{\vec{v}}(t)H_{0}U_{\vec{v}}^{-1}(t))U_{\vec{v}}(t)|\Psi_{p}(t)\rangle = i\hbar\left(U_{\vec{v}}(t)\partial_{t}U_{\vec{v}}^{-1}(t)\right)U_{\vec{v}}(t)|\Psi_{p}(t)\rangle$$
$$+ i\hbar\partial_{t}(U_{\vec{v}}(t)|\Psi_{p}(t)\rangle) \ .$$

Noting from (1.3) and (1.2) that

$$i\hbar U_{\vec{v}}(t)\partial_{t}U_{\vec{v}}^{-1}(t) = \frac{1}{2}m\vec{v}^{2} + \vec{v}\cdot\vec{P}$$

we obtain, using (1.6),

$$[U_{\vec{v}}(t)H_{0}U_{v}^{-1}(t) - (\frac{1}{2}m\vec{v}^{2} + \vec{v}\cdot\vec{P})]|\Psi_{\vec{p}-m\vec{v}}(t)\rangle = i\hbar\partial_{t}|\Psi_{\vec{p}-m\vec{v}}(t)\rangle \ ,$$

or, recalling (1.7),

$$\frac{\vec{P}^{2}}{2m}|\Psi_{\vec{p}-m\vec{v}}(t)\rangle = i\hbar\partial_{t}|\Psi_{\vec{p}-m\vec{v}}(t)\rangle.$$

This equation expresses on operator level the covariance of the free-particle equation of motion: If $|\psi_{\vec{p}}(t)\rangle$ is a solution of the equations of motion, then $|\psi_{\vec{p}-m\vec{v}(t)}\rangle$ is also a solution.

Group-property

As Eq. (1.4) shows, a Galilei transformation is represented by the unitary operator

$$U_{\vec{v}}(t) = e^{i\vec{v}\cdot\mathbf{K}}$$

with

$$\vec{K} = (t\vec{P} - m\vec{R})$$

the generators of *boosts*. We have

$$[K_i, K_j] = [tP_i - mR_i, tP_j - mR_j] = 0 \ ,$$

so that different "boosts" commute with each other. The Galilei transformations thus correspond to an *abelian Lie group*. In particular

$$U_{\vec{v}'}(t)U_{\vec{v}}(t) = U_{\vec{v}'+\vec{v}}(t) \ .$$

Boosts

Let S and S' be two inertial frames whose clocks are synchronized in such a way, that their respective origins coincide at time $t = 0$. Then we have for an eigenstate of the momentum operator, as seen by observers O and O' in S and S',

$$O : |\vec{p}\rangle$$
$$O' : U_{\vec{v}}(0)|\vec{p}\rangle = e^{-\frac{i}{\hbar}m\vec{v}\cdot\vec{R}}|\vec{p}\rangle = |\vec{p} - m\vec{v}\rangle$$

respectively. In particular, for a particle at rest in system S we obtain, from the point of view of O',

$$O' : U_{\vec{v}}(0)|\vec{0}\rangle = |-m\vec{v}\rangle \ .$$

Define

$$U_{-\vec{v}}(0) =: U[B(\vec{p})] = e^{\frac{i}{\hbar}\vec{p}\cdot\vec{R}} \ ,$$

where $B(\vec{p})$ stands for a Galilei transformation taking $\vec{p} = 0 \to \vec{p} = m\vec{v}$ and $E = 0 \to E = \frac{p^2}{2m} = \frac{1}{2}mv^2$. $U[B(\vec{p})]$ is thus an operator which takes a particle at rest into a particle with momentum $\vec{p}$. One refers to this as a "boost" (active point of view).

We have the following property of Galilei transformations not shared by Lorentz transformations (compare with (2.23)): boosts and rotations *separately* form a group. Indeed one easily checks that

$$[K_i, K_j] = 0 \ , \quad [K_i, L_j] = 0 \ .$$

(4) *Causality*

We next want to show that NRQM violates the principle of causality. We have for any interacting theory,

$$|\Psi(t')\rangle = e^{-\frac{i}{\hbar}(t'-t)H}|\Psi(t)\rangle.$$

Let $|E_n\rangle$ be a complete set of eigenstates of H:[3]

$$H|E_n\rangle = E_n|E_n\rangle$$

$$\sum_n |E_n\rangle\langle E_n| = 1 \ .$$

Then

$$\langle\vec{r}'|\Psi(t')\rangle = \sum_n \langle\vec{r}'|E_n\rangle\langle E_n|\Psi(t)\rangle e^{-\frac{i}{\hbar}(t'-t)E_n}$$

$$= \sum_n \int d^3r \langle\vec{r}'|E_n\rangle\langle E_n|\vec{r}\rangle\langle\vec{r}|\Psi(t)\rangle e^{-\frac{i}{\hbar}(t'-t)E_n} \ .$$

Define

$$\langle\vec{r}|E_n\rangle = \varphi_n(\vec{r}), \quad \sum_n \varphi_n(\vec{r}')\varphi_n^*(\vec{r}) = \delta^3(\vec{r}' - \vec{r})$$

$$\langle\vec{r}|\Psi(t)\rangle = \psi(\vec{r}, t) \ ,$$

as well as

$$\sum_n \varphi_n(\vec{r}')\varphi_n^*(\vec{r})e^{-\frac{i}{\hbar}E_n(t'-t)} := K(\vec{r}', t'; \vec{r}, t). \tag{1.8}$$

The kernel $K(\vec{r}, t; \vec{r}_0, t_0)$ satisfies a heat-like equation:

$$(i\hbar\partial_t - H)K(\vec{r}, t; \vec{r}_0, t_0) = 0$$

$$K(\vec{r}, t; \vec{r}_0, t_0)|_{t=t_0} = \delta^3(\vec{r} - \vec{r}_0). \tag{1.9}$$

In terms of this kernel we have from above,

$$\psi(\vec{r}', t') = \int d^3r \ K(\vec{r}', t'; \vec{r}, t)\psi(\vec{r}, t) \ . \tag{1.10}$$

We now specialize to the case of a free point-like particle. In that case

$$H_0 = \frac{\mathbf{P}^2}{2m} \Rightarrow \begin{cases} E_n \to E(\vec{p}) = \frac{\vec{p}^2}{2m} \\ \sum_n \to d^3p \\ \langle\vec{r}|\varphi_n\rangle \to \varphi_p(\vec{r}) = \frac{1}{(2\pi\hbar)^{3/2}} e^{\frac{i}{\hbar}\vec{p}\cdot\vec{r}} \end{cases}$$

and correspondingly we have with (1.8),

$$K_0(\vec{r}', t'; \vec{r}, t) = \frac{1}{(2\pi\hbar)^3} \int d^3p \, e^{\frac{i}{\hbar}\vec{p}\cdot(\vec{r}'-\vec{r})} e^{-\frac{i}{\hbar}\frac{\vec{p}^2}{2m}(t'-t)}$$

$$= \left(\frac{m}{2\pi\hbar i(t'-t)}\right)^{3/2} e^{\frac{i}{\hbar}\frac{m}{2}\frac{(\vec{r}'-\vec{r})^2}{t'-t}} \ . \tag{1.11}$$

---

[3] For notational simplicity we suppose the spectrum to be discrete.

Notice that the kernel K_0 satisfies the desired initial condition (1.9).

From (1.10), for the initial condition $\psi(\vec{r}, t_0) = \delta^3(\vec{r} - \vec{r}_0)$, we get

$$\psi(\vec{r}, t) = \left(\frac{m}{2\pi i \hbar (t - t_0)} \right)^{3/2} e^{\frac{i}{\hbar} \frac{m}{2} \frac{(\vec{r} - \vec{r}_0)^2}{(t - t_0)}} \ ,$$

or in particular

$$\mathcal{P}(\vec{r}, t) = |\psi(\vec{r}, t)|^2 = \left(\frac{m}{2\pi \hbar (t - t_0)} \right)^3 \ ,$$

that is, for an infinitesimal time after t_0 one already finds the particle with equal probability anywhere in space; this violates obviously the principle of relativity, as well as causality.

Chapter 2

Lorentz Group and Hilbert Space

In this chapter we first discuss the realization of the homogeneous Lorentz transformations in four-dimensional space-time, as well as the corresponding Lie algebra. From here we obtain all finite dimensional representations, and in particular the explicit form of the matrices representing the boosts for the case of spin $= 1/2$, which will play a fundamental role in Chapter 3. The Lorentz transformation properties of massive and zero-mass 1-particle in Hilbert space (and their explicit realization in Chapter 9) lie at the heart of the Fock space representation (second quantization) in Chapter 9. It is assumed that the reader is already familiar with the essentials of the Special Theory of Relativity and of Group Theory.

2.1   Defining properties of Lorentz transformations

Homogeneous Lorentz transformations are linear transformations on the space-time coordinates,[1]

$$x^\mu \to x'^\mu = \Lambda^\mu{}_\nu x^\nu, \quad (x^\mu) = (ct, \vec{r}), \quad \mu = 0, 1, 2, 3 \tag{2.1}$$

leaving the quadratic form

$$ds^2 = c^2 dt^2 - d\vec{r}^2 \tag{2.2}$$

invariant. Any 4-tuplet transforming like the coordinates in (2.1) is called a contravariant 4-vector. In particular, energy and momentum of a particle are components of a 4-vector

$$(p^\mu) = (\omega(\vec{p}), \vec{p}) = (\sqrt{\vec{p}^2 + m^2 c^2}, \vec{p}), \quad \vec{p} = \gamma(v) m \vec{v} \tag{2.3}$$

---

[1] The inhomogeneous Lorentz transformations including the space-time translations will be discussed in Chapter 9. We adopt the convention that repeated upper and lower indices are to be summed over.

with

$$\gamma(v) = \frac{1}{\sqrt{1 - \frac{v^2}{c^2}}} \, .$$

The same transformation law defines 4-vector *fields* at a given *physical* point:

$$V'^{\mu}(x) = (\Lambda^{-1})^{\mu}{}_{\nu} V^{\nu}(\Lambda x) \, . \tag{2.4}$$

Note that the quadruple $(\Lambda x)^{\mu}$ referred to $\mathcal{S}$ is the same point as the quadruple x^{μ} referred to $\mathcal{S}'$. Thus, alternatively

$$V'^{\mu}(x') = \Lambda^{\mu}{}_{\nu} V^{\nu}(x) \, . \tag{2.5}$$

Notice that (2.4) and (2.5) represent inverse transformations of the reference frame, respectively. Examples are provided by the 4-vector current $j^{\mu} = (c\rho, \vec{j})$ and the vector potential $A^{\mu} = (\phi, \vec{A})$ of electrodynamics in a Lorentz-covariant gauge.

The differential element dx^{μ} transforms like

$$dx'^{\mu} = \left(\frac{\partial x'^{\mu}}{\partial x^{\nu}} \right) dx^{\nu}.$$

Hence it also transforms like a *contravariant* 4-vector, since

$$\frac{\partial x'^{\mu}}{\partial x^{\nu}} = \Lambda^{\mu}{}_{\nu} \, .$$

The partial derivative $\frac{\partial}{\partial x^{\mu}}$, on the other hand, transforms differently. The usual chain rule of differentiation gives

$$\frac{\partial}{\partial x'^{\mu}} = \left(\frac{\partial x^{\nu}}{\partial x'^{\mu}} \right) \frac{\partial}{\partial x^{\nu}} \, .$$

From the inversion of (2.1) it follows that

$$\left(\frac{\partial x^{\nu}}{\partial x'^{\mu}} \right) = (\Lambda^{-1})^{\nu}{}_{\mu} \, . \tag{2.6}$$

Hence for the partial derivative $\partial_{\mu} = \frac{\partial}{\partial x^{\mu}}$ we have the transformation law

$$\partial'_{\mu} = \partial_{\nu} (\Lambda^{-1})^{\nu}{}_{\mu} \, . \tag{2.7}$$

Four-tuples which transform like the partial derivative are called *covariant* 4-vectors. Contravariant and covariant 4-vectors are obtained from each other by raising and lowering the indices with the aid of the metric tensors $g^{\mu\nu}$ and $g_{\mu\nu}$, defined by[2]

$$\|g_{\mu\nu}\| = \|g^{\mu\nu}\| = \begin{pmatrix} 1 & & & \\ & -1 & & \\ & & -1 & \\ & & & -1 \end{pmatrix}, \quad g^2 = 1$$

---

[2]This chapter is largely based on the papers by S. Weinberg in *Physical Review*. Note that Weinberg uses the metric $g^{\mu\nu} = (-1, 1, 1, 1)$.

respectively, in terms of which the invariant element of length (2.2) can be written in the form

$$ds^2 = g_{\mu\nu} dx^\mu dx^\nu .$$

The requirement that ds^2 be a Lorentz invariant

$$g_{\mu\nu} dx'^\mu dx'^\nu = g_{\lambda\rho} dx^\lambda dx^\rho$$

now implies

$$g_{\mu\nu} \Lambda^\mu_{\ \lambda} \Lambda^\nu_{\ \rho} = g_{\lambda\rho} . \tag{2.8}$$

Thus the metric $g_{\mu\nu}$ is said to be a Lorentz-invariant tensor. It is convenient to write this equation in matrix notation by grouping the elements $\Lambda^\mu_{\ \nu}$ into a matrix as follows:

$$\Lambda = \begin{pmatrix} \Lambda^0_{\ 0} & \Lambda^0_{\ 1} & \Lambda^0_{\ 2} & \Lambda^0_{\ 3} \\ \Lambda^1_{\ 0} & \cdot & \cdot & \cdot \\ \cdot & \cdot & \cdot & \cdot \\ \cdot & \cdot & \cdot & \Lambda^3_{\ 3} \end{pmatrix} .$$

Defining the elements of the transpose matrix Λ^T by

$$(\Lambda^T)^\mu_{\ \nu} = \Lambda^\nu_{\ \mu}$$

we can write (2.8) as follows:

$$\Lambda^T g \Lambda = g .$$

From here we obtain for the inverse Λ^{-1},

$$\Lambda^{-1} = g \Lambda^T g ,$$

or in terms of components

$$(\Lambda^{-1})^\mu_{\ \nu} = g^{\mu\lambda} \Lambda^\rho_{\ \lambda} g_{\rho\nu} = \Lambda_\nu^{\ \mu} . \tag{2.9}$$

Define the *dual* to a contravariant 4-vector v^μ by

$$v_\mu = g_{\mu\nu} v^\nu . \tag{2.10}$$

Thus $g_{\mu\nu} (g^{\mu\nu})$ serve to lower (raise) the Lorentz indices. In particular $g^{\mu\lambda} g_{\lambda\nu} = g^\mu_{\ \nu} = \delta^\mu_{\ \nu}$. We have after a Lorentz transformation, upon using (2.9)

$$v'_\mu = v_\nu (\Lambda^{-1})^\nu_{\ \mu} \tag{2.11}$$

or we conclude that v_μ defined by (2.10) does indeed transform like a *covariant* 4-vector. In particular we see that the following 4-tuplets transform like *covariant* and *contravariant* 4-vectors, respectively:

$$(\partial_\mu) = (\partial_0, \vec{\nabla}) , \quad (\partial^\nu) = (\partial_0, -\vec{\nabla}) ,$$

where

$$\vec{\nabla} = \left(\frac{\partial}{\partial x^1}, \frac{\partial}{\partial x^2}, \frac{\partial}{\partial x^3} \right) . \tag{2.12}$$

2.2   Classification of Lorentz transformations

The Lorentz invariance of the scalar product $v_\mu v^\mu = v^2$ allows us to divide the 4-vectors into three classes which cannot be transformed into each other by a Lorentz transformation:

(a) v^μ time-like $(v^2 > 0)$
(b) v^μ space-like $(v^2 < 0)$
(c) v^μ light-like $(v^2 = 0)$

This means in particular that space-time separates, as far as Lorentz transformations are concerned, into three disconnected regions referring to the *interior* and *exterior* of the light cone $x^2 = 0$, as well as to the *surface* of the light cone itself. The trajectory of a point particle localized at the origin of the light cone at time $t = 0$ lies within the *forward* light cone; Moreover, if we attach a light cone to the particle at the point where it is momentarily localized, the tangent to the trajectory at that point does not intersect the surface of that light cone.

The Lorentz invariant

$$d\tau = \sqrt{\frac{ds^2}{c^2}} = \sqrt{dt^2 - \frac{d\vec{r}^2}{c^2}} = dt\sqrt{1 - \frac{v^2}{c^2}}$$

taken along the trajectory of the particle is just the *proper time*, measured in the rest frame of the particle.

From (2.8) follow two important properties of Lorenz transformations:

(i) $\det \Lambda = \pm 1$

(ii) $(\Lambda^0{}_0)^2 - \sum_i (\Lambda^i{}_0)^2 = 1$,

implying

$$\Lambda^0{}_0 \geq 1 \quad \text{or} \quad \Lambda^0{}_0 \leq -1 \ .$$

Note that this allows for four types of transformations which cannot be smoothly connected by varying continuously the parameter labelling the transformation. We thus have four possibilities characterizing the Lorentz invariance of the differential element (2.2):

(a) $L^\uparrow_+$: $\det \Lambda = 1$, $\Lambda^0{}_0 \geq 1$ (*proper, orthochrone LT*)

(b) $L^\uparrow_-$: $\det \Lambda = -1$, $\Lambda^0{}_0 \geq 1$ (*improper, orthochrone LT*)

(c) $L^\downarrow_+$: $\det \Lambda = 1$, $\Lambda^0{}_0 \leq -1$ (*proper, non-orthochrone LT*)

(d) $L^\downarrow_-$: $\det \Lambda = -1$, $\Lambda^0{}_0 \leq -1$ (*improper, non-orthochrone LT*)

Only the first set of transformations is smoothly connected to the identity and hence form a Lie group. The remaining transformations do not have the group property.

They are obtained by adjoining to the transformations in $L_+^\uparrow$ *space reflections*, *space-time reflections* and *time inversion*, respectively, as represented by the matrices

$$\Lambda_P = \begin{pmatrix} 1 & 0 \\ 0 & -1 \end{pmatrix}, \quad \Lambda_{PT} = \begin{pmatrix} -1 & 0 \\ 0 & -1 \end{pmatrix}, \quad \Lambda_T = \begin{pmatrix} -1 & 0 \\ 0 & 1 \end{pmatrix}.$$

Only $L_+^\uparrow$ represents an exact symmetry of nature.

General form of a Lorentz boosts

From (2.3) one has for a boost in the z-direction, taking the particle from rest to a momentum $\vec{p} = \gamma(v)mv\hat{e}_3$,[3]

$$L(mv\hat{e}_3) = \begin{pmatrix} \gamma(v) & 0 & 0 & \beta\gamma(v) \\ 0 & 1 & 0 & 0 \\ 0 & 0 & 1 & 0 \\ \beta\gamma(v) & 0 & 0 & \gamma(v) \end{pmatrix}, \tag{2.13}$$

where[4]

$$\gamma(v) = \frac{1}{\sqrt{1-\beta^2}} = \frac{\omega(\vec{p})}{mc}, \quad \beta = \frac{v}{c} = \frac{|\vec{p}|}{\omega(\vec{p})}.$$

Since

$$\gamma^2 - (\beta\gamma)^2 = 1$$
$$\omega(\vec{p}) = \sqrt{\vec{p}^2 + m^2c^2},$$

we may parametrize γ and $\beta\gamma$ as follows:

$$\gamma(v) = \frac{\omega(\vec{p})}{mc} = \cosh\theta, \quad \beta\gamma = \frac{|\vec{p}|}{mc} = \sinh\theta, \quad \tanh\theta = \frac{|\vec{p}|}{\omega(\vec{p})} \tag{2.14}$$

or (2.13) now reads

$$L(|p|\hat{e}_3) = \begin{pmatrix} \cosh\theta & 0 & 0 & \sinh\theta \\ 0 & 1 & 0 & 0 \\ 0 & 0 & 1 & 0 \\ \sinh\theta & 0 & 0 & \cosh\theta \end{pmatrix}. \tag{2.15}$$

---

[3]Lorentz transformations representing a boost to momentum $\vec{p}$ we denote by $L(\vec{p})$.
[4]We have

$$\frac{\omega(p)}{mc} = \frac{\sqrt{p^2 + m^2c^2}}{mc} = \frac{\sqrt{\gamma^2(v)m^2v^2 + m^2c^2}}{mc}$$

$$= \sqrt{\gamma^2(v)\beta^2 + 1} = \sqrt{\frac{\beta^2}{1-\beta^2} + 1} = \frac{1}{\sqrt{1-\beta^2}} = \gamma(v)$$

$$\frac{|\vec{p}|}{\omega(p)} = \frac{|\vec{p}|}{mc\gamma(v)} = \frac{\gamma(v)m|\vec{v}|}{mc\gamma(v)} = \frac{v}{c}$$

Note that this is a hermitian matrix! For a boost in an arbitrary direction one can show that the corresponding matrix elements are given by (note that $g^i{}_j = \delta_{ij}$, $\hat{p}_j = -\hat{p}^j$)

$$L^0{}_0(\vec{p}) = \cosh\theta$$
$$L^i{}_0(\vec{p}) = L^0{}_i(\vec{p}) = \hat{p}^i \sinh\theta \qquad (2.16)$$
$$L^i{}_j(\vec{p}) = g^i{}_j - \hat{p}^i\hat{p}_j[\cosh\theta - 1]$$

2.3 Lie algebra of the Lorentz group

Consider an infinitesimal Lorentz transformation in 3+1 dimensions. It is customary to parametrize its matrix elements as follows. In the case of $L^\uparrow_+$ we are dealing with a six-parameter group parametrized by the "velocities" associated with boosts, and the Euler angles associated with the rotations. This is just the number of independent components of an antisymmetric second rank tensor. In analogy to the rotation group it is customary to write for the matrix elements of an infinitesimal Lorentz transformation

$$\Lambda^\mu{}_\nu \simeq g^\mu{}_\nu + \omega^\mu{}_\nu , \qquad \omega_{\mu\nu} = -\omega_{\nu\mu} .$$

Using the metric tensor as raising and lowering operators for the indices, we further have

$$\omega^i{}_j = -\omega^j{}_i = -\omega_{ij}$$
$$\omega^i{}_0 = \omega^0{}_i = -\omega_{i0} = \omega_{0i} .$$

We may rewrite this transformation in matrix form as follows

$$\Lambda \simeq 1 + \frac{i}{2}\omega_{\mu\nu}\mathcal{M}^{\mu\nu} , \qquad (2.17)$$

where $\mathcal{M}^{\mu\nu}$ denote the generators associated with the transformations in $L^\uparrow_+$, with the property

$$\mathcal{M}^{\mu\nu} = -\mathcal{M}^{\nu\mu}$$

and the matrix elements[5]

$$(\mathcal{M}^{\mu\nu})^\lambda{}_\rho = -i\left(g^{\mu\lambda}g^\nu{}_\rho - g^{\nu\lambda}g^\mu{}_\rho\right) . \qquad (2.18)$$

Here again it is implied that Lorentz indices can be raised and lowered with the aid of the metric tensor $g^{\mu\nu} = g_{\mu\nu}$. Expression (2.17) is just the infinitesimal expansion of

$$\Lambda = e^{\frac{i}{2}\omega_{\mu\nu}\mathcal{M}^{\mu\nu}} . \qquad (2.19)$$

_____

[5]$\Lambda^\lambda{}_\rho \simeq g^\lambda{}_\rho + \frac{\omega_{\mu\nu}}{2}\left(g^{\mu\lambda}g^\nu{}_\rho - g^{\nu\lambda}g^\mu{}_\rho\right) \simeq g^\lambda{}_\rho + \frac{1}{2}(\omega^\lambda{}_\rho - \omega_\rho{}^\lambda) = g^\lambda{}_\rho + \omega^\lambda{}_\rho.$

Explicitly we have

$$\mathcal{M}^{01} = -i \begin{pmatrix} 0\,1\,0\,0 \\ 1\,0\,0\,0 \\ 0\,0\,0\,0 \\ 0\,0\,0\,0 \end{pmatrix}, \quad \mathcal{M}^{02} = -i \begin{pmatrix} 0\,0\,1\,0 \\ 0\,0\,0\,0 \\ 1\,0\,0\,0 \\ 0\,0\,0\,0 \end{pmatrix}, \quad \mathcal{M}^{03} = -i \begin{pmatrix} 0\,0\,0\,1 \\ 0\,0\,0\,0 \\ 0\,0\,0\,0 \\ 1\,0\,0\,0 \end{pmatrix}.$$

Hence for a pure Lorentz transformation in the $-z$-direction (boost in the $+z$ direction)

$$\Lambda \simeq 1 + \theta \begin{pmatrix} 0\,0\,0\,1 \\ 0\,0\,0\,0 \\ 0\,0\,0\,0 \\ 1\,0\,0\,0 \end{pmatrix}, \tag{2.20}$$

where we have set

$$\omega_{03} = \theta .$$

Equation (2.20) just represents the first two leading terms of the expansion of the finite Lorentz boost (2.15) in the x^3-direction around $\theta = 0$. We thus identify $\mathcal{M}^{0i}$ with the generators for pure velocity transformations along the z-axis. Notice that these matrices are anti-hermitian. On the other hand, the *hermitian* matrices $\mathcal{M}^{ij}$ generate rotations in the (ij)-plane:

$$\mathcal{M}^{12} = -i \begin{pmatrix} 0\ 0\ 0\,0 \\ 0\ 0\ 1\,0 \\ 0\,-1\,0\,0 \\ 0\ 0\ 0\,0 \end{pmatrix}, \quad \mathcal{M}^{23} = -i \begin{pmatrix} 0\,0\ 0\ 0 \\ 0\,0\ 0\ 0 \\ 0\,0\ 0\ 1 \\ 0\,0\,-1\,0 \end{pmatrix}, \quad \mathcal{M}^{31} = -i \begin{pmatrix} 0\,0\,0\ 0 \\ 0\,0\,0\,-1 \\ 0\,0\,0\ 0 \\ 0\,1\,0\ 0 \end{pmatrix}$$

the generators of rotations about the z-, x- and y-axis respectively. One verifies from (2.18) that the generators of Lorentz transformations satisfy the Lie algebra

$$[\mathcal{M}^{\mu\nu}, \mathcal{M}^{\lambda\rho}] = -i(g^{\mu\lambda}\mathcal{M}^{\nu\rho} + g^{\nu\rho}\mathcal{M}^{\mu\lambda} - g^{\nu\lambda}\mathcal{M}^{\mu\rho} - g^{\mu\rho}\mathcal{M}^{\nu\lambda}) . \tag{2.21}$$

Explicitly

$$[\mathcal{M}^{ij}, \mathcal{M}^{kl}] = -i(g^{ik}\mathcal{M}^{jl} + g^{jl}\mathcal{M}^{ik} - g^{il}\mathcal{M}^{jk} - g^{jk}\mathcal{M}^{il})$$

$$[\mathcal{M}^{i0}, \mathcal{M}^{j0}] = -i\mathcal{M}^{ij}$$

$$[\mathcal{M}^{j0}, \mathcal{M}^{kl}] = -i(g^{jk}\mathcal{M}^{0l} - g^{jl}\mathcal{M}^{0k}) .$$

This algebra simplifies if we define the generators

$$J_i = \frac{1}{2}\epsilon_{ijk}\mathcal{M}^{jk} , \quad K_i = \mathcal{M}^{i0} , \quad \vec{J}^\dagger = \vec{J} , \quad \vec{K}^\dagger = -\vec{K} \tag{2.22}$$

which now satisfy the simple commutation relations[6,7]

$$[J_i, J_j] = i\epsilon_{ijk}J_k$$
$$[J_i, K_j] = i\epsilon_{ijk}K_k \qquad (2.23)$$
$$[K_i, K_j] = -i\epsilon_{ijk}J_k$$

reminiscent of the rotation group in four dimensions (except for the minus sign). This is not surprising, since the Lorentz transformations are connected to the four-dimensional rotation group by an analytic continuation in the parameters parametrizing the boosts, to *pure imaginary* values. The underlying Minkowski character of the space-time manifold is hidden in the anti-hermiticity of the operators $\mathcal{M}^{i0}$ generating the boosts. Correspondingly we have for a *finite* transformation $L_+^\uparrow$ in space-time

$$\Lambda = e^{i\vec{\alpha}\cdot\vec{J} - i\vec{\theta}\cdot\vec{K}}, \quad \Lambda^\dagger \neq \Lambda^{-1}, \qquad (2.24)$$

where we have made the identifications

$$\omega_{ij} = \epsilon_{ijk}\alpha_k, \quad \omega_{0i} = \theta\hat{p}^i. \qquad (2.25)$$

with θ the angle labelling the boost (2.16). For a pure rotation in the ij-plane,

$$\mathcal{M}^{ij} = i \begin{pmatrix} 0 & 0 & 0 & 0 \\ 0 & 0 & 0 & 0 \\ 0 & & \epsilon_{ijk}J_k & \\ 0 & & & \end{pmatrix}, \qquad (2.26)$$

where J_k is the generator of rotations around the k-axis (i, j, k taken cyclicly), with the explicit realization

$$J_1 = -i \begin{pmatrix} 0 & 0 & 0 \\ 0 & 0 & 1 \\ 0 & -1 & 0 \end{pmatrix}, \quad J_2 = -i \begin{pmatrix} 0 & 0 & -1 \\ 0 & 0 & 0 \\ 1 & 0 & 0 \end{pmatrix}$$

$$J_3 = -i \begin{pmatrix} 0 & 1 & 0 \\ -1 & 0 & 0 \\ 0 & 0 & 0 \end{pmatrix}.$$

Next, it will be our aim to obtain higher dimensional representations of the generators of the Lorentz group.

---

[6]These commutation relations do not fix the sign of the generator K_i of boosts. We follow in (2.22) the convention adopted by S. Weinberg.

[7]The minus sign in the last commutation relations reflects the fact that the Lorentz group $SO(3,1)$ can be considered as the *complexification* of the rotation group in four dimensions, $SO(4)$. The fact that the commutator of two generators of the boost is given as a linear combination of the generators of rotations is related to the phenomenon of the *Thomas precession*. From the group theoretic point of view it expresses the fact that, if we perform the sequence of infinitesimal boosts $g(-\delta\theta_2)g(-\delta\theta_1)g(\delta\theta_2)g(\delta\theta_1)$ with $g(\delta\theta) = 1 + i\delta\vec{\theta}\cdot\vec{K} - \frac{1}{2}(\delta\vec{\theta}\cdot\vec{K})^2 + \cdots$, the effect is just an infinitesimal rotation in the frame we started from.

2.4   Finite irreducible representation of $L_+^\uparrow$

In order to obtain a characterization of the irreducible representation $L_+^\uparrow$, we define the new operators

$$\vec{A} = \frac{1}{2}(\vec{J} + i\vec{K}) , \quad \vec{A} = (A_1, A_2, A_3) ,$$
$$\vec{B} = \frac{1}{2}(\vec{J} - i\vec{K}) , \quad \vec{B} = (B_1, B_2, B_3) . \tag{2.27}$$

Note that $A_i^\dagger = A_i$ and $B_i^\dagger = B_i$ (see (2.22)). It then follows from the commutation relations (2.23) that

$$[A_i, A_j] = i\epsilon_{ijk} A_k$$
$$[B_i, B_j] = i\epsilon_{ijk} B_k$$
$$[A_i, B_j] = 0 . \tag{2.28}$$

The operators $\vec{A}$ and $\vec{B}$ thus satisfy two Lie algebras which are decoupled from each other. Correspondingly, there exist two Casimir operators, $\vec{A}^2$ and $\vec{B}^2$, which commute with all elements of the group. Their eigenvalues thus serve to classify the irreducible representations of $L_+^\uparrow$. In the irreducible basis, the matrix elements of the generators $\vec{A}$ and $\vec{B}$ are evidently given by the well-known expressions known from the rotation group:

$$\langle A, a'; B, b' | \vec{A} | A, a; B, b \rangle = \delta_{b'b} \vec{J}_{a'a}^{(A)}$$

$$\langle A, a'; B, b' | \vec{B} | A, a; B, b \rangle = \delta_{a'a} \vec{J}_{b'b}^{(B)} ,$$

where a and b take the values

$$a = -A, -A+1, ..., A ; \quad A = 0, \frac{1}{2}, 1, ...$$
$$b = -B, -B+1, ..., B ; \quad B = 0, \frac{1}{2}, 1, ...$$

In particular, we have

$$(J_3^{(A)})_{a'a} = a\delta_{a'a}, \quad (J_3^{(B)})_{b'b} = b\delta_{b'b}$$
$$(J_1^{(A)} \pm iJ_2^{(A)})_{a'a} = \sqrt{A(A+1) - a(a \pm 1)} \, \delta_{a',a\pm 1}$$
$$(J_1^{(B)} \pm iJ_2^{(B)})_{b'b} = \sqrt{B(B+1) - b(b \pm 1)} \, \delta_{b',b\pm 1} .$$

For the operators $\vec{J}$ and $i\vec{K}$ these matrix elements read from (2.27)

$$\langle A, a'; B, b' | \vec{J} | A, a; B, b \rangle =: \vec{J}_{a'b';ab}^{(A,B)} = \delta_{b'b} \vec{J}_{a'a}^{(A)} + \delta_{a'a} \vec{J}_{b'b}^{(B)} .$$
$$\langle A, a'; B, b' | i\vec{K} | A, a; B, b \rangle =: i\vec{K}_{a'b';ab}^{(A,B)} = \delta_{b'b} \vec{J}_{a'a}^{(A)} - \delta_{a'a} \vec{J}_{b'b}^{(B)} .$$

From these matrix elements we then obtain the corresponding representations of finite Lorentz transformations $L_+^\uparrow$ by simple exponentiation, in a way analogous to (2.24),

$$D_{a'b',ab}^{(A,B)}[\Lambda] = < A, a'; B, b' | e^{i\vec{\alpha}\cdot\vec{J} - i\vec{\theta}\cdot\vec{K}} | A, a; B, b > ,$$

or compactly

$$D^{(A,B)}[\Lambda] = \exp\{i\vec{\alpha}\cdot\vec{J}^{(A,B)} - i\vec{\theta}\cdot\vec{K}^{(A,B)}\} ,$$

with

$$\vec{\theta} = \theta\hat{p} ,$$

where $\hat{p}$ denotes the direction of the boost. The matrices $D^{(A,B)}$ exhaust all possible finite dimensional representations of $L_+^\uparrow$. They are *non-unitary* except for the trivial representation $D^{(0,0)}$.

Example 1

Consider the representation $(A, B) = (j, 0)$. In that representation the operator $\vec{B}$ is realized by *zero* or $\vec{K} = -i\vec{J}$. This possibility is allowed by the commutation relations (2.28). One has

$$D^{(j,0)}[\Lambda] = e^{i(\vec{\alpha}+i\vec{\theta})\cdot\vec{J}^{(j,0)}} \tag{2.29}$$

with

$$\vec{J}_{a'a}^{(j,0)} = \langle j, a'; 0, 0 | \vec{J} | j, a; 0, 0 \rangle .$$

Example 2

Consider the representation $(A, B) = (0, j)$. In that case $\vec{A} = 0$ or $\vec{K} = i\vec{J}$, and the Lorentz transformations are realized by

$$D^{(0,j)}[\Lambda] = e^{i(\vec{\alpha}-i\vec{\theta})\cdot\vec{J}^{(0,j)}} \tag{2.30}$$

with the corresponding matrix elements for $\vec{J}$:

$$\vec{J}_{b'b}^{(0,j)} = \langle 0, 0; j, b' | \vec{J} | 0, 0; j, b \rangle .$$

The above representations will play a central role in the chapters to follow. We observe that

$$D^{(j,0)}[\Lambda]^\dagger = D^{(0,j)}[\Lambda^{-1}] .$$

We introduce the following notation[8]

$$D^{(j)}[\Lambda] := D^{(j,0)}[\Lambda] ,$$
$$\bar{D}^{(j)}[\Lambda] := D^{(0,j)}[\Lambda] . \tag{2.31}$$

We then have in particular

$$\bar{D}^{(j)}[L(\vec{p})] = D^{(j)}[L(-\vec{p})] , \tag{2.32}$$

---

[8]We follow closely the notation of S. Weinberg, *Phys. Rev.* **133** (1964) B1318.

as well as

$$\bar{D}^{(j)}[\Lambda] = D^{(j)}[\Lambda^{-1}]^\dagger . \tag{2.33}$$

For a pure rotation

$$\bar{D}^{(j)}[R] = D^{(j)}[R] . $$

The case $j = \frac{1}{2}$

For the case of $j = 1/2$ it is easy to give an explicit expression for the matrices $D^{(\frac{1}{2})}[L(\vec{p})]$ and $\bar{D}^{(\frac{1}{2})}[L(\vec{p})]$ representing a boost, since in that case the matrices representing $\vec{J}$ are just one half of the *Pauli matrices* (4.4):

$$D^{(\frac{1}{2})}[L(\vec{p})] = e^{-\frac{\theta}{2}\hat{p}\cdot\vec{\sigma}} , \quad \bar{D}^{(\frac{1}{2})}[L(\vec{p})] = e^{\frac{\theta}{2}\hat{p}\cdot\vec{\sigma}} . \tag{2.34}$$

Using

$$(\hat{p} \cdot \vec{\sigma})^2 = 1$$

we have

$$e^{-\theta\hat{p}\cdot\vec{\sigma}} = \cosh\theta - (\hat{p}\cdot\vec{\sigma})\sinh\theta , \tag{2.35}$$

or recalling (2.14), we may write this also as

$$e^{-\theta\hat{p}\cdot\vec{\sigma}} = \frac{\omega(\vec{p}) - \vec{p}\cdot\vec{\sigma}}{mc} .$$

We thus conclude that

$$e^{-\frac{\theta}{2}\hat{p}\cdot\vec{\sigma}} = (e^{-\theta\hat{p}\cdot\vec{\sigma}})^{1/2} = \left(\frac{\omega(\vec{p}) - \vec{p}\cdot\vec{\sigma}}{mc}\right)^{1/2} .$$

Defining

$$t^\mu = (1, \vec{\sigma}) , \quad \bar{t}^\mu = (1, -\vec{\sigma})$$

we may thus write the matrices (2.34) in the compact form

$$D^{(\frac{1}{2})}[L(\vec{p})] = \left(\frac{t \cdot p}{mc}\right)^{1/2} , \quad \bar{D}^{(\frac{1}{2})}[L(\vec{p})] = \left(\frac{\bar{t} \cdot p}{mc}\right)^{1/2} . \tag{2.36}$$

These matrices will play a central role in our discussion of the Dirac equation in Chapter 4. Their explicit form is most easily obtained by returning to (2.35) and noting that

$$e^{-\frac{\theta}{2}\hat{p}\cdot\vec{\sigma}} = \cosh\frac{\theta}{2} - (\hat{p}\cdot\vec{\sigma})\sinh\frac{\theta}{2} .$$

Now,

$$\cosh\frac{\theta}{2} = \sqrt{\frac{1}{2}(\cosh\theta + 1)} = \sqrt{\frac{\omega(\vec{p}) + mc}{2mc}}$$

$$\sinh\frac{\theta}{2} = \sqrt{(\cosh^2\frac{\theta}{2} - 1)} = \sqrt{\frac{\omega(\vec{p}) - mc}{2mc}} = \sqrt{\frac{\omega(\vec{p}) + mc}{2mc}}\frac{|\vec{p}|}{\omega(\vec{p}) + mc} .$$

Hence

$$D^{(\frac{1}{2})}[L(\vec{p})] = \sqrt{\frac{\omega(p) + mc}{2mc}} \left(1 - \frac{\vec{p} \cdot \vec{\sigma}}{\omega(p) + mc} \right) . \tag{2.37}$$

Similarly

$$\bar{D}^{(\frac{1}{2})}[L(\vec{p})] = \sqrt{\frac{\omega(p) + m}{2m}} \left(1 + \frac{\vec{p} \cdot \vec{\sigma}}{\omega(p) + mc} \right) . \tag{2.38}$$

These explicit expressions will prove useful in our discussion of the Dirac equation in Chapter 4.

For the rest of these lectures we set $c = 1$.

2.5 Transformation properties of massive 1-particle states

In analogy to the Galilei transformations discussed in Chapter 1, we take $U[L(\vec{p})]$ to be the unitary operator taking the state $|s, \sigma >$ of a particle of spin $s, s_z = \sigma$ at rest into a 1-particle state of momentum $\vec{p}$.[9]

$$|\vec{p}, s, \sigma\rangle = \sqrt{\frac{m}{\omega(\vec{p})}} U[L(\vec{p})]|s, \sigma\rangle, \quad \omega(\vec{p}) = \sqrt{\vec{p}^2 + m^2} \tag{2.39}$$

where

$$|s, \sigma\rangle := |\vec{p} = 0, s, \sigma\rangle \tag{2.40}$$

with normalization

$$\langle s, \sigma'|s, \sigma\rangle = \delta_{\sigma'\sigma} . \tag{2.41}$$

Note that the spin of a particle at rest is a well defined quantity, whereas for a moving relativistic particle this is not the case. The kinematical factor introduced in (2.39) compensates for the non-covariant normalization of the 1-particle states:

$$\langle \vec{p}, s, \sigma'|\vec{p}, s, \sigma\rangle = \delta_{\sigma'\sigma}\delta^3(\vec{p}\,' - \vec{p}). \tag{2.42}$$

The form of this kinematical factor can be motivated in the following way: The normalization (2.42) of the 1-particle states corresponds to the completeness relation

$$\sum_\sigma \int d^3p |\vec{p}, s, \sigma\rangle\langle\vec{p}, s, \sigma| = 1 .$$

Now, d^3p is not a relativistically invariant integration measure, whereas $d^3p/\omega(\vec{p})$ is. Indeed, making use of the usual properties of the Dirac delta-function we have

$$\int \frac{d^3p}{2\omega(\vec{p})} f(\omega(\vec{p}), \vec{p}) = \int d^3p \int dp^0 \theta(p^0)\delta((p^0)^2 - \omega^2(\vec{p}))f(p^0, \vec{p}) .$$

---

[9]We follow again closely the notation of S. Weinberg, *Phys. Rev.* **133** (1964) B1318 and *Phys. Rev.* **134** (1964) B882. It is to be kept in mind that, unlike us, S. Weinberg uses the metric $g_{\mu\nu} = (-1, 1, 1, 1)$.

The delta-function insures the proper energy momentum relation for a free particle,

$$p^2 = m^2 \, ,$$

while the theta-function insures that the vector p^μ is time-like, that is, the particle has positive energy. Both properties are preserved by Lorentz transformations in $L_+^\uparrow$. Furthermore, d^4p is a Lorentz-invariant measure since

$$d^4p' = \left| \frac{\partial p'^\mu}{\partial p^\nu} \right| d^4p$$

and

$$\left| \frac{\partial p'^\mu}{\partial p^\nu} \right| = \det \Lambda = 1 \, .$$

We thus conclude that

$$\frac{d^3p'}{2\omega(\vec{p}\,')} = \frac{d^3p}{2\omega(\vec{p})} \, . \tag{2.43}$$

This explains roughly the origin of the kinematical factor in (2.39).[10] Now let $U[\Lambda]$ be the unitary operator inducing a Lorentz transformation on the 1-particle state $|\vec{p}, s, \sigma'\rangle$. Using the group property of Lorentz transformations $L_+^\uparrow$, we have

$$U[\Lambda]|\vec{p}, s, \sigma\rangle = \left(\frac{m}{\omega(\vec{p})} \right)^{1/2} U[\Lambda] U[L(\vec{p})]|s, \sigma\rangle$$

$$= \left(\frac{m}{\omega(\vec{p})} \right)^{1/2} U[L(\vec{\Lambda p})] U[L^{-1}(\vec{\Lambda p}) \Lambda L(\vec{p})]|s, \sigma\rangle \, . \tag{2.44}$$

It is easy to see that the matrix

$$R_W = L^{-1}(\vec{\Lambda p}) \Lambda L(\vec{p}) \tag{2.45}$$

is not equal to *one* unless Λ represents a pure boost colinear with $\vec{p}$. In general R_W represents a pure rotation — the so-called *Wigner rotation* — in the rest frame of the particle. We may thus make use of the completeness relation

$$\sum_\sigma |s, \sigma\rangle\langle s, \sigma| = 1 \, ,$$

valid in the rest frame of the particle in order to write (2.44) in the form

$$U[\Lambda]|\vec{p}, s, \sigma\rangle = \left(\frac{m}{\omega(\vec{p})} \right)^{1/2} \sum_{\sigma'} U[L(\vec{\Lambda p})]|s, \sigma'\rangle\langle s, \sigma'|U[R_W]|s, \sigma\rangle$$

$$= \left(\frac{m}{\omega(\vec{p})} \right)^{1/2} \sum_{\sigma'} U[L(\vec{\Lambda p})]|s, \sigma'\rangle D_{\sigma'\sigma}^{(s)}[R_W] \, ,$$

---

[10] For a more detailed analysis see E. P. Wigner, *Ann. Phys. Math.* **40** (1939) 149.

where we have made the identification

$$\langle s, \sigma' | R_W | s, \sigma \rangle = D^{(s)}_{\sigma'\sigma}[R_W]$$

with $D^{(s)}[R_W]$ a $(2s+1)$-dimensional irreducible representation of the rotation group. We thus finally have

$$U[\Lambda]|\vec{p}, s, \sigma\rangle = \sqrt{\frac{\omega(\vec{\Lambda p})}{\omega(\vec{p})}} \sum_{\sigma'} |\vec{\Lambda p}, s, \sigma'\rangle D^{(s)}_{\sigma'\sigma}[R_W] \,. \tag{2.46}$$

At this point we can now firmly establish the correctness of our choice of normalization factor in (2.46). To this end we start from the completeness relation

$$\sum_{\sigma=-s}^{\sigma=s} \int d^3p \, |\vec{p}, s, \sigma\rangle\langle\vec{p}, s, \sigma| = 1 \,, \tag{2.47}$$

and multiply this relation from the left with $U[\Lambda]$, and from the right with $U^{-1}[\Lambda]$:

$$\sum_{\sigma=-s}^{\sigma=s} \int d^3p \, U[\Lambda]|\vec{p}, s, \sigma\rangle\langle\vec{p}, s, \sigma| U^{-1}[\Lambda] = 1 \,.$$

We now make use of (2.46) in order to rewrite this relation as

$$\sum_{\sigma', \sigma''} \int d^3p \, \frac{\omega(\vec{\Lambda p})}{\omega(\vec{p})} |\vec{\Lambda p}, s, \sigma'\rangle \sum_{\sigma=-s}^{s} D^{(s)}_{\sigma'\sigma}[R_W] D^{(s)\star}_{\sigma''\sigma}[R_W] \langle\vec{\Lambda p}, s, \sigma''| = 1 \,.$$

Making use of the unitarity of the matrix representation of the rotation group, we have

$$\sum_{\sigma} D^{(s)}_{\sigma'\sigma}[R_W] D^{(s)\star}_{\sigma''\sigma}[R_W] = \delta_{\sigma'\sigma''} \,.$$

Hence we obtain from above

$$\sum_{\sigma} \int d^3p \, \frac{\omega(\vec{\Lambda p})}{\omega(\vec{p})} |\vec{\Lambda p}, s, \sigma\rangle\langle\vec{\Lambda p}, s, \sigma| = 1 \,.$$

Recalling the transformation property of the integration measure, Eq. (2.43), the above expression reduces to

$$\sum_{\sigma} \int d^3(\vec{\Lambda p}) |\vec{\Lambda p}, s, \sigma\rangle\langle\vec{\Lambda p}, s, \sigma| = 1 \,,$$

showing that our choice of normalization is consistent with the Lorentz covariance of the completeness relation.

2.6   Transformation properties of zero-mass 1-particle states

We are now in the position of discussing the Lorentz transformation properties of zero-mass particle states. The transformation rules have been completely worked out by E. Wigner.[11]

In the case of zero mass particles we can no longer go into the rest frame of the particle to define a general state in terms of a Lorentz boost. In fact, it is well known that a massless particle of spin j is polarized either along or opposite to its direction of motion, corresponding to two possible *helicity* states. If parity is not conserved, there may exist but one helicity state, as is exemplified by the neutrino (anti-neutrino) with *negative (positive)* helicity. Correspondingly we expect these helicity states to transform under a one-dimensional representation, independent of the spin of the particle. Following Wigner, we choose for our "standard" state a particle moving in the positive z-direction with four-momentum $\tilde{k}^\mu = |\kappa, 0, 0, \kappa >$, $(\tilde{k}^2 = 0)$ and helicity $\lambda : |\tilde{k}, \lambda\rangle$. These states replace the states $|s, \sigma\rangle$ in the massive case. Whereas the states $|s, \sigma\rangle$ belong to a representation of the *rotation group*, the helicity states $|\tilde{k}, \lambda\rangle$ furnish a representation of the *little* group, a subgroup of the Lorentz group consisting of all homogeneous proper Lorentz transformations leaving our standard 4-vector $\tilde{k}$ invariant.

In analogy to the massive case, we define the state of a massless particle of arbitrary momentum $\vec{p}$ by "boosting" the standard state $|\tilde{k}, \lambda >$ into the desired new state:

$$|\vec{p}, \lambda\rangle = \sqrt{\frac{\mu}{|\vec{p}|}} U[\mathcal{L}(\vec{p})]|\tilde{k}, \lambda\rangle \ , \tag{2.48}$$

where $U[\mathcal{L}(\vec{p})]$ is the unitary operator corresponding to the Lorentz transformation $\mathcal{L}(p)$ which takes our standard four-momentum $\tilde{k}^\mu$ into p^μ,

$$p^\mu = \mathcal{L}^\mu{}_\nu(\vec{p})\tilde{k}^\nu \ , \tag{2.49}$$

and μ in (2.48) is an arbitrary parameter with the dimensions of a mass. There are various ways of defining $\mathcal{L}(\vec{p})$; we shall make the choice[12]

$$\mathcal{L}^\mu{}_\nu(\vec{p}) = R^\mu{}_\lambda(\hat{p})B^\lambda{}_\nu(|\vec{p}|) \ . \tag{2.50}$$

Here $B^\lambda{}_\nu(|p|)$ is a "boost" along the z-axis with non-zero components (compare with (2.16)

$$B^1{}_1(|\vec{p}|) = B^2{}_2(|\vec{p}|) = 1$$
$$B^3{}_3(|\vec{p}|) = B^0{}_0(|\vec{p}|) = \cosh\phi(|\vec{p}|)$$
$$B^3{}_0(|\vec{p}|) = B^0{}_3(|\vec{p}|) = \sinh\phi(|\vec{p}|) \ .$$

---

[11] E.P. Wigner, *Theoretical Physics* (International Atomic Energy Vienna, 1963) p. 59.
[12] We follow again the notation of S. Weinberg, *Phys. Rev.* **134** (1964) B882.

To determine $\phi(|\vec{p}|)$ we observe that

$$B^{\mu}_{\nu}(|\vec{p}|)\tilde{k}^{\nu} = \kappa e^{\phi}\begin{pmatrix}1\\0\\0\\1\end{pmatrix} = \begin{pmatrix}|\vec{p}|\\0\\0\\|\vec{p}|\end{pmatrix}$$

so that

$$\phi(|\vec{p}|) = \ln\left(\frac{|\vec{p}|}{\kappa}\right). \tag{2.51}$$

We choose $R(\hat{p})$ as the rotation (say, in the plane containing $\vec{p}$ and the z-axis) into the unit vector $\hat{p}$. The kinematic factor $(\mu/|\vec{p}|)^{1/2}$ in (2.48) is inserted because of our choice of non-relativistic normalization of the states

$$\langle\vec{p}\,',\lambda'|\vec{p},\lambda\rangle = \delta^3(\vec{p}-\vec{p}\,')\delta_{\lambda\lambda'}.$$

In order to obtain the transformation law of the states (2.48) under a general Lorentz transformation, we now proceed in a way analogous to that followed in Section 2.5. We thus have

$$U[\Lambda]|\vec{p},\lambda> = \sqrt{\frac{\mu}{|\vec{p}|}}U[\Lambda]U[\mathcal{L}(\vec{p})]|\tilde{k},\lambda\rangle \tag{2.52}$$

$$= \sqrt{\frac{\mu}{|\vec{p}|}}U[\mathcal{L}(\vec{\Lambda p})]U[\mathcal{L}^{-1}(\vec{\Lambda p})\Lambda\mathcal{L}(\vec{p})]|\tilde{k},\lambda\rangle.$$

But the transformation $\mathcal{L}^{-1}(\vec{\Lambda p})\Lambda\mathcal{L}(\vec{p})$ leaves our standard vector $(\kappa,0,0,\kappa)$ invariant and hence belongs to the *little* group. Indeed,

$$\mathcal{L}^{-1}(\vec{\Lambda p})\Lambda\mathcal{L}(\vec{p})\tilde{k} = \mathcal{L}^{-1}(\vec{\Lambda p})\Lambda p = \tilde{k}. \tag{2.53}$$

Hence the four-by-four matrix

$$\tilde{\Lambda} = \mathcal{L}^{-1}(\vec{\Lambda p})\Lambda\mathcal{L}(\vec{p}) \tag{2.54}$$

belongs to the "little group" of the Lorentz Group. The corresponding unitary operator $U[\tilde{\Lambda}]$ thus also does not change the momentum of the state $|\tilde{k},\lambda\rangle$, and must induce the following linear transformation on the massless 1-particle state $|\tilde{k},\lambda>$:

$$U[\tilde{\Lambda}]|\tilde{k},\lambda\rangle = \sum_{\lambda'}|\tilde{k},\lambda'\rangle d_{\lambda'\lambda}[\tilde{\Lambda}], \quad \tilde{\Lambda}^{\mu}_{\ \nu}\tilde{k}^{\nu} = \tilde{k}^{\mu},$$

where $d[\tilde{\Lambda}]$ is an irreducible representation of the *little* group (compare with (2.46)). Correspondingly we have from (2.52)

$$U[\Lambda]|\vec{p},\lambda> = \sqrt{\frac{\mu}{|\vec{p}|}}\sum_{\lambda'}|\vec{\Lambda p},\lambda'> d_{\lambda'\lambda}[\tilde{\Lambda}]. \tag{2.55}$$

In order to obtain the representation matrices of the *little* group, we must examine the nature of the transformations, leaving our standard vector $\tilde{k}^{\mu} = (\kappa,0,0,\kappa)$

invariant. Since the *little* group is a subgroup of the Lorentz group, the representation matrices $d[\tilde{\Lambda}]$ are obtained as a special case of the representation matrices for a general Lorentz transformation, discussed in Section 2.4.

It suffices to look at an infinitesimal transformation of the form

$$\tilde{\Lambda}^{\mu}_{\ \nu} \approx g^{\mu}_{\ \nu} + \tilde{\omega}^{\mu}_{\ \nu} \quad , \quad \tilde{\omega}^{\mu\nu} = -\tilde{\omega}^{\nu\mu} \ ,$$

where the infinitesimal parameters are now required to satisfy

$$\tilde{\omega}^{\mu}_{\ \nu} \tilde{k}^{\nu} = 0 \ . \tag{2.56}$$

Inspection of (2.56) shows that $\Omega^{\mu}_{\ \nu}$ is in general a function of three parameters θ, χ_1 and χ_2 with the *non-zero* components given by

$$\tilde{\omega}_{12} = \theta$$
$$\tilde{\omega}_{01} = -\tilde{\omega}_{31} = \chi_1 \ , \ \tilde{\omega}_{02} = \tilde{\omega}_{23} = \chi_2 \tag{2.57}$$

or

$$(\tilde{\Lambda}^{\mu}_{\ \nu}) \approx \begin{pmatrix} 0 & \chi_1 & \chi_2 & 0 \\ \chi_1 & 0 & -\theta & -\chi_1 \\ \chi_2 & \theta & 0 & -\chi_2 \\ 0 & \chi_1 & \chi_2 & 0 \end{pmatrix} \ . \tag{2.58}$$

The unitary operator acting on the states is correspondingly given by

$$U[\tilde{\Lambda}] = e^{\frac{i}{2}\tilde{\omega}_{\mu\nu}\mathcal{M}^{\mu\nu}} \ ,$$

where $\mathcal{M}^{\mu\nu}$ are the generators of Lorentz transformations satisfying the Lie algebra (2.21). Using (2.22) we then have for an infinitesimal transformation,

$$U[\tilde{\Lambda}] \approx 1 + i\theta J_3 + i\chi_1(K_1 - J_2) + i\chi_2(K_2 + J_1) \ ,$$

or using (2.27),

$$U[\tilde{\Lambda}] \approx 1 + i\theta J_3 - (\chi_1 + i\chi_2)A_- + (\chi_1 - i\chi_2)B_+ \ , \tag{2.59}$$

where $J_3 = A_3 + B_3$, and where A_- and B_+ stand for

$$A_- = A_1 - iA_2 \quad , \quad B_+ = B_1 + iB_2 \ .$$

Since

$$[J_3, A_-] = -A_- \quad , \quad [J_3, B_+] = B_+$$

we see that B_+ and A_- act as *raising* and *lowering* operators for the eigenvalues of J_3, respectively. The eigenvalues of the helicity operator $J_3 = A_3 + B_3$ of a general state $|A, a; B, b>$ are $\lambda = a + b$. In nature a massless spin j particle only exists in two helicity states with helicity $\lambda = +j$ (right handed) or $\lambda = -j$ (left handed). For a transformation of the little group not to change this helicity we therefore demand for such an helicity state

$$A_-|\lambda> = 0 \ , \quad B_+|\lambda> = 0 \ .$$

This leads to the identification

$$|\lambda> = |A, -A; B, B> \quad \text{with} \quad \lambda = B - A .$$

A spin-j massless particle thus transforms under the $(0, j)$, $(\frac{1}{2}, j + \frac{1}{2})$, $\cdots$ (right handed) or $(j, 0)$, $(j + \frac{1}{2}, j), \cdots$ (left handed) representation of the little group. It then follows from (2.59) by exponentiation, that (A_- and B_+ annihilate the state)

$$d_{\lambda\lambda'}[\tilde{\Lambda}] = \langle \tilde{k}, \lambda' | U[\mathcal{L}^{-1}(\vec{\Lambda p})\Lambda^{-1}\mathcal{L}(\vec{p})] | \tilde{k}, \lambda \rangle = \delta_{\lambda'\lambda} \exp\{i\lambda\Theta[\mathcal{L}^{-1}(\vec{\Lambda p})\Lambda\mathcal{L}(\vec{p})]\} ,$$
$$(2.60)$$

the phase $\Theta(\theta, \chi_1, \chi_2)$ being some more or less complicated function of the *little group parameters*, which reduces to θ in (2.57) for infinitesimal transformations. It must satisfy the group property

$$\Theta[\tilde{\Lambda}_1 \tilde{\Lambda}_2] = \Theta[\tilde{\Lambda}_1] + \Theta[\tilde{\Lambda}_2] . \tag{2.61}$$

Hence finally we have from (2.55),

$$U[\Lambda]|\vec{p}, \lambda\rangle = \sqrt{\frac{|\vec{\Lambda p}|}{|\vec{p}|}} e^{i\lambda\Theta[\tilde{\Lambda}]}|\vec{\Lambda p}, \lambda\rangle, \quad \tilde{\Lambda} = \mathcal{L}^{-1}(\vec{\Lambda p})\Lambda\mathcal{L}(\vec{p})) . \tag{2.62}$$

This result will play an important role when we proceed to discuss the quantization of the electromagnetic field in Chapter 7.

Chapter 3

Search for a Relativistic Wave Equation

In this chapter we engage in the search for a relativistic wave equation for a *spin zero* particle moving in an external potential, reducing to the ordinary Schrödinger equation in the non-relativistic limit. We begin by looking at the case of a free particle. We adhere to the familiar quantum mechanical principles, as long as we can. We shall encounter a number of difficulties which will lead us to eventually abandon the usual probability interpretation.

3.1 A relativistic Schrödinger equation

As discussed in Chapter 1, the time evolution of a quantum mechanical state is governed by the equation of motion ($\hbar = c = 1$)

$$i\partial_t |\Psi(t) >= H|\Psi(t) > ,$$

where H is the Hamiltonian operator of the system obtained from the corresponding classical Hamiltonian by representing the canonically conjugate dynamical variables q and p in $H(q,p)$ by operators satisfying canonical commutation relations. In this and the following two sections we consider the case of a free particle. Since in a relativistic theory the relation between the energy of a free particle and its momentum is given by $E = \sqrt{\vec{p}^2 + m^2}$, it is natural to take for the Hamiltonian

$$H = \sqrt{\vec{p}^2 + m^2} .$$

In the coordinate representation, the canonical commutation relations are as usual realized by

$$\vec{p} \to \frac{\hbar}{i}\vec{\nabla} .$$

Defining the wave function associated with the state $|\Psi >$ by $\phi(\vec{r},t)$, one arrives thus at the wave equation

$$i\partial_t \phi(\vec{r},t) = \sqrt{-\vec{\nabla}^2 + m^2}\phi(\vec{r},t) . \tag{3.1}$$

This equation in turn implies that $\phi(\vec{r}, t)$ is also a solution of the Klein–Gordon (KG) equation

$$(\Box + m^2)\phi(\vec{r}, t) = 0 \ , \tag{3.2}$$

where $\Box$ denotes the D'Alembert operator. Since this operator is manifestly Lorentz-invariant, covariance of the physical laws demand that under a Lorentz transformation the wave function transforms like a scalar:

$$\phi'(\vec{r}', t') = \phi(\vec{r}, t) \ . \tag{3.3}$$

This is also implied by Eq. (3.1), since *on the space of solutions* of the Klein–Gordon equation (3.2) the operator appearing on the left- and right-hand sides of Eq. (3.1) transform in the same way:

$$\sqrt{-\vec{\nabla}^2 + m^2} = \sqrt{(\Box + m^2)^2 - \partial_0^2} \to i\partial_0 \ .$$

Note that in contrast to the non-relativistic case, where the wave function does not transform like a scalar, but rather picks up a phase under Galilei transformations, such a phase is absent in the relativistic case, rendering the covariance of the equation of motion manifest in this case.

The operator appearing on the rhs of Eq. (3.1) is defined in terms of its Taylor expansion:

$$\sqrt{-\vec{\nabla}^2 + m^2} = m - \frac{\vec{\nabla}^2}{2m} + \cdots$$

Hence its eigenfunctions are given by the plane waves $e^{i\vec{p}\cdot\vec{r}}$, with the corresponding eigenvalues $\sqrt{\vec{p}^2 + m^2}$. As a consequence, the action of this operator on any absolute integrable function $f(\vec{r}, t)$ in $\mathbf{R}^3$ is defined via its Fourier transform

$$\phi(\vec{r}, t) = \frac{1}{(2\pi)^{3/2}} \int \frac{d^3p}{2\omega(\vec{p})} \tilde{\phi}(\vec{p}, t) e^{i\vec{p}\cdot\vec{r}}$$

as

$$\sqrt{-\vec{\nabla}^2 + m^2}\phi(\vec{r}, t) = \int \frac{d^3p}{2\omega(\vec{p})} \sqrt{\vec{p}^2 + m^2}\tilde{\phi}(\vec{p}, t) e^{i\vec{p}\cdot\vec{r}} \ .$$

The factor $1/2\omega(\vec{p})$ has been included in the measure to make it relativistically invariant. In order for $\phi(\vec{r}, t)$ to be a solution of Eq. (3.1), we must further have

$$\tilde{\phi}(\vec{p}, t) = a(\vec{p})e^{-i\omega(\vec{p})t}$$

where $\omega(p) = \sqrt{\vec{p}^2 + m^2}$. Hence, the most general solution to Eq. (3.2) has the form

$$\phi(\vec{r}, t) = \frac{1}{(2\pi)^{3/2}} \int \frac{d^3p}{2\omega(\vec{p})} a(\vec{p})e^{i(\vec{p}\cdot\vec{r}-\omega(\vec{p})t)} \ . \tag{3.4}$$

In order for $\phi(\vec{r}, t)$ to be a Lorentz *scalar*, the Fourier coefficients will themselves have to be Lorentz scalars:

$$a'(\vec{p}') = a(\vec{p}) \ . \tag{3.5}$$

In the non-relativistic limit

$$\omega(\vec{p}) = m + \frac{\vec{p}^2}{2m} + O(\vec{p}^4) \ ,$$

so that we recover the Schrödinger 1-particle wave function of the non-relativistic theory. Since the argument of the exponential appearing in the Fourier integral is not Galilei-invariant, the corresponding non-relativistic field $\phi_{nr}(\vec{r},t)$ does not transform as a scalar. In fact, each Fourier component will pick up a phase such as witnessed in Eq. (1.1) of Chapter 1. We therefore see that in this respect the transformation law in the relativistic case is simpler than in the case of Galilei transformations.

3.2 Difficulties with the wave equation

As we now show, the wave equation (3.1) presents a number of problems which will eventually lead us to abandon it.

Locality

The wave equation (3.1) has an unwanted property: In order to determine the change in the solution at the point $\vec{r}$ in an infinitesimal time interval $(t, t + dt)$, we must know the function for time t at *all* points $\vec{r}$. This property is contained in the Taylor expansion of the Hamilton operator in powers of the Laplacian and is referred to as *non-locality*. It is also evident from the form of the solution (3.4) by noting that

$$\phi(\vec{r}, t + dt) \simeq \phi(\vec{r}, t) - \frac{idt}{(2\pi)^{3/2}} \int \frac{d^3 p}{2\omega(\vec{p})} \sqrt{\vec{p}^2 + m^2} a(\vec{p}) e^{i(\vec{p}\cdot\vec{r} - \omega(\vec{p})t)} + \cdots$$

which upon using the Taylor expansion in p^2 becomes

$$\phi(\vec{r}, t + dt) \simeq \phi(\vec{r}, t) - idt \left[m - \frac{\vec{\nabla}^2}{2m} + \cdots \right] \phi(\vec{r}, t) + \cdots$$

Probability interpretation

In order for the wave function to have the interpretation of a probability amplitude, it should have at least the following properties:

(i) Its *normalization* with respect to some suitable integration measure should be independent of time, as well as of the choice of inertial frame.

(ii) The associated probability *density* should be *positive semi-definite*.

The second condition is satisfied by defining the probability density by

$$\mathcal{P}(\vec{r}, t) = |\phi(\vec{r}, t)|^2 \ .$$

Furthermore, adopting the following definition for the scalar product

$$(\varphi, \chi)_t = \int d^3 r \varphi^*(\vec{r}, t) \chi(\vec{r}, t) ,$$

and making use of the equation of motion (3.1), one finds that the normalization of the wave function is preserved in time:

$$\frac{d}{dt}(\phi, \phi)_t = 0 .$$

The probability of finding a particle anywhere in space should, however, also be independent of the choice of inertial frame. Since the wave function $\phi(\vec{r}, t)$ transforms like a scalar (see Eq. (3.3)) this is not the case due to the Lorentz non-invariance of the measure $d^3 r$.

For the case of a *free* particle this defect is easily repaired by adopting a new definition for the scalar product:

$$(\varphi, \chi) = i \int d^3 r (\varphi^* \partial_0 \chi - \partial_0 \varphi^* \cdot \chi) =: i \int d^3 r \varphi^* \overset{\leftrightarrow}{\partial_0} \chi . \tag{3.6}$$

Indeed, making use of the Fourier decomposition (3.4) for a free particle wave function, one computes

$$(\phi, \phi) = \int \frac{d^3 p}{2\omega(\vec{p})} |a(\vec{p})|^2 .$$

Because of the invariance property (2.43) of the integration measure, and the scalar property (3.5) of the Fourier amplitude, we conclude that the normalization of the wave function is a Lorentz invariant with respect to the scalar product (3.6). In fact, one easily shows that it has all the properties expected from a scalar product:

$$(\varphi, \chi)_t = (\varphi, \chi)_t^*$$
$$(\varphi_1 + \varphi_2, \chi)_t = (\varphi_1, \chi)_t + (\varphi_2, \chi)_t ,$$

and

$$(\phi, \phi)_t > 0$$

for solutions of the equation of motion. The scalar product $(\phi, \phi)_t$ is also time-independent, as one easily checks using the equation of motion (3.2):

$$\begin{aligned}
\frac{d}{dt}(\phi, \phi) &= i \int_V d^3 r \left(\phi^* \partial_0^2 \phi - (\partial_0^2 \phi^*) \phi \right) \\
&= i \int_V d^3 r \left(\phi^* \vec{\nabla}^2 \phi - (\vec{\nabla}^2 \phi^*) \phi \right) \\
&= i \int_V d^3 r \vec{\nabla} \cdot \left(\phi^* \vec{\nabla} \phi - (\vec{\nabla} \phi^*) \phi \right) \\
&= i \int_{\partial V} d\vec{f} \cdot \left(\phi^* \vec{\nabla} \phi - (\vec{\nabla} \phi^*) \phi \right) \to 0
\end{aligned}$$

for $\partial V \to \infty$ and $\phi \to 0$ at infinity. In fact, the probability density satisfies a continuity equation analogous to that in non-relativistic Quantum Mechanics:

$$\partial_0 \mathcal{P}(\vec{r}, t) + \vec{\nabla} \cdot \vec{j}(\vec{r}, t) = 0 \ ,$$

where

$$\mathcal{P} = i\phi^* \overset{\leftrightarrow}{\partial_0} \phi, \quad \vec{j} = -i(\phi^* \vec{\nabla}\phi - (\vec{\nabla}\phi^*)\phi) \ .$$

We may in fact collect $\mathcal{P}(\vec{r}, t)$ and $j(\vec{r}, t)$ into a 4-vector as follows:

$$j^\mu(\vec{r}, t) = (\mathcal{P}(\vec{r}, t), \vec{j}(\vec{r}, t)) = i(\phi^* \overset{\leftrightarrow}{\partial^\mu} \phi)$$

where $\overset{\leftrightarrow}{\partial^\mu} = \vec{\partial}^\mu - \overleftarrow{\partial}^\mu$. Indeed, under a Lorentz transformation we have (compare with (2.5))

$$j'^\mu(x') = i\phi'^*(x') \overset{\leftrightarrow}{\partial'^\mu} \phi'(x')$$
$$= i\phi^*(x)\Lambda^\mu_{\ \nu} \overset{\leftrightarrow}{\partial^\nu} \phi(x) = \Lambda^\mu_{\ \nu} j^\nu(x) \ .$$

Although we have succeeded in satisfying the requirement (i), this is not the case as far as requirement (ii) is concerned. As a simple example shows, the probability *density* is not a positive semi-definite quantity. To see this, consider the superposition of two plane waves for two free particles of mass m:

$$\phi = ce^{-ik\cdot x} + de^{-iq\cdot x}, \quad k^2 = q^2 = m^2$$

with $k \cdot x = k_\mu x^\mu$, etc.

Let c and d be real. One then has

$$\phi^* i \overset{\leftrightarrow}{\partial^0} \phi = 2c^2\omega_k + 2d^2\omega_q + 2cd(\omega_k + \omega_q)\cos((k - q) \cdot x) \ .$$

Now, for every 4-vector $(k - q)^\mu$ there exists a vector x^μ such that $(k - q) \cdot x = 0$. For this vector

$$x^0 = \frac{(\vec{k} - \vec{q}) \cdot \vec{x}}{\omega_k - \omega_q} \ .$$

We may then rewrite the above expression as

$$\phi^* i \overset{\leftrightarrow}{\partial^0} \phi = (c, d) \begin{pmatrix} 2\omega_k & \omega_k + \omega_q \\ \omega_k + \omega_q & 2\omega_q \end{pmatrix} \begin{pmatrix} c \\ d \end{pmatrix} \ .$$

The matrix appearing here is hermitian, and may thus be diagonalized, its eigenvalues being

$$\lambda_\pm = (\omega_k + \omega_q) \pm \sqrt{(\omega_k - \omega_q)^2 + (\omega_k + \omega_q)^2} \ .$$

Hence, though the *total* probability is positive, the density in space-time is not. Since $\lambda_- < 0$, there always exist coefficients such that $\phi^* i \overset{\leftrightarrow}{\partial_0} \phi < 0$, which proves our claim.

3.3 The Klein–Gordon equation

One tentative way out to save Lorentz covariance in the presence of interaction would be to treat space and time from the outset on equal footing by working with a differential equation of second order in space as well as time. This equation should nevertheless contain the solution discussed previously. Noting that

$$(i\partial_t + \sqrt{-\vec{\nabla}^2 + m^2})(-i\partial_t + \sqrt{-\vec{\nabla}^2 + m^2}) = \Box + m^2 \tag{3.7}$$

one is thus led in the absence of interaction, to the Klein–Gordon equation

$$(\Box + m^2)\phi = 0 . \tag{3.8}$$

The general solution of this equation is now given by

$$\phi(x) = \phi^{(+)}(x) + \phi^{(-)}(x) \tag{3.9}$$

where

$$\phi^{(\pm)}(\vec{r},t) = \frac{1}{(2\pi)^{3/2}} \int \frac{d^3k}{2\omega_k} a^{(\pm)}(\vec{k}) e^{i(\vec{k}\cdot\vec{r}\mp\omega_k t)} .$$

The solution (3.9) thus represents in general two wave packets moving away from each other with time. If we choose the Fourier amplitudes $a^{(+)}(k)$ and $a^{(-)}(k)$ to be concentrated around $\vec{k} = \vec{k}_0$, then these two wave packets will separate from each other with twice the group velocity

$$v_g = \left(\frac{\partial\omega(\vec{k})}{\partial|\vec{k}|} \right)_{\vec{k}=\vec{k}_0} = \frac{|\vec{k}_0|}{\omega(\vec{k}_0)} ,$$

which could be interpreted as a two-particle state. Since

$$i\partial_t\phi^{(-)} = -H_0\phi^{(-)}, \quad H_0 = \sqrt{-\vec{\nabla}^2 + m^2} ,$$

we may regard the solution $\phi^{(-)}$ as solutions of the Schrödinger equation for *negative* energy. These negative energy solutions do not fit into our probabilistic interpretation, since with the scalar product (3.6),

$$(\phi^{(-)}, \phi^{(-)}) = \int d^3r \phi^{(-)*}(\vec{r},t) \overset{\leftrightarrow}{\partial}_0 \phi^{(-)}(\vec{r},t) = -\int \frac{d^3k}{2\omega_k} |a^{(-)}(k)|^2 < 0 .$$

The negative energy solutions of the KG equation thus carry negative norm with respect to the scalar product (3.6). In the free case we may nevertheless ignore their existence, since they satisfy the orthogonality property

$$(\phi^{(+)}, \phi^{(-)}) = 0$$

and as a result we have no mixing of positive and negative energy solutions:

$$(\phi, \phi) = \int \frac{d^3k}{2\omega_k} (|a^{(+)}(\vec{k})|^2 - |a^{(-)}(\vec{k})|^2) .$$

Thus we may restrict ourselves to the *positive energy sector* of the theory. This will, however, no longer be true if we allow for interactions with an external potential, which will induce transitions between positive and negative energy states.

3.4   KG equation in the presence of an electromagnetic field

The interaction of a charge q with an external electromagnetic field is introduced in the Klein–Gordon equation by the usual minimal substitution

$$\partial_\mu \to \partial_\mu + iqA_\mu, \tag{3.10}$$

where $A^\mu = (\Phi, \vec{A})$ is related to the electric field $\mathbf{E}$ in the usual way:

$$\vec{E} + \partial_t \vec{A} = -\vec{\nabla}\Phi \;.$$

This leads us to consider the equation of motion

$$(D_\mu D^\mu + m^2)\phi = 0 \tag{3.11}$$

with the covariant derivative

$$D_\mu = \partial_\mu + iqA_\mu. \tag{3.12}$$

It is important to realize that unlike the free particle case, this equation can no longer be factorized in the form (3.7):

$$(D_\mu D^\mu + m^2) \neq (H + i\partial_0)(H - i\partial_0) \;,$$

where H is the Hamiltonian for a relativistic particle moving in an external electromagnetic field:

$$H = \sqrt{-(\vec{\partial} + iq\vec{A})^2 + m^2} + q\Phi$$

Eq. (3.11) is covariant under the following gauge transformation

$$A_\mu \to A'_\mu = A_\mu + \partial_\mu \Lambda \;, \tag{3.13}$$

where $\Lambda(x)$ denotes an arbitrary function of x. Indeed, under this transformation

$$D_\mu \to D'_\mu = \partial_\mu + iqA_\mu + iq\partial_\mu \Lambda \;,$$

or equivalently

$$D_\mu \to D'_\mu = e^{-iq\Lambda(x)} D_\mu e^{iq\Lambda(x)} \;. \tag{3.14}$$

In particular

$$D'^\mu D'_\mu = e^{-iq\Lambda(x)} D^\mu D_\mu e^{iq\Lambda(x)} \;.$$

Hence defining the gauge-transformed wave function $\phi'(x)$ by

$$\phi(x) \to \phi'(x) = e^{-iq\Lambda(x)}\phi(x) \tag{3.15}$$

Eq. (3.14) implies

$$(D'^\mu D'_\mu + m^2)\phi'(x) = 0 \;. \tag{3.16}$$

The transformation law (3.15) can be restated in the following way: The wave function $\phi(x)$ is a *functional* of the vector potential $A^\mu(x)$:

$$\phi(x) = \phi(x; A^\mu) .$$

The transformation law (3.15) for the covariant derivative then implies that under the gauge transformation (3.13) the functional $\phi(x; A^\mu)$ transforms as follows:

$$\phi(x; A_\mu + \partial_\mu \Lambda) = e^{-iq\Lambda(x)} \phi(x; A_\mu) . \tag{3.17}$$

The gauge covariance of the equation of motion (3.11) allows us to choose in particular the covariant Lorentz gauge $\partial \cdot A = 0$. In this gauge the 4-tuplet $A^\mu = (A^0, \vec{A})$ transforms like a 4-vector. This demonstrates the manifest Lorentz covariance of the equation of motion (3.16) in the *Lorentz gauge*.

Negative energy solutions and antiparticles

Consider the case where the vector potential is independent of time. In that case there exist stationary solutions

$$\phi^{(\pm)}(x) = \varphi^{(\pm)}(\vec{r}) e^{\mp iEt}, \quad E > 0$$

suggesting again a two-particle interpretation for a general wave packet. Labelling these solutions by the charge q appearing in the covariant derivative (3.12), substitution into Eq. (3.11) leads to the equations

$$[m^2 - E^2 - \vec{\nabla}^2 + 2qEA^0 + iq(\vec{\nabla} \cdot \vec{A}) + 2iq\vec{A} \cdot \vec{\nabla} - q^2 A_\mu A^\mu] \varphi^{(+)}(\vec{r}) = 0$$
$$[m^2 - E^2 - \vec{\nabla}^2 - 2qEA^0 + iq(\vec{\nabla} \cdot \vec{A}) + 2iq\vec{A} \cdot \vec{\nabla} - q^2 A_\mu A^\mu] \varphi^{(-)}(\vec{r}) = 0 ,$$

where we have set $\Phi = A^0$. From here we see that

$$\phi^{(-)}(\vec{r}; q) = \phi^{(+)*}(\vec{r}; -q) . \tag{3.18}$$

This suggests the identification of the "negative energy" solution $\phi^{(-)}$ with the respective antiparticle. The transformation (3.18) for scalar fields is referred to as *charge conjugation*.

Probability interpretation

In the free-particle case, we attempted to identify the probability density with the time component of a conserved 4-vector-current. It was found to satisfy all requirements, provided we restricted ourselves to positive energy solutions from the outset. As we now show, this will no longer be possible in the presence of an electromagnetic interaction, which will invariably lead to transitions to states involving negative energy solutions.

Following the general line of approach adopted in the free particle case, we note to begin with that we can define again a conserved current by

$$j_\mu(x; A) = i[\phi^*(x) D_\mu \phi(x) - (D_\mu \phi(x))^* \phi(x)] .$$

This current is gauge-invariant

$$j_\mu(x; A + \partial A) = i[\phi'^*(x) D'_\mu \phi'(x) - (D'_\mu \phi'(x))^* \phi'(x)]$$
$$= j_\mu(x; A)$$

and thus defines a Lorentz covariant *observable*.

With the aid of the equation of motion (3.11) one easily checks that this current is conserved:

$$\partial_\mu j^\mu(x; A) = 0 \ .$$

Since j^μ transforms like a 4-vector density, we take its zero component to define the probability density:

$$\mathcal{P}(x; A) \equiv j^0(x; A) = i\phi^*(x)(\overset{\leftrightarrow}{\partial^0} + 2iqA^0)\phi(x) \ . \tag{3.19}$$

Unfortunately this definition of the probability density already violates positivity for "positive energy" solutions; indeed, consider the stationary wave function

$$\phi(x) = \varphi^{(+)}(\vec{r})e^{-iEt}.$$

Substitution into (3.19) yields

$$\mathcal{P}(x; A) = 2(E - q\Phi)|\varphi^{(+)}(\vec{r})|^2 \ .$$

Hence, even if $E > 0$, this density is not positive semi-definite, since the sign of the Coulomb potential can be either positive or negative.

The above considerations lead us to abandon at this stage our search for a relativistic scalar wave equation conforming to the principles of non-relativistic quantum mechanics. We shall, however, return to the field equation (3.11) after having learned in Chapter 7 to interpret $\phi(x)$ as an *operator-valued* field acting on a Hilbert space of *Fock states*.

Chapter 4

The Dirac Equation

We begin this chapter by obtaining the relativistic "Schrödinger" equation for a free spin-1/2 field by following first the historical approach, and then presenting a derivation based on Lorentz covariance and space-time parity alone. This leads us to a four-component wave equation which is first-order in space and time coordinates. We present the solution of this equation for three choices of basis: The *Dirac*, *Weyl* (or chiral), and *Majorana* representations. The latter representation is shown to be particularly useful for the case of *Majorana fermions*, i.e. fermions which are their own anti-particles. We show that the Dirac equation allows for the notion of a probability density after suitable interpretation of the negative energy states.

4.1 Dirac spinors in the Dirac and Weyl representations

In this section we present the derivation of the Dirac equation by following the historical path, as well as a purely group-theoretical approach relying on Lorentz transformation properties alone. We then obtain the general solution of these equations in terms of the four independent Dirac spinors.

Dirac equation: historical derivation

Since in a manifestly Lorentz-covariant wave equation, space and time variables should appear on equal footing, Dirac demanded that the hamiltonian in the equation

$$\partial_t \psi(\vec{r}, t) = H\psi(\vec{r}, t) \tag{4.1}$$

should depend linearly on the momentum $\vec{p}$ canonically conjugate to $\vec{r}$. This led him to the Ansatz[1]

$$i\partial_t \psi = (-i\vec{\alpha} \cdot \vec{\nabla} + m\beta)\psi . \tag{4.2}$$

---

[1]P.A.M. Dirac, *The Principles of Quantum Mechanics*, 4th edn. (Oxford University Press, Oxford 1958).

The triple $\vec{\alpha} = (\alpha_1, \alpha_2, \alpha_3)$ and β in this equation cannot be just numbers, since this would already be inconsistent with rotational covariance. Hence they are expected to be given by matrices. These matrices must be hermitian, in order to warrant the hermiticity of the Hamilton operator. Furthermore, Eq. (4.2) should lead to the correct relation between energy and momentum for free particles. In order to see what this implies, we differentiate Eq. (4.2) with respect to time, thus obtaining

$$-\partial_t^2 \psi = [-(\vec{\alpha} \cdot \vec{\nabla})^2 + m^2\beta^2 - im(\beta\alpha_i + \alpha_i\beta)\partial_i]\psi$$
$$= [-\frac{1}{2}\{\alpha_i, \alpha_j\}\partial_i\partial_j + m^2\beta^2 - im\{\beta, \alpha_i\}\partial_i]\psi .$$

Here the bracket $\{A, B\}$ denotes the anticommutator of two objects:

$$\{A, B\} = AB + BA .$$

In order to get the desired energy momentum relation, this equation has to reduce to the Klein–Gordon equation, which is the case if

$$\{\alpha_i, \alpha_j\} = 2\delta_{ij} , \quad \{\beta, \alpha_i\} = 0, \quad \beta^2 = 1 . \tag{4.3}$$

From here we deduce the following properties of the matrices:

Tracelessness

Since $\beta^2 = \alpha_i^2 = 1$, it follows from $\{\beta, \alpha_i\} = 0$ that

$$tr\alpha_i = tr\beta^2\alpha_i = tr(\beta\alpha_i\beta) = -tr\alpha_i ,$$
$$tr\beta = tr\alpha_i^2\beta = tr(\alpha_i\beta\alpha_i) = -tr\beta ,$$

or

$$tr\,\alpha_i = 0, \quad tr\,\beta = 0.$$

Dimensionality

Since $\alpha_i^2 = \beta^2 = 1$, the eigenvalues of α_i and β are either $+1$ or -1. From the tracelessness of the matrices it then follows that the dimension of the matrices must be *even*.

Minimal dimension

The Pauli matrices

$$\sigma_1 = \begin{pmatrix} 0 & 1 \\ 1 & 0 \end{pmatrix} , \quad \sigma_2 = \begin{pmatrix} 0 & -i \\ i & 0 \end{pmatrix} , \quad \sigma_3 = \begin{pmatrix} 1 & 0 \\ 0 & -1 \end{pmatrix} , \tag{4.4}$$

together with the identity matrix **1** represent a complete basis for 2×2 *hermitian* matrices. Of these, the Pauli matrices satisfy the first of the conditions (4.3); however, the identity matrix cannot be identified with β, since $tr\beta = 0$. Since the dimension of the matrices must be even, we conclude that the dimension of these matrices must be at least *four*.

The following 4×4 matrices satisfy all the requirements (4.3):

$$\alpha_i = \begin{pmatrix} 0 & \sigma_i \\ \sigma_i & 0 \end{pmatrix} \quad , \quad \beta = \begin{pmatrix} 1 & \\ & -1 \end{pmatrix} . \tag{4.5}$$

The same applies of course to matrices obtained from the above ones via a *unitary* transformation (unitary, in order to preserve the hermiticity of the matrices). For the choice of basis (4.5), the equation reads

$$i\partial_t \psi = \begin{pmatrix} m & -i\vec{\sigma} \cdot \vec{\nabla} \\ -i\vec{\sigma} \cdot \vec{\nabla} & -m \end{pmatrix} \psi .$$

We can compactify the notation by introducing the definitions

$$\gamma_D^0 = \beta, \quad \gamma_D^i = \beta\alpha_i, \quad \{\gamma_D^\mu, \gamma_D^\nu\} = 2g^{\mu\nu} , \tag{4.6}$$

where the subscript D stands for "Dirac representation". Explicitly we have

$$\gamma_D^0 = \begin{pmatrix} 1 & 0 \\ 0 & -1 \end{pmatrix}, \quad \gamma_D^i = \begin{pmatrix} 0 & \sigma_i \\ -\sigma_i & 0 \end{pmatrix}, \quad i = 1, 2, 3. \tag{4.7}$$

We may collect these matrices into a 4-tuplet γ_D^μ. This notation is justified since we shall show later that these matrices "transform" (in a sense to be made precise later) under Lorentz transformations as a "4-vector". In terms of the matrices (4.6) the Dirac equation takes the compact form[2]

$$i\gamma^0 \partial_0 \psi = (m - i\gamma^i \partial_i)\psi ,$$

$$(i\gamma^\mu \partial_\mu - m)\psi = 0, \quad \{\gamma^\mu, \gamma^\nu\} = 2g^{\mu\nu} . \tag{4.8}$$

This equation implies that $\psi(r,t)$ is also a solution of the Klein–Gordon equation (3.8). We thus have the following Fourier decomposition into positive and negative energy solutions,

$$\psi_\alpha(x) = \frac{1}{(2\pi)^{3/2}} \int d^3p \sqrt{\frac{m}{\omega(\vec{p})}} \sum_\sigma [U_\alpha(\vec{p}, \sigma)a(\vec{p}, \sigma)e^{-ip\cdot x} + V_\alpha(\vec{p}, \sigma)b^*(\vec{p}, \sigma)e^{ip\cdot x}] , \tag{4.9}$$

where $p^\mu = (\omega(\vec{p}), \vec{p})$, and the sum in σ extends over the two spin orientations in the rest-frame of the particle, as we shall see. The reason for displaying explicitly the factor $\sqrt{\frac{m}{\omega(\vec{p})}}$ will become clear from the transformation (7.15) and canonical normalization (7.40) in Chapter 7.

For $\psi(x)$ to be a solution of the Dirac equation (4.8), the (*positive* and *negative* energy) *Dirac spinors* $U(p, \sigma)$ and $V(p, \sigma)$ must satisfy the equations

$$(\gamma^\mu p_\mu - m)U(\vec{p}, \sigma) = 0$$

$$(\gamma^\mu p_\mu + m)V(\vec{p}, \sigma) = 0 . \tag{4.10}$$

---

[2]Here and in what follows: in formulae which hold generally, without reference to a particular basis such as the Dirac representation, we omit the subscript D.

Recalling the explicit form (4.7) of the γ_D^μ-matrices, we obtain for the independent solutions in the Dirac representation,

$$U^{(D)}(\vec{p}, \sigma) = N_+ \begin{pmatrix} \chi(\sigma) \\ \frac{\vec{p} \cdot \vec{\sigma}}{\omega + m} \chi(\sigma) \end{pmatrix}, \quad V^{(D)}(\vec{p}, \sigma) = N_- \begin{pmatrix} \frac{\vec{p} \cdot \vec{\sigma}}{\omega + m} \chi^c(\sigma) \\ \chi^c(\sigma) \end{pmatrix}, \quad (4.11)$$

where $\omega(\vec{p}) = \omega$, and $N_\pm$ are normalization constants to be determined below. $\chi(\sigma)$ and $\chi^c(\sigma)$ denote the spinor and its *conjugate* in the rest frame of the particle,

$$\chi\left(\tfrac{1}{2}\right) = \begin{pmatrix} 1 \\ 0 \end{pmatrix}, \quad \chi\left(-\tfrac{1}{2}\right) = \begin{pmatrix} 0 \\ 1 \end{pmatrix}, \quad \chi^c(\sigma) = c^{-1}\chi(\sigma),$$

with c the "charge conjugation" matrix defined by

$$\chi^c(\sigma) = c^{-1}\chi(\sigma) = (-1)^{\frac{1}{2}+\sigma}\chi(-\sigma). \quad (4.12)$$

Notice that

$$c = \begin{pmatrix} 0 & -1 \\ 1 & 0 \end{pmatrix} = -i\sigma_2, \quad c^2 = -1.$$

The matrix c has the fundamental property[3]

$$c\vec{\sigma}c^{-1} = -\vec{\sigma}^*. \quad (4.13)$$

The following algebraic relations will turn out to be useful:

$$C\gamma^\mu C^{-1} = -\gamma^{\mu*}, \quad (4.14)$$

where

$$C = C\gamma^5\gamma^0 = i\gamma^2, \quad C^2 = 1, \quad C^\dagger = C \quad (4.15)$$

with

$$C = \begin{pmatrix} c & 0 \\ 0 & c \end{pmatrix}, \quad C^2 = -1, \quad C^\dagger = -C \quad (4.16)$$

and

$$\gamma^5 = i\gamma^0\gamma^1\gamma^2\gamma^3.$$

We further have

$$C\gamma^\mu C^{-1} = \gamma^{\mu T} \quad C\gamma^{0*}C^{-1} = \gamma^0, \quad C\gamma^{i*}C^{-1} = -\gamma^i. \quad (4.17)$$

In the Dirac representation, γ^5 is the *off-diagonal* 4×4 matrix

$$\gamma_D^5 = \begin{pmatrix} 0 & 1 \\ 1 & 0 \end{pmatrix}. \quad (4.18)$$

In order to fix the normalization constants in (4.11), we need to choose a scalar product. To this end we observe that

$$\gamma_\mu^\dagger = \gamma^0\gamma_\mu\gamma^0.$$

---

[3]The reason for introducing c will become clear in Chapter 7, Eq. (7.15).

Hence the Dirac operator $i\gamma^\mu \partial_\mu - m$ is hermitian with respect to the "Dirac" scalar product

$$(\psi, \psi') = \int d^4 x \bar{\psi} \psi', \quad \bar{\psi} = \psi^\dagger \gamma^0 . \tag{4.19}$$

Correspondingly we normalize the Dirac spinors by requiring[4]

$$\bar{U}(\vec{p}, \sigma) U(\vec{p}, \sigma') = \delta_{\sigma\sigma'}, \quad \bar{U} = U^\dagger \gamma^0 ,$$
$$\bar{V}(\vec{p}, \sigma) V(\vec{p}, \sigma') = -\delta_{\sigma\sigma'}, \quad \bar{V} = V^\dagger \gamma^0 , \tag{4.20}$$

which finally leads in the *Dirac representation* to the normalized Dirac spinors

$$U^{(D)}(\vec{p}, \sigma) = \sqrt{\frac{\omega + m}{2m}} \begin{pmatrix} \chi(\sigma) \\ \frac{\vec{p}\cdot\vec{\sigma}}{\omega+m}\chi(\sigma) \end{pmatrix}, \quad V^{(D)}(\vec{p}, \sigma) = \sqrt{\frac{\omega + m}{2m}} \begin{pmatrix} \frac{\vec{p}\cdot\vec{\sigma}}{\omega+m}\chi^c(\sigma) \\ \chi^c(\sigma) \end{pmatrix}. \tag{4.21}$$

Dirac spinors in the Weyl representation

We now present a derivation of the Dirac equation based on group-theoretical arguments alone. Our fundamental requirement will be that the solution of the "relativistic Schrödinger equation" should belong to a representation of the Lorentz group. In particular consider the irreducible representations $(1/2,0)$ and $(0,1/2)$ in (2.28). Denoting the wave functions in the respective representations by $\varphi(x)$ and $\bar{\varphi}(x)$, this means that in analogy to (2.4), under Lorentz transformations,

$$\varphi_\sigma(x) \longrightarrow \varphi'_\sigma(x) = \sum_{\sigma'} D^{(\frac{1}{2})}_{\sigma\sigma'}[\Lambda^{-1}] \varphi_{\sigma'}(\Lambda x) \tag{4.22}$$

$$\bar{\varphi}_\sigma(x) \longrightarrow \bar{\varphi}'_\sigma(x) = \sum_{\sigma'} \bar{D}^{(\frac{1}{2})}_{\sigma\sigma'}[\Lambda^{-1}] \bar{\varphi}_{\sigma'}(\Lambda x) . \tag{4.23}$$

The operators acting on these fields are thus given by 2×2 matrices. A complete set of such matrices is given by the identity and the three Pauli matrices. As we next show the set of four matrices[5]

$$t^\mu = (1, \vec{\sigma}) , \quad \bar{t}^\mu = (1, -\vec{\sigma}) \tag{4.24}$$

transform as a "4-vector" in the following sense:

$$D^{(\frac{1}{2})}[\Lambda^{-1}] t^\mu \bar{D}^{(\frac{1}{2})}[\Lambda] = \Lambda^\mu_{\ \nu} t^\nu \tag{4.25}$$

$$\bar{D}^{(\frac{1}{2})}[\Lambda^{-1}] \bar{t}^\mu D^{(\frac{1}{2})}[\Lambda] = \Lambda^\mu_{\ \nu} \bar{t}^\nu . \tag{4.26}$$

Note that these transformation laws cannot be interpreted as a change in basis. They are easily verified for an infinitesimal Lorentz transformation

$$\Lambda^\mu_{\ \nu} \simeq g^\mu_{\ \nu} + \omega^\mu_{\ \nu} , \quad \omega^0_{\ i} = \omega^i_{\ 0} = \theta_i; \quad \omega_{ij} = -\epsilon_{ijk}\alpha_k ,$$

---

[4]The minus sign is a consequence of our Dirac scalar product.
[5]We follow in general the notation of S. Weinberg, *Phys. Rev.* **133** (1964) B1318.

and the corresponding expression for the $(1/2,0)$ and $(0,1/2)$ representations (2.29) and (2.30) with

$$\vec{J}^{(\frac{1}{2},0)} = \vec{J}^{(0,\frac{1}{2})} = \frac{\vec{\sigma}}{2} \ .$$

We have

$$D^{(\frac{1}{2})}[\Lambda] \simeq 1 + i(\vec{\alpha} + i\vec{\theta}) \cdot \frac{\vec{\sigma}}{2}$$

$$\bar{D}^{(\frac{1}{2})}[\Lambda] \simeq 1 + i(\vec{\alpha} - i\vec{\theta}) \cdot \frac{\vec{\sigma}}{2}.$$

Notice that the rotational part does not care about which of the two representations we are in. Keeping only terms linear in the parameters we have

(i) for $t^0 = 1$,

$$D^{(\frac{1}{2})}[\Lambda^{-1}]t^0 \bar{D}^{(\frac{1}{2})}[\Lambda] \simeq 1 + i[(\vec{\alpha} - i\vec{\theta}) - i(\vec{\alpha} + i\vec{\theta})] \cdot \frac{\vec{\sigma}}{2}$$

$$\simeq 1 + \vec{\theta} \cdot \vec{\sigma} + \cdots$$

$$\simeq 1 + \theta_i t^i + \cdots$$

(ii) *for* $t^i = \sigma_i$,

$$D^{(\frac{1}{2})}[\Lambda^{-1}]t^i \bar{D}^{(\frac{1}{2})}[\Lambda] \simeq \sigma_i + \frac{i}{2}\sigma_i\sigma_j(\alpha_j - i\theta_j) - \frac{i}{2}\sigma_j\sigma_i(\alpha_j + i\theta_j)$$

$$\simeq \sigma_i + \frac{i}{2}[\sigma_i, \sigma_j]\alpha_j + \frac{1}{2}\{\sigma_i, \sigma_j\}\theta_j$$

$$\simeq \sigma_i - \epsilon_{ijk}\alpha_j\sigma_k + \theta_i$$

$$\simeq t^i - \epsilon_{ijk}\alpha_j t^k + \theta_i \ .$$

On the other hand we have for an infinitesimal Lorentz transformation $\Lambda^\mu{}_\nu$,

$$\Lambda^0{}_\nu t^\nu \simeq t^0 + \omega^0{}_i t^i$$

$$\Lambda^i{}_\nu t^\nu \simeq t^i + \omega^i{}_j t^j + \omega^i{}_0 t^0 \ .$$

In accordance with our previous parametrization we have

$$\Lambda^0{}_\nu t^\nu \simeq 1 + \theta_i t^i$$

$$\Lambda^i{}_\nu t^\nu \simeq t^i - \epsilon_{ijk}\alpha_j t^k + \theta_i \ ,$$

where we have set $t^0 = 1$. This establishes our claim (4.25) for t^μ. In a similar way one demonstrates the transformation law (4.26). It now follows from (4.25) and (4.26) that the equations

$$-m\varphi(x) + it^\mu \partial_\mu \bar{\varphi}(x) = 0 \tag{4.27}$$

$$-m\bar{\varphi}(x) + i\bar{t}^\mu \partial_\mu \varphi(x) = 0 \tag{4.28}$$

transform covariantly under Lorentz transformations.[6] Indeed, multiplying the first equation from the left with $D^{(\frac{1}{2})}[\Lambda^{-1}]$ and making the replacement $x \to x' = \Lambda x$, one has

$$-mD^{(\frac{1}{2})}[\Lambda^{-1}]\varphi(x') + (D^{(\frac{1}{2})}[\Lambda^{-1}]it^\mu \partial'_\mu \bar{D}[\Lambda])\bar{D}[\Lambda^{-1}]\bar{\varphi}(x') = 0 \,,$$

or recalling that $\partial'_\mu = \partial_\nu (\Lambda^{-1})^\nu{}_\mu$, we have from (4.25)

$$D[\Lambda^{-1}]t^\mu \partial'_\mu \bar{D}[\Lambda] = t^\nu \partial_\mu (\Lambda^{-1})^\mu{}_\nu = t^\nu \partial_\nu \,.$$

Together with (4.22) this implies

$$m\varphi'(x) + it^\mu \partial_\mu \bar{\varphi}'(x) = 0 \,,$$

which proves the covariance of Eq. (4.27). In the same way, one also proves the covariance of the second equation.

Equations (4.27), (4.28) represent a coupled set of equations, which only decouple in the case of zero-mass fermions. They may be collected into a single equation by defining the 4×4 matrices

$$\gamma^\mu_W = \begin{pmatrix} 0 & t^\mu \\ \bar{t}^\mu & 0 \end{pmatrix} \,, \tag{4.29}$$

where the subscript W stands for "Weyl".[7] One explicitly checks that they satisfy the anticommutation relations

$$\{\gamma^\mu, \gamma^\nu\} = 2g^{\mu\nu} \,. \tag{4.30}$$

Writing $\psi(x)$ in the form

$$\psi(x) := \begin{pmatrix} \varphi(x) \\ \bar{\varphi}(x) \end{pmatrix} \,, \tag{4.31}$$

the above coupled set of equations takes the form

$$(i\gamma^\mu_W \partial_\mu - m)\psi = 0 \,.$$

Multiplying this equation from the left with the operator $(i\gamma^\mu \partial_\mu + m)$ and using the anticommutation relations (4.30) we see that ψ is also a solution of the Klein–Gordon equation:

$$(\Box + m^2)\psi(x) = 0$$

describing the propagation of a free particle with the correct energy-momentum relation. By further defining the 4×4 matrices (in the Weyl-representation)

$$\mathcal{D}^{(\frac{1}{2})} = \begin{pmatrix} D^{(\frac{1}{2})} & 0 \\ 0 & \bar{D}^{(\frac{1}{2})} \end{pmatrix} \,, \tag{4.32}$$

---

[6]Substituting (4.27) into (4.28) yields $(m^2 + t^\mu \bar{t}^\nu \partial_\mu \partial_\nu)\varphi = 0$. Noting that with our definitions (4.24) for t^μ and $\bar{t}^\mu$, we have $t^\mu \bar{t}^\nu + t^\nu \bar{t}^\mu = 2g^{\mu\nu}$, we recover the Klein–Gordon equation, as desired.
[7]H. Weyl, *The Theory of Groups and Quantum Mechanics* (Dover Publications, Inc. New York, 1931).

the transformation laws (4.25), (4.26) can be collected to read

$$\mathcal{D}^{(\frac{1}{2})}[\Lambda^{-1}]\gamma_W^\mu \mathcal{D}^{(\frac{1}{2})}[\Lambda] = \Lambda^\mu_{\ \nu}\gamma_W^\nu \tag{4.33}$$

and

$$\psi(x) \to \psi'(x) = \mathcal{D}[\Lambda^{-1}]\psi(\Lambda x). \tag{4.34}$$

On this level we now have manifest Lorentz covariance of the Dirac equation (4.8). Note also that the *inverse* of the matrix $\mathcal{D}$ is now *equivalent* to the corresponding "Dirac" adjoint (recall (2.33))

$$\mathcal{D}^{(\frac{1}{2})}[\Lambda^{-1}] = \gamma_W^0 \mathcal{D}^{(\frac{1}{2})}[\Lambda]^\dagger \gamma_W^0 \ . \tag{4.35}$$

This will play an important role when we come to define scalar products.

We now decompose the solution to the Dirac equation as in (4.9). For $U_\alpha^{(W)}(\vec{p}, \sigma)$ and $V_\alpha^{(W)}(\vec{p}, \sigma)$ the Dirac equation then reads

$$(\gamma_W \cdot p - m)U^{(W)}(\vec{p}, \sigma) = 0 \ , \quad (\gamma_W \cdot p + m)V^{(W)}(\vec{p}, \sigma) \ . \tag{4.36}$$

Recalling the explicit form of the (1/2,0) and (0,1/2) representations (2.36) of boosts, we conclude that

$$\gamma_W \cdot p = \begin{pmatrix} 0 & t \cdot p \\ \bar{t} \cdot p & 0 \end{pmatrix} = \begin{pmatrix} 0 & m(D^{(\frac{1}{2})}[L(\vec{p})])^2 \\ m(\bar{D}^{(\frac{1}{2})}[L(\vec{p})])^2 & 0 \end{pmatrix}. \tag{4.37}$$

Recalling from (2.32) that $\bar{D}[L(\vec{p})] = D^{-1}[L(\vec{p})]$, we can solve the set of algebraic equations (4.36) for the four independent Dirac spinors, to give

$$U^{(W)}(\vec{p}, \sigma) = \frac{1}{\sqrt{2}} \begin{pmatrix} D^{(\frac{1}{2})}[L(\vec{p})]\chi(\sigma) \\ \bar{D}^{(\frac{1}{2})}[L(\vec{p})]\chi(\sigma) \end{pmatrix}$$

$$V^{(W)}(\vec{p}, \sigma) = \frac{1}{\sqrt{2}} \begin{pmatrix} D^{(\frac{1}{2})}[L(\vec{p})]\chi^c(\sigma) \\ -\bar{D}^{(\frac{1}{2})}[L(\vec{p})]\chi^c(\sigma) \end{pmatrix} \ , \tag{4.38}$$

where $U^{(W)}(\vec{p}, \sigma)$ and $V^{(W)}(\vec{p}, \sigma)$ have been normalized with respect to the scalar product (4.19).

Making use of the explicit form (2.37) and (2.38) of the 2×2 matrices representing the boosts, one can rewrite the expressions (4.38) in the explicit form

$$U^{(W)}(\vec{p}, \sigma) = \frac{1}{\sqrt{2}}\sqrt{\frac{\omega + m}{2m}} \begin{pmatrix} (1 - \frac{\vec{p}\cdot\vec{\sigma}}{\omega+m})\chi(\sigma) \\ (1 + \frac{\vec{p}\cdot\vec{\sigma}}{\omega+m})\chi(\sigma) \end{pmatrix},$$

$$V^{(W)}(\vec{p}, \sigma) = \frac{1}{\sqrt{2}}\sqrt{\frac{\omega + m}{2m}} \begin{pmatrix} (1 - \frac{\vec{p}\cdot\vec{\sigma}}{\omega+m})\chi^c(\sigma) \\ -(1 + \frac{\vec{p}\cdot\vec{\sigma}}{\omega+m})\chi^c(\sigma) \end{pmatrix} \ . \tag{4.39}$$

Comparing with (4.21), we seem to be arriving at different results. In fact, these results can be shown to be unitarily equivalent. Indeed, the γ^μ-matrices (4.29) and (4.7) are related by the unitary transformation

$$\gamma_D^\mu = S\gamma_W^\mu S^{-1}$$

with

$$S = \frac{1}{\sqrt{2}} \begin{pmatrix} 1 & 1 \\ -1 & 1 \end{pmatrix}, \quad S^{-1} = S^{\dagger}. \tag{4.40}$$

Correspondingly we have for the Dirac spinors

$$\begin{aligned} U^{(D)}(\vec{p}, \sigma) &= S U^{(W)}(\vec{p}, \sigma) \\ V^{(D)}(\vec{p}, \sigma) &= S V^{(W)}(\vec{p}, \sigma), \end{aligned} \tag{4.41}$$

which are readily seen to coincide with the spinors (4.21).

The basis in which the γ-matrices take the form (4.7) is referred to as the *Dirac representation*. The basis in which the γ-matrices take the form (4.29) is referred to as the *Weyl representation*. The same applies to the Dirac spinors (4.21) and (4.39), respectively.

The choice of representation is a matter of taste and depends on the specific problem and question one wants to address. Thus, to discuss the non-relativistic limit of the Dirac equation, it is convenient to work in the Dirac representation. If one is dealing with massless charged fermions, it is more convenient to work in the Weyl representation, since the Dirac equation reduces to two uncoupled equations in this case. We shall have the opportunity to work in still another basis, the so-called *Majorana representation*, which turns out to be particularly suited if the fermions are massless and *charge neutral* (neutrinos, for example).

4.2   Properties of the Dirac spinors

One easily proves the following results for both representations:

(a) Orthogonality relations

$$\begin{aligned} \bar{U}(\vec{p}, \sigma') U(\vec{p}, \sigma) &= \chi^{\dagger}(\sigma') \chi(\sigma) = \delta_{\sigma'\sigma}, \\ \bar{V}(\vec{p}, \sigma') V(\vec{p}, \sigma) &= -\chi^{\dagger c}(\sigma') \chi^{c}(\sigma) = -\delta_{\sigma'\sigma}. \end{aligned} \tag{4.42}$$

The *positive (negative)* energy spinors are seen to have *positive (negative)* norm and to be orthogonal, respectively. One furthermore has

$$\bar{U}(\vec{p}, \sigma') V(\vec{p}, \sigma) = 0, \quad \bar{V}(\vec{p}, \sigma') U(\vec{p}, \sigma) = 0.$$

We thus conclude that the "positive" and "negative" energy solutions[8] for half-integral spin are also mutually orthogonal with respect to the "Dirac scalar product".

(b) Projectors on positive and negative energy states

According to (a) the matrices

$$\Lambda_{+}(p)_{\alpha\beta} = \sum_{\sigma} U_{\alpha}(\vec{p}, \sigma) \bar{U}_{\beta}(\vec{p}, \sigma),$$

$$\Lambda_{-}(p)_{\alpha\beta} = -\sum_{\sigma} V_{\alpha}(\vec{p}, \sigma) \bar{V}_{\beta}(\vec{p}, \sigma)$$

---

[8]See Chapter 5 for this terminology.

have the properties of projectors on the positive and negative energy solutions, respectively. In particular, the property

$$\Lambda_+(p) + \Lambda_-(p) = 1$$

follows from the completeness relation

$$\sum_{\sigma=\pm\frac{1}{2}} (U_\alpha(\vec{p},\sigma)\bar{U}_\beta(\vec{p},\sigma) - V_\alpha(\vec{p},\sigma)\bar{V}_\beta(\vec{p},\sigma)) = \delta_{\alpha\beta}$$

for the spinors. We have for both representations

$$\Lambda_\pm(p) = \frac{m \pm \gamma \cdot p}{2m} . \tag{4.43}$$

4.3 Properties of the γ-matrices

We next list some useful properties of the γ-matrices which are independent of the choice of representation.

(a) The trace of an odd number of γ-matrices vanishes

Proof:

$$\mathrm{tr}[\gamma^{\mu_1}...\gamma^{\mu_\ell}] = \mathrm{tr}[\gamma^5\gamma^{\mu_1}...\gamma^{\mu_\ell}\gamma^5] = (-1)^\ell \,\mathrm{tr}[\gamma^{\mu_1}...\gamma^{\mu_\ell}] , \tag{4.44}$$

where we have used the cyclic property of the trace, as well as $\gamma_5^2 = 1$.

(b) Reduction of the trace of a product of γ-matrices

In general it follows, by repeated use of the anticommutator (4.30) of γ-matrices, that

$$\mathrm{tr}(\gamma_{\mu_1}...\gamma_{\mu_{2n}}) = 2g_{\mu_1\mu_2}\,\mathrm{tr}(\gamma_{\mu_3}...\gamma_{\mu_{2n}}) - 2g_{\mu_1\mu_3}\,\mathrm{tr}(\gamma_{\mu_2}\gamma_{\mu_4}...\gamma_{\mu_{2n}}) \tag{4.45}$$
$$+ ... + 2(-1)^{2n} g_{\mu_1\mu_{2n}}\,\mathrm{tr}(\gamma_{\mu_2}\gamma_{\mu_3}...\gamma_{\mu_{2n-1}}) - \mathrm{tr}(\gamma_{\mu_1}...\gamma_{\mu_{2n}}) ,$$

or

$$\mathrm{tr}(\gamma_{\mu_1}...\gamma_{\mu_{2n}}) = g_{\mu_1\mu_2}\mathrm{tr}(\gamma_{\mu_3}...\gamma_{\mu_{2n}}) - g_{\mu_1\mu_3}\,\mathrm{tr}(\gamma_{\mu_2}\gamma_{\mu_4}...\gamma_{\mu_{2n}})$$
$$+ ... + (-1)^{2n} g_{\mu_1\mu_{2n}}\,\mathrm{tr}(\gamma_{\mu_2}\gamma_{\mu_3}...\gamma_{\mu_{2n-1}}) .$$

As a Corollary to this we have the "contraction" identity

$$\mathrm{tr}[\gamma^\lambda\gamma^\mu\gamma^\nu\gamma^\rho\gamma_\lambda] = -2tr[\gamma^\rho\gamma^\nu\gamma^\mu] \tag{4.46}$$

as well as

$$\mathrm{tr}(\gamma^\mu\gamma^\nu) = 4g^{\mu\nu} , \quad \mathrm{tr}(\slashed{a}\slashed{b}) = 4a \cdot b$$
$$\mathrm{tr}(\gamma^\mu\gamma^\nu\gamma^\rho\gamma^\sigma) = 4(g^{\mu\nu}g^{\rho\sigma} - g^{\mu\rho}g^{\nu\sigma} + g^{\mu\sigma}g^{\nu\rho}) , \tag{4.47}$$

where we followed the Feynman convention of writing

$$\slashed{a} := \gamma^\mu a_\mu \,.$$

Notice that the factor 4 arises from $\mathrm{tr}\,1 = 4$, the dimension of space-time. We further have the *contraction identities*

$$\gamma^\lambda \gamma^\mu \gamma^\nu \gamma_\lambda = 4g^{\mu\nu}$$
$$\gamma^\lambda \gamma^\mu \gamma^\nu \gamma^\rho \gamma_\lambda = \gamma^\mu \gamma^\nu \gamma^\rho \tag{4.48}$$
$$\gamma^\lambda \gamma^\mu \gamma^\nu \gamma^\rho \gamma^\sigma \gamma_\lambda = 2(\gamma^\sigma \gamma^\mu \gamma^\nu \gamma^\rho + \gamma^\rho \gamma^\nu \gamma^\mu \gamma^\sigma)$$

which will prove useful in Chapters 15 and 16.

(c) The γ^5-matrix

In the Weyl representation the upper and lower components of the Dirac spinors are referred to as the positive and negative chiality components, corresponding to the eigenvalues of the matrices[9]

$$\gamma_D^5 = \begin{pmatrix} 0 & 1 \\ 1 & 0 \end{pmatrix}, \quad \gamma_W^5 = \begin{pmatrix} -1 & 0 \\ 0 & 1 \end{pmatrix}.$$

As one easily convinces oneself, one has (from here on we follow the convention of Itzykson and Zuber and of most other authors, and choose $\epsilon^{0123} = 1$)

$$\gamma^5 = -\frac{i}{4!}\epsilon_{\mu\nu\lambda\rho}\gamma^\mu \gamma^\nu \gamma^\lambda \gamma^\rho = i\gamma^0 \gamma^1 \gamma^2 \gamma^3 \,, \quad \gamma_5^2 = 1 \,. \tag{4.49}$$

This expression defines the γ_5 matrix in *both* representations.

(d) Lorentz transformation properties of γ^5

For Λ a Lorentz transformation, we have the *algebraic* property

$$\mathcal{D}[\Lambda^{-1}]\epsilon_{\mu\nu\lambda\rho}\gamma^\mu \gamma^\nu \gamma^\lambda \gamma^\rho \mathcal{D}[\Lambda] = \epsilon_{\mu\nu\lambda\rho}\Lambda^\mu{}_{\mu'}\Lambda^\nu{}_{\nu'}\Lambda^\lambda{}_{\lambda'}\Lambda^\rho{}_{\rho'}\gamma^{\mu'}\gamma^{\nu'}\gamma^{\lambda'}\gamma^{\rho'} \,.$$

Now

$$\epsilon_{\mu\nu\lambda\rho}\Lambda^\mu{}_{\mu'}\Lambda^\nu{}_{\nu'}\Lambda^\lambda{}_{\lambda'}\Lambda^\rho{}_{\rho'} = \det\Lambda \cdot \epsilon_{\mu'\nu'\lambda'\rho'} \,.$$

Hence we conclude that γ^5 "transforms" in particular like a pseudoscalar under *space reflections*, and in general as

$$\mathcal{D}[\Lambda^{-1}]\gamma^5 \mathcal{D}[\Lambda] = \gamma^5 \det\Lambda \,.$$

---

[9]Note that

$$S\gamma_D^5 S^{-1} = \gamma_W^5 \,, \quad S\gamma_D^0 S^{-1} = \begin{pmatrix} 0 & 1 \\ 1 & 0 \end{pmatrix} = \gamma_W^0$$

where S is given by (4.40). This agrees with the usual $V - A$ coupling of neutrinos in the weak interactions.

(e) Traces involving γ^5

$$\text{tr}(\gamma^5\gamma^\mu\gamma^\nu) = 0 \ ,$$
$$\text{tr}(\gamma^5\gamma^\mu\gamma^\nu\gamma^\lambda\gamma^\rho) = -4i\epsilon^{\mu\nu\lambda\rho} = 4i\epsilon_{\mu\nu\lambda\rho} \ .$$

Here the first relation follows from the fact that there exists no Levi–Civita tensor with two indices in four dimensions. The second relation follows from the fact that the right-hand side should be a Lorentz invariant pseudotensor of rank four, for which $\epsilon_{\mu\nu\lambda\rho}$ is the only candidate, and choosing the indices as in (4.49) to fix the constant.

4.4   Zero-mass, spin $=\frac{1}{2}$ fields

In the extreme relativistic limit we expect the mass of the fermion to be negligible. For $m = 0$ the Dirac *Hamiltonian* operator *commutes* with γ_5. Hence we may classify the eigenfunctions of the Hamiltonian by the eigenvalues of γ_5. It is thus desirable to work in the *Weyl* representation, where γ_5 is diagonal. In this representation the Dirac operator becomes *off-diagonal* in the large momentum limit, and the Weyl equations (4.27) and (4.28) reduce to the form

$$it^\mu\partial_\mu\bar{\varphi}(x) = 0 \ , \tag{4.50}$$
$$i\bar{t}^\mu\partial_\mu\varphi(x) = 0 \ , \tag{4.51}$$

or

$$(i\partial_0 + i\vec{\sigma}\cdot\nabla)\bar{\varphi}(x) = 0 \ , \tag{4.52}$$
$$(i\partial_0 - i\vec{\sigma}\cdot\nabla)\varphi(x) = 0 \ . \tag{4.53}$$

The 2×2 matrix $\vec{\sigma}\cdot\hat{n}/2$ represents the projection of the angular momentum operator on the direction of motion of the particle and is called the *helicity operator*. Correspondingly one refers to its eigenvalues $\pm1/2$ as *helicity*.

Equations (4.52) and (4.53) are just the Weyl equations for a *massless* particle. If *parity* is not conserved, we may confine ourselves to either one of the two equations, that is to either particles polarized in the direction of motion (positive helicity) or opposite to the direction of motion (negative helicity). This is the case for neutrinos (antineutrinos) participating in the parity-violating weak interactions, which carry helicity $-1/2$ $(+1/2)$. If parity is conserved, both helicity states must exist.

The fact that the massive Dirac equation turns into Weyl equations in the "infinite momentum frame" shows that at high energies massive particles are polarized "parallel" or "anti-parallel" to the direction of motion. However, whereas the *helicity* of a *massless* particle is a Lorentz invariant, this is not the case for a massive particle: If a massive particle is polarized in the direction of motion in one inertial frame, its polarization will be a superposition of all possible spin projections in a different inertial system. Phrased in a different way: If the particle is massive one can always catch up with it and ultrapass it, so that the particle appears moving "backwards", while continuing to be polarized in the original direction. With a zero mass particle you can never catch up since it is moving at the speed of light.

Chirality

As the last argument above shows, the $m = 0$ case has to be treated separately, and cannot be obtained as the zero-mass limit of massive case discussed so far, which was based on the existence of a rest frame of the particle. According to our discussion in the chapter on Lorentz transformations, zero-mass 1-particle states indeed transform quite differently from the massive ones.

In the zero mass case, the Dirac equations for the U and V spinors reduce to one and the same equation:

$$\gamma \cdot p U(\vec{p}, \lambda) = 0, \quad \gamma \cdot p V(\vec{p}, \lambda) = 0. \tag{4.54}$$

Let us define the "spin" operator

$$\vec{\Sigma} = \begin{pmatrix} \vec{\sigma} & 0 \\ 0 & \vec{\sigma} \end{pmatrix} . \tag{4.55}$$

In terms of the Gamma matrices (Dirac or Weyl basis) this operator reads

$$\vec{\Sigma} = \gamma^5 \gamma^0 \vec{\gamma} . \tag{4.56}$$

The Dirac equation may then be written in the form

$$(\vec{\Sigma} \cdot \hat{n}) U(\vec{p}, \lambda) = \gamma^5 U(\vec{p}, \lambda)$$

where $\hat{n} = \frac{\vec{p}}{|\vec{p}|}$. Thus $\frac{1}{2} \vec{\Sigma} \cdot \hat{n}$ is just the helicity operator in the 4-component representation.

The helicity operator (4.56) commutes with the free Dirac Hamiltonian. The same applies to γ_5, if the mass of the particle is *zero*. Since furthermore

$$[\gamma_5, \vec{\Sigma} \cdot \hat{n}] = 0 ,$$

we may classify the eigenstates of the zero-mass Dirac Hamiltonian according to their *helicity* and *chirality*, the latter being defined as the corresponding eigenvalue ± 1 of γ^5. Such states are obtained from the solutions U to the Dirac equation with the aid of the projection operator

$$P_\pm = \frac{1}{2}(1 \pm \gamma_5), \quad P_\pm^2 = P_\pm .$$

We have

$$\gamma_5 U_\pm = \pm U_\pm, \quad \frac{\vec{\Sigma} \cdot \hat{n}}{2} U_\pm = \pm \frac{1}{2} U_\pm$$

where

$$U_\pm = P_\pm U, \quad U(\vec{p}) = \begin{pmatrix} \omega(\vec{p}, -\frac{1}{2}) \\ \bar{\omega}(\vec{p}, \frac{1}{2}) \end{pmatrix} .$$

Recalling that in the Weyl representation

$$\gamma_W^5 = \begin{pmatrix} -1 & 0 \\ 0 & 1 \end{pmatrix}, \quad P_+^{(W)} = \begin{pmatrix} 0 & 0 \\ 0 & 1 \end{pmatrix}, \quad P_-^W = \begin{pmatrix} 1 & 0 \\ 0 & 0 \end{pmatrix}$$

we have

$$U_+^{(W)}(\vec{p}) = \begin{pmatrix} 0 \\ \bar{\omega}(\vec{p}, \frac{1}{2}) \end{pmatrix}, \quad U_-^{(W)}(\vec{p}) = \begin{pmatrix} \omega(\vec{p}, -\frac{1}{2}) \\ 0 \end{pmatrix}.$$

The eigenvalue of γ_5 thus coincides with twice the eigenvalue of the helicity operator: *particles* of positive (negative) chirality, carry helicity $+1/2(-1/2)$.

Solution of Weyl equations

Experiment shows that neutrinos (antineutrinos) only occur with negative (positive) helicity. One thus refers to ν ($\bar{\nu}$) as being left (right) handed. This is reflected by the so-called $V - A$ (vector minus axial vector) coupling of the neutrino sector. Since parity is violated, the absence of right-handed neutrinos and left-handed antineutrinos is admissible. The 4-component Dirac field (4.9) of the massive case is thus replaced in this case by

$$\psi_\alpha(x) = \begin{pmatrix} \varphi_\sigma(x) \\ \bar{\varphi}_{\bar\sigma}(x) \end{pmatrix},$$

with

$$\varphi_\sigma(x) = \frac{1}{(2\pi)^{2/3}} \int d^3p \sqrt{\frac{m}{|\vec{p}|}} \, \omega_\sigma(\vec{p}, -\frac{1}{2}) \left[a(\vec{p}, -\frac{1}{2}) e^{-ip\cdot x} + b^*(\vec{p}, \frac{1}{2}) e^{ip\cdot x} \right], \quad (4.57)$$

$$\bar{\varphi}_{\bar\sigma}(x) = \frac{1}{(2\pi)^{2/3}} \int d^3p \sqrt{\frac{m}{|\vec{p}|}} \, \bar{\omega}_{\bar\sigma}(\vec{p}, \frac{1}{2}) \left[a(\vec{p}, \frac{1}{2}) e^{-ip\cdot x} + b^*(\vec{p}, -\frac{1}{2}) e^{ip\cdot x} \right], \quad (4.58)$$

where the spin projection now refers to helicity. The Weyl equations thus reduce to solving the eigenvalue problems

$$\frac{\vec{\sigma} \cdot \hat{n}}{2} w\left(\vec{p}, -\frac{1}{2}\right) = -\frac{1}{2} w\left(\vec{p}, -\frac{1}{2}\right), \quad \frac{\vec{\sigma} \cdot \hat{n}}{2} \bar{w}\left(\vec{p}, \frac{1}{2}\right) = \frac{1}{2} \bar{w}\left(\vec{p}, \frac{1}{2}\right). \quad (4.59)$$

For the momentum pointing in the z-direction, the eigenvalue problems are solved by

$$w\left(\tilde{k}, -\frac{1}{2}\right) = \begin{pmatrix} 0 \\ 1 \end{pmatrix}, \quad \bar{w}\left(\tilde{k}, \frac{1}{2}\right) = \begin{pmatrix} 1 \\ 0 \end{pmatrix} \quad (4.60)$$

with $\tilde{k} = (\kappa, 0, 0, \kappa)$. Now perform the following transformation: First boost the momentum $\tilde{k}^\mu$ to the momentum $p^\mu = (|\vec{p}|, 0, 0, |\vec{p}|)$ with the matrix (see (2.15))

$$B(|\vec{p}|) = \begin{pmatrix} \cosh\theta & 0 & 0 & \sinh\theta \\ 0 & 1 & 0 & 0 \\ 0 & 0 & 1 & 0 \\ \sinh\theta & 0 & 0 & \cosh\theta \end{pmatrix}.$$

Hence

$$B^\mu{}_\nu(|\vec{p}|)\tilde{k}^\nu = \kappa e^\theta \begin{pmatrix} 1 \\ 0 \\ 0 \\ 1 \end{pmatrix} = \begin{pmatrix} |\vec{p}| \\ 0 \\ 0 \\ |\vec{p}| \end{pmatrix} .$$

This determines θ as a function of $|\vec{p}|$:

$$\theta = \ln\left(\frac{|\vec{p}|}{\kappa}\right) .$$

We now rotate the vector p^μ thus obtained in the desired direction of the final vector p^μ with the rotation matrix $\mathcal{R}(\hat{p})^\mu{}_\nu$,

$$\mathcal{R}(\hat{p}) = \begin{pmatrix} 0\,0 & 0 & 0 \\ 0 & & \\ 0 & R(\hat{p}) & \\ 0 & & \end{pmatrix} .$$

The result is

$$p^\mu = \mathcal{L}^\mu{}_\nu(\vec{p})\tilde{k}^\nu$$

where $\mathcal{L}^\mu_\nu(\vec{p})$ is the matrix (2.50):

$$\mathcal{L}(\vec{p})^\mu{}_\nu = (R(\hat{p})B(|\vec{p}|))^\mu{}_\nu = \left(\frac{|\vec{p}|}{\kappa}\right)R(\hat{p})^\mu{}_\nu .$$

Here $\mathcal{L}(\vec{p})$ plays the role of the boost $L(\vec{p})$ in the massive case. Correspondingly we have from (4.60) and (2.34) for the 2-component spinors

$$\bar{\omega}_{\bar{\sigma}}(\vec{p}, \frac{1}{2}) = D^{(\frac{1}{2})}[\mathcal{L}(\vec{p})]_{\bar{\sigma}, -\frac{1}{2}} = \left(\frac{|\vec{p}|}{\kappa}\right)^{\frac{1}{2}} D^{(\frac{1}{2})}[R(\hat{p})]_{\bar{\sigma}, -\frac{1}{2}} . \qquad (4.61)$$

Similarly we have

$$\bar{\omega}_{\bar{\sigma}}(\vec{p}, \frac{1}{2}) = D^{(\frac{1}{2})}[\mathcal{L}(\vec{p})]_{\bar{\sigma}, \frac{1}{2}} = \left(\frac{|\vec{p}|}{\kappa}\right)^{\frac{1}{2}} D^{(\frac{1}{2})}[R(\hat{p})]_{\bar{\sigma}, \frac{1}{2}} . \qquad (4.62)$$

The fields (4.57) and (4.58) now take the form

$$\varphi_\sigma(x) = \frac{1}{(2\pi)^{2/3}} \int d^3p \sqrt{\frac{m}{|\vec{p}|}} D^{(\frac{1}{2})}[\mathcal{L}(\vec{p})]_{\sigma, -\frac{1}{2}} \left[a\left(\vec{p}, -\frac{1}{2}\right)e^{-ip\cdot x} + b^*\left(\vec{p}, \frac{1}{2}\right)e^{ip\cdot x}\right]$$

$$\qquad (4.63)$$

$$\bar{\varphi}_{\bar{\sigma}}(x) = \frac{1}{(2\pi)^{2/3}} \int d^3p \sqrt{\frac{m}{|\vec{p}|}} \bar{D}^{(\frac{1}{2})}[\mathcal{L}(\vec{p})]_{\bar{\sigma}, \frac{1}{2}} \left[a\left(\vec{p}, \frac{1}{2}\right)e^{-ip\cdot x} + b^*\left(\vec{p}, -\frac{1}{2}\right)e^{ip\cdot x}\right] .$$

In this form, the Fourier decomposition resembles closely that of a massive field except for the fact that in the massless case $U = V$. Alternatively we have using

(4.61) and (4.62),

$$\varphi_\sigma(x) = \frac{1}{(2\pi)^{2/3}} \int d^3p D^{(\frac{1}{2})} [R(\vec{p})]_{\sigma,-\frac{1}{2}} \left[a\left(\vec{p}, -\frac{1}{2}\right) e^{-ip\cdot x} + b^* \left(\vec{p}, \frac{1}{2}\right) e^{ip\cdot x} \right]$$

$$(4.64)$$

$$\bar{\varphi}_{\bar\sigma}(x) = \frac{1}{(2\pi)^{2/3}} \int d^3p \bar{D}^{(\frac{1}{2})} [R(\vec{p})]_{\bar\sigma,\frac{1}{2}} \left[a\left(\vec{p}, \frac{1}{2}\right) e^{-ip\cdot x} + b^* \left(\vec{p}, -\frac{1}{2}\right) e^{ip\cdot x} \right]$$

where we have chosen $\kappa = m$.

4.5 Majorana fermions

So far we have considered the *Dirac* representation, particularly suited for discussing the non-relativistic limit as we shall see, and the Weyl (or chiral) representation particularly suited for discussing the relativistic limit, or the case of zero mass particles. There exists another choice of basis for the Gamma matrices called the *Majorana* representation which is particularly suited for the case of charge self-conjugate fermions, referred to as *Majorana fermions*.

Spin zero, charge neutral particles are called "self-conjugate", and are described by real fields satisfying the Klein–Gordon equation, which itself is real. In the case of "self-conjugate" spin one-half particles, the analogon is provided by the "Majorana" representation.

The Majorana representation is, however, also useful in the case where fermions and anti-fermions are distinct particles if the symmetry group in question is for instance the *orthogonal group* $O(N)$ rather than the unitary group $U(N)$. The reason is that in the Majorana representation the Dirac equation is *real*. We therefore discuss separately the notion of Majorana *representation* and Majorana *fermions*.

Majorana representation

There exists a choice of basis in which all Dirac matrices are purely imaginary. This is called the *Majorana representation* of the Gamma-matrices. They are obtained from the corresponding ones in the Dirac representation via the unitary transformation

$$\gamma_M^\mu = S\gamma_D^\mu S^{-1}$$

with

$$S = \frac{1}{\sqrt{2}} \begin{pmatrix} 1 & \sigma_2 \\ \sigma_2 & -1 \end{pmatrix}$$

$$(4.65)$$

and have the property

$$\gamma_M^{\mu*} = -\gamma_M^\mu .$$

One explicitly computes

$$\gamma_M^0 = \begin{pmatrix} 0 & \sigma_2 \\ \sigma_2 & 0 \end{pmatrix}, \quad \gamma_M^1 = \begin{pmatrix} i\sigma_3 & 0 \\ 0 & i\sigma_3 \end{pmatrix}, \quad \gamma_M^2 = \begin{pmatrix} 0 & -\sigma_2 \\ \sigma_2 & 0 \end{pmatrix},$$

$$\gamma_M^3 = \begin{pmatrix} -i\sigma_1 & 0 \\ 0 & -i\sigma_1 \end{pmatrix}, \quad \gamma_M^5 = \begin{pmatrix} \sigma_2 & 0 \\ 0 & -\sigma_2 \end{pmatrix}.$$

Now, the Dirac wave function and its charge conjugate (in the Dirac and Weyl representations) are related by (see Chapter 7, Eq. (7.48)),

$$\psi_c(x) = \mathcal{C}\psi^*(x) . \tag{4.66}$$

Applying the unitary operator (4.65) to both sides of the equation, we obtain for the Dirac wave function in the *Majorana representation*

$$\psi_c^{(M)} = \Gamma\psi^{(M)*}$$

where

$$\Gamma = S\mathcal{C}S^{-1*} = -i.$$

We thus conclude that

$$\psi_c^{(M)} = -i\psi^{(M)*}. \tag{4.67}$$

Summarizing we have in the *Dirac, Weyl* and *Majorana* representations

$$\psi_c = \Gamma\psi^* \tag{4.68}$$

where

$$\Gamma = \begin{cases} i\gamma^2 \ (\text{Dirac, Weyl}) \\ -i \ (\text{Majorana}) \end{cases} . \tag{4.69}$$

Majorana spinors

Applying the unitary operator (4.65) to the Dirac spinors in the *Dirac representation* (4.21) (we use $c^{-1} = i\sigma_2$)

$$U^{(D)}(\vec{p},\sigma) = \sqrt{\frac{\omega+m}{2m}} \begin{pmatrix} \chi(\sigma) \\ \frac{\vec{\sigma}\cdot\vec{p}}{\omega+m}\chi(\sigma) \end{pmatrix}$$

$$V^{(D)}(\vec{p},\sigma) = \sqrt{\frac{\omega+m}{2m}} \begin{pmatrix} \frac{\vec{p}\cdot\vec{\sigma}}{\omega+m}i\sigma_2\chi(\sigma) \\ i\sigma_2\chi(\sigma) \end{pmatrix} ,$$

we obtain for the corresponding spinors in the *Majorana representation*

$$U^{(M)}(\vec{p},\sigma) = \frac{1}{\sqrt{2}}\sqrt{\frac{\omega+m}{2m}} \begin{pmatrix} (1+\sigma_2\frac{\vec{\sigma}\cdot\vec{p}}{\omega+m})\chi(\sigma) \\ (\sigma_2 - \frac{\vec{\sigma}\cdot\vec{p}}{\omega+m})\chi(\sigma) \end{pmatrix} \tag{4.70}$$

$$V^{(M)}(\vec{p},\sigma) = \frac{i}{\sqrt{2}}\sqrt{\frac{\omega+m}{2m}} \begin{pmatrix} (\frac{\vec{p}\cdot\vec{\sigma}}{\omega+m}\sigma_2 + 1)\chi(\sigma) \\ (\sigma_2\frac{\vec{p}\cdot\vec{\sigma}}{\omega+m}\sigma_2 - \sigma_2)\chi(\sigma) \end{pmatrix} . \tag{4.71}$$

Recalling that

$$\sigma_2\vec{\sigma}\sigma_2 = -\vec{\sigma}^*$$

we see from (4.70) and (4.71) that

$$U^{(M)*}(\vec{p},\sigma) = iV^{(M)}(\vec{p},\sigma) . \tag{4.72}$$

In the case of Majorana fermions, it is convenient to redefine the phase of the Dirac spinors via the replacements

$$U^{(M)} \longrightarrow e^{-\frac{i\pi}{4}}U^{(M)}, \quad V^{(M)} \longrightarrow e^{-\frac{i\pi}{4}}V^{(M)} .$$

For this new choice of phase, Eq. (4.72) is replaced by

$$U^{(M)*}(\vec{p}, \sigma) = V^{(M)}(\vec{p}, \sigma) . \tag{4.73}$$

Self-conjugate Dirac fields

Fields describing fermions which are their own anti-particles are said to be self-conjugate and have the Fourier representation

$$\psi^{(M)}(x) = \frac{1}{(2\pi)^{3/2}} \int \frac{d^3p}{\sqrt{2\omega_p}} \sum_\sigma \left\{ a(\vec{p}, \sigma)U^{(M)}(\vec{p}, \sigma)e^{-ip\cdot x} + +a^*(\vec{p}, \sigma)V^{(M)}(\vec{p}, \sigma)e^{ip\cdot x} \right\},$$

with the property (4.73). These fields are real,

$$\psi^{(M)}(x) = \psi^{(M)*}(x) ,$$

and play the role of real scalar fields in the case of spin $1/2$ fields.

Chapter 5

The Free Maxwell Field

As is well known, the electromagnetic field can be interpreted on the quantum level as a flux of quanta, called *photons*. In fact, this interpretation first arose in connection with Planck's formula describing the spectrum of black-body radiation. As Maxwell's equations show, these quanta propagate with the velocity of light in all inertial frames, so there exists no rest frame we can associate with them. Photons can thus be viewed as "massless" particles. According to our discussion in Section 6 of Chapter 2, the respective 1-particle *states* should thus transform according to a one-dimensional representation of the *little group*.

The *Aharonov Bohm effect* shows that it is the *vector potential A^μ* which plays the fundamental role in quantum mechanics. This vector potential however transforms under the $D^{(\frac{1}{2},\frac{1}{2})}$ representation of the Lorentz group and involves a priori *four* degrees of freedom. Of these, A^0 is associated with the "Coulomb potential" and thus corresponds to *non-radiative* degrees of freedom which are only present if there is a *matter source*. This leaves us with three degrees of freedom. One of these is not observable (on classical level) as a result of the underlying *gauge invariance* of physical quantities. For a pure radiation field one is thus left with only two degrees of freedom, corresponding to the two helicity states of a photon. These statements become obvious in the *Coulomb gauge*, which is thus also called the "physical" gauge. We are, however, not limited to this choice of gauge which in practical calculations complicates matter considerably, due to the fact that it breaks *manifest* Lorentz covariance. We shall thus review the solutions of Maxwell's equations in two different gauges — the *Lorentz* gauge and the (non-covariant) Coulomb gauge.

5.1 The radiation field in the Lorentz gauge

In the absence of a source, the Maxwell equations (Coulomb's and Ampère's law) become

$$\partial_\mu F^{\mu\nu} = 0 \, , \tag{5.1}$$

with the usual identification

$$B^i = -\frac{1}{2}\epsilon_{ijk}F^{jk} \, , \quad E^i = F^{i0} \tag{5.2}$$

for the magnetic and electric fields, respectively.

The electromagnetic field tensor $F_{\mu\nu}$ can be written in the form

$$F_{\mu\nu} = \partial_\mu A_\nu - \partial_\nu A_\mu$$

and is evidently invariant under the *gauge transformation*

$$A_\mu \to A_\mu + \partial_\mu \Lambda \ . \tag{5.3}$$

In terms of the vector potential A_μ the homogeneous Maxwell equations (5.1) read

$$\Box A^\mu - \partial^\mu \partial_\nu A^\nu = 0 \ . \tag{5.4}$$

By choosing Λ in (5.3) to be given by

$$\Lambda(x) = -\int d^4 y\, G(x-y)\partial_\mu A^\mu(y) \ , \quad \Box G(x) = \delta^4(x)$$

we arrive at the Lorentz gauge

$$\partial_\mu A^\mu = 0 \ . \tag{5.5}$$

In this gauge the equations of motion (5.4) read

$$\Box A^\mu(x) = 0 \ .$$

The general solution of this equation is well known from the course in electrodynamics and is given by

$$A^\mu(x) = \frac{1}{(2\pi)^{3/2}} \int \frac{d^3 k}{\sqrt{2|\vec{k}|}} \sum_{\nu=0}^{3} \left[\epsilon^\mu(\vec{k},\nu)a(\vec{k},\nu)e^{-ik\cdot x} + \epsilon^{\mu *}(\vec{k},\nu)a^*(\vec{k},\nu)e^{ik\cdot x} \right] \ ,$$

where $a(k,\nu), \nu = 0...3$ are the Fourier coefficients and ϵ^μ are the corresponding *polarization tensors* playing a role analogous to the Dirac spinors. In the Lorentz gauge (5.5) we must have

$$k \cdot \epsilon(\vec{k},\nu) = 0, \quad k^2 = 0, \quad \nu = 1,2,3,4 \ .$$

With A^μ real, we choose the polarization tensors to be real. The choice of the Fourier coefficients is then dictated by the *reality* of the electromagnetic field. In particle language it corresponds to the fact that the photon is its own antiparticle!

The Fourier coefficients $a^*(\vec{k},\nu)$ ($a(\vec{k},\nu)$) will eventually be identified with the creation (destruction) operators of 1-particle states. However, only two of these states can correspond to photons of helicity $+1$ and -1. In the Coulomb gauge this becomes manifest. This gauge, thus often referred to as the "physical gauge", has however the drawback of not being manifestly Lorentz invariant.

5.2   The radiation field in the Coulomb gauge

The Coulomb gauge

$$\vec{\nabla} \cdot \vec{A} = 0 \tag{5.6}$$

can be reached by performing the gauge transformation (5.3) with

$$\Lambda(\vec{r}, t) = \int d^3 r' \frac{-1}{4\pi |\vec{r} - \vec{r}\,'|} \vec{\nabla}' \cdot \vec{A}(\vec{r}\,', t) \ .$$

In this gauge, the free Maxwell equations for the vector potential read:

$$\Box A^\mu - \partial^\mu \partial_0 A^0 = 0 \ . \tag{5.7}$$

Setting $\mu = 0$, it follows from here that

$$\vec{\nabla}^2 A^0 = 0 \ .$$

This is *Laplace's* equation; it only has the trivial solution if we require that the vector potential tends to zero at infinity. In the Coulomb gauge we thus have in the absence of sources, $A^0 = 0$. This shows that in this gauge the vector potential possesses only *two* degrees of freedom, corresponding to a *radiation* field, in agreement with our general considerations in Section 6 of Chapter 2, showing that a zero mass particle can exist only in two helicity states. Correspondingly we have for the general solution of (5.7) in the gauge (5.6)

$$\vec{A}(\vec{r}, t) = \frac{1}{(2\pi)^{3/2}} \int \frac{d^3 k}{\sqrt{2|\vec{k}|}} \sum_{\lambda = \pm 1} \vec{\epsilon}(\vec{k}, \lambda) [a(\vec{k}, \lambda) e^{-ik \cdot x} + a^*(\vec{k}, -\lambda) e^{ik \cdot x}] \tag{5.8}$$

with $\vec{k} \cdot \vec{\epsilon}(\vec{k}, \lambda) = 0$, and λ the helicity, where we have used (5.13). We are allowing for both helicity states, since parity is conserved. We now show how to choose these polarization tensors for $\vec{A}$ a real field.

Define the two orthogonal vectors

$$(\epsilon_\pm^\mu) = \frac{1}{\sqrt{2}}(0, 1, \pm i, 0) \ . \tag{5.9}$$

Introducing as in Section 6 of Chapter 2 the light-like *standard* 4-momentum $\tilde{k}$ by

$$\tilde{k}^\mu = |\kappa| n^\mu, \quad (n^\mu) = (1, 0, 0, 1) = (1, \hat{n}) \tag{5.10}$$

describing the motion of a photon in z-direction with energy $|\vec{k}|$, we see that the standard polarization tensors (5.9) are in fact eigenvectors of the helicity operator

$$\vec{\mathcal{J}} \cdot \hat{n} = \begin{pmatrix} 0 & 0 \\ 0 & J_3 \end{pmatrix}, \quad J_3 = -i \begin{pmatrix} 0 & 1 & 0 \\ -1 & 0 & 0 \\ 0 & 0 & 0 \end{pmatrix} \ .$$

Indeed,

$$\vec{\mathcal{J}} \cdot \hat{n} \ \epsilon_\pm^\mu = \pm \epsilon_\pm^\mu \ .$$

Inspired by our considerations in Section 6 of Chapter 2 on the *Little group*, we now define the polarization tensors for a photon with general momentum $\vec{k}$ by

$$\epsilon^\mu(\vec{k}, \pm 1) = \mathcal{L}^\mu{}_\nu(\vec{k}) \epsilon^\nu_\pm , \tag{5.11}$$

where $\mathcal{L}^\mu{}_\nu(k)$ are the elements of the Little group taking the standard vector $\tilde{k}^\mu$ into the final vector k^μ (see Chapter 2),

$$\mathcal{L}^\mu{}_\nu(\vec{k}) = \mathcal{R}^\mu_\lambda(\vec{k}) B^\lambda_\nu(|\vec{k}|).$$

The *boost* matrix $B^\mu{}_\nu$ represents a boost of $\tilde{k}$ along the 3-direction to $k^\mu = (|\kappa|, 0, 0, |\kappa|)$, and leaves the standard polarization tensors unchanged since it only affects the 0 and 3 elements, leaving the $(1, 2)$ elements unchanged. Hence

$$\epsilon^\mu(\vec{k}, \pm 1) = \mathcal{L}^\mu{}_\nu(\vec{k}) \epsilon^\nu_\pm = \mathcal{R}^\mu{}_\nu(\hat{k}) \epsilon^\nu_\pm = \begin{pmatrix} 0 \\ R^i{}_j(\hat{k}) \epsilon^j_\pm \end{pmatrix} , \tag{5.12}$$

where we have used

$$\mathcal{R} = \begin{pmatrix} 1 & 0 \\ 0 & R \end{pmatrix} , \qquad B^\mu{}_\nu \epsilon^\nu_\pm = \epsilon^\nu_\pm ,$$

with R the matrix rotating the standard momentum into the direction of $\vec{k}$. Noting that

$$(R^{-1})^\mu{}_\nu = R^\nu{}_\mu ,$$

we easily verify the following properties of the polarization tensors:

$$\epsilon^*_\mu(\vec{k}, \lambda') \epsilon^\mu(\vec{k}, \lambda) = \delta_{\lambda\lambda'} , \qquad \epsilon^0(\vec{k}, \lambda) = 0 ,$$

$$\epsilon^{*\mu}(\vec{k}, \lambda) = \epsilon^\mu(\vec{k}, -\lambda), \qquad k \cdot \epsilon(\vec{k}, \lambda) = 0 . \tag{5.13}$$

The last property follows from the following manipulations:

$$\tilde{k}_\mu \epsilon^\mu_\pm = 0 \Rightarrow \tilde{k}_\lambda (\mathcal{R}^{-1}\mathcal{R})^\lambda{}_\nu \epsilon^\nu_\pm = 0 \Rightarrow \tilde{k}_\lambda (\mathcal{R}^{-1})^\lambda{}_\rho \mathcal{R}^\rho{}_\nu \epsilon^\nu_\pm = 0 \Rightarrow k \cdot \epsilon(\vec{k}, \lambda) = 0 ,$$

where we have used the Lorentz transformation property (2.6) of a covariant 4-vector. The polarization tensor satisfies the following completeness relation:

$$\sum_{\lambda = \pm 1} \epsilon^i(\vec{k}, \lambda) \epsilon^{*j}(\vec{k}, \lambda) + \frac{k^i k^j}{\vec{k}^2} = \delta^{ij} . \tag{5.14}$$

Chapter 6

Quantum Mechanics of Dirac Particles

In the case of the scalar field we have seen that there was no possibility for a probability interpretation in the sense of non-relativistic Quantum Mechanics. In the case of Dirac spin one-half particles, such a probabilistic interpretation is possible if their electromagnetic interaction is restricted to external electromagnetic fields.

6.1 Probability interpretation

In order for $P(\vec{r}, t)$ to describe a probability density, we must require

$$\frac{d}{dt} \int d^3r \, P(\vec{r}, t) = 0 .$$

This property is satisfied if $P(\vec{r}, t)$ satisfies a continuity equation

$$\partial_t P(\vec{r}, t) + \vec{\nabla} \cdot \vec{j}(\vec{r}, t) = 0 . \tag{6.1}$$

Now, in order that the normalization

$$\int d^3r \, P(\vec{r}, t) = 1$$

be independent of the choice of the inertial system, $d^3r \, P(\vec{r}, t)$ must transform like a scalar.

Claim:

If $P(\vec{r}, t)$ is the zero component of a *conserved* 4-vector field $j^\mu(\vec{r}, t)$,

$$j^\mu(x) = \begin{pmatrix} P(x) \\ \vec{j}(x) \end{pmatrix}, \qquad \partial_\mu j^\mu(x) = 0 , \tag{6.2}$$

then

$$\int d^3r\, P(\vec{r}, t) = \int d^3r'\, P'(\vec{r}', t') \,, \tag{6.3}$$

where x and x' are related by a Lorentz transformation: $x'^\mu = \Lambda^\mu{}_\nu x^\nu$.

Proof:

Since the probability density satisfies a continuity equation (6.1), it follows that the integrals in (6.3) are independent of time. It is therefore convenient to choose $t = t' = 0$.

We now convert the three-dimensional integrals in (6.3) into four-dimensional integrals by noting that

$$\int_{x^0=0} d^3x\, P(x) = \int d^4x\, \delta(x^0) j^0(x)$$
$$= \int d^4x\, \delta(n(0) \cdot x)(n(0) \cdot j(x)) \,, \tag{6.4}$$

where $n^\mu(0)$ is the unit time-like vector

$$n^\mu(0) = n_\mu(0) = (1, 0, 0, 0) \,.$$

The integration in (6.4) is thus restricted to the plane "perpendicular" to the unit 4-vector $n^\mu(0)$.

We now rewrite the above integral in terms of Lorentz-transformed variables. By assumption we have under a Lorentz transformation

$$x^\mu \to x'^\mu = \Lambda^\mu{}_\nu x^\nu$$

the transformation law (see (2.5))

$$j'^\mu(x') = \Lambda^\mu{}_\nu j^\nu(x) \,.$$

Making use of the Lorentz invariance of the integration measure d^4x and *defining* the Lorentz-transformed unit 4-vector n'^μ by

$$n(0) \cdot x = n'(0) \cdot x' \,,$$

we have

$$\int d^3r\, P(\vec{r}, 0) = \int d^4x'\, \delta(n'(0) \cdot x')(n'_\mu(0) j'^\mu(x')) \,. \tag{6.5}$$

We next want to convert the right-hand side of this equation again into a three-dimensional integral with respect to the Lorentz-transformed variables. In order to achieve this, we shall have to do some work. Note that the right-hand side involves an integral in t' over the hyperplane $n' \cdot x' = 0$ where

$$n'_\mu = n_\nu(\Lambda^{-1})^\nu{}_\mu \,.$$

Noting that x^μ can also be written in the form

$$x^\mu = \sum_\rho x^\rho n^\mu(\rho) \,,$$

where $n^\mu(\rho), \rho = 0, 1, 2, 3$ are the 4-unit basis 4-vectors

$$n^\mu(\rho) = \delta^\mu_{\ \rho} \ ,$$

we now do the analogous thing with x'^μ. Expanding x'^μ in terms of the corresponding Lorentz-transformed basis vectors,

$$n'^\mu(\rho) = \Lambda^\mu_{\ \nu} n^\nu(\rho)$$

we have

$$x'^\mu = \sum_\rho \xi^\rho n'^\mu(\rho) \ .$$

Since

$$n'(\rho) \cdot n'(\lambda) = g^\lambda_{\ \rho}$$

we also have

$$x' \cdot n'(0) = \xi^0, \quad x' \cdot n'(i) = \xi^i \ . \tag{6.6}$$

Now

$$J = \left| \frac{\partial x'^\mu}{\partial \xi^\nu} \right| = \det\Lambda = 1 \ .$$

Hence we have from (6.5),

$$\int d^3r P(\vec{r}, t) = \int d^4\xi \delta(\xi^0)(n'_\mu(0)j'^\mu(x'(\xi))$$

$$= \int_{\xi^0=0} d^3\xi (n'_\mu(0)j'^\mu(x'(\xi)) \ . \tag{6.7}$$

Because of (6.6), the $d^3\xi$ integration extends over the surface σ' "perpendicular" to $n'^\mu(0)$. We want to convert this integral into an integral over the surface "perpendicular" to $n^\mu(0)$. We thus need some kind of "Gauß theorem" in Minkowski space. Such a theorem exists and states

$$\int_V d^4x \partial_\mu F^\mu(x) = \int_{S(V)} d\sigma(x) n_\mu(x) F^\mu(x) \ , \tag{6.8}$$

where $S(V)$ is the closed *oriented* surface composed of $\xi^0 = 0$ and $x^0 = 0$, closed at infinity, bounding the volume V. According to Gauß' law (6.8) we have

$$\int_{\xi^0=0} d^3\xi n'_\mu(0)j'^\mu(x'(\xi)) - \int_{x'^0=0} d^3x' n_\mu(0)j'^\mu(x') = \int_V d^4x' \partial'_\mu j'^\mu(x') = 0$$

where we have used current conservation, and n'^μ and n^μ are *oriented* unit vectors "orthogonal" to the surfaces $\xi^0 = 0$ and $x^0 = 0$, respectively. Hence

$$\int d^3\xi n'_\mu(0)j'^\mu(x'(\xi)) = \int d^3r' j'^0(\vec{r}', t') \ .$$

From (6.7) we thus conclude that (6.3) is satisfied. This establishes the independence of the normalization integrals (6.3) of·the choice of Lorentz frame.

It remains to find a suitable candidate for the probability density. Now, since the Dirac field transforms as (see (4.34) and (4.35))

$$\psi(x) \longrightarrow \psi'(x) = \mathcal{D}_{\alpha\beta}[\Lambda^{-1}]\psi_\beta(\Lambda x) \ ,$$

$$\bar\psi(x) \longrightarrow \bar\psi'(x) = \bar\psi_\beta(\Lambda x)\mathcal{D}_{\beta\alpha}[\Lambda]$$

under a Lorentz transformation Λ, it follows, using (4.33), that the bilinear

$$j^\mu(x) = \bar\psi(x)\gamma^\mu\psi(x)$$

transforms like a 4-vector field. One also easily establishes that it is conserved. Indeed, from the Dirac equation

$$(i\slashed{\partial} - m)\psi = 0 \tag{6.9}$$

follows

$$\bar\psi(-i \overleftarrow{\slashed{\partial}} - m) = 0 \ , \tag{6.10}$$

where $\overleftarrow{\partial}$ is defined by

$$\bar\psi \overleftarrow{\partial_\mu} := \partial_\mu\bar\psi \ ,$$

and use has been made of

$$\beta\gamma^{\mu\dagger}\beta = \gamma^\mu \quad (\beta = \gamma^0) \ .$$

Multiplying (6.9) with $\bar\psi$ from the left and (6.10) with ψ from the right, and subtracting one equation from the other, we finally have

$$\partial_\mu(\bar\psi\gamma^\mu\psi) = 0 \ ,$$

which proves our claim. We thus identify the probability density with

$$P(x) = \bar\psi(x)\gamma^0\psi(x) \ . \tag{6.11}$$

6.2 Non-relativistic limit

In order to discuss the non-relativistic limit of the Dirac equation, one preferably works in the *Dirac representation* of the Gamma matrices. Of course such a limit can only be discussed for massive particles. From (4.21) we have[1]

$$U(\vec p, \sigma) = \begin{pmatrix} U^{(1)}(\vec p, \sigma) \\ U^{(2)}(\vec p, \sigma) \end{pmatrix} = \sqrt{\frac{\omega + m}{2m}}\begin{pmatrix} \chi(\sigma) \\ \frac{\vec p \cdot \vec\sigma}{\omega + m}\chi(\sigma) \end{pmatrix}. \tag{6.12}$$

Noting that

$$\frac{\vec p}{\omega + m} \simeq \frac{\vec p}{2m} \simeq \frac{1}{2}v \ ,$$

---

[1] J.D. Bjorken and S.D. Drell, *Relativistic Quantum Fields* (McGraw-Hill, 1965).

we see from (6.12) that

$$U^{(2)}(\vec{p}, \sigma) \simeq \frac{\vec{v} \cdot \vec{\sigma}}{2} U^{(1)}(\vec{p}, \sigma) .$$

Hence one refers to the upper (lower) components of the positive energy spinor $U(p, \sigma)$ as the "large" ("small") components. For the case of the negative energy spinor $V(p, \sigma)$ the situation is just the reverse. Hence in the non-relativistic limit the positive and negative energy solutions have the form

$$\psi^{(+)}(x) \sim e^{-imt} e^{i\vec{p} \cdot \vec{x} - i\frac{p^2}{2m}t} \begin{pmatrix} \chi(\sigma) \\ 0 \end{pmatrix} ,$$

$$\psi^{(-)}(x) \sim e^{imt} e^{-i\vec{p} \cdot \vec{x} + i\frac{p^2}{2m}t} \begin{pmatrix} 0 \\ \chi^c(\sigma) \end{pmatrix} . \tag{6.13}$$

They evidently satisfy

$$i\partial_t \psi^{(\pm)}(x)_{n.r.} = \pm H_{n.r.} \psi^{(\pm)}(x)_{n.r.} ,$$

with

$$H_{n.r.} = -\frac{\vec{\nabla}^2}{2m} + m . \tag{6.14}$$

This equation shows that only the solutions of positive energy lead to the Schrödinger equation in the non-relativistic limit. The mass dependent exponential in (6.13) is irrelevant, and merely corresponds to a shift in the "zero point energy"; it accounts for the rest-mass energy mc^2 of a particle at rest.

6.3   Negative-energy solutions and localization

If at time $t = 0$ we have a *Gaussian* wave packet localized within a distance $2d$ around $r = 0$, such as

$$\psi(\vec{r}, t = 0) = \frac{1}{(\pi d^2)^{3/4}} e^{-\frac{r^2}{2d^2}} U\left(0, \frac{1}{2}\right), \quad U\left(0, \frac{1}{2}\right) = \begin{pmatrix} \chi(\frac{1}{2}) \\ 0 \end{pmatrix} , \tag{6.15}$$

then this corresponds to a Gaussian momentum distribution with uncertainty

$$\Delta p \simeq \hbar/d$$

in the momentum. To make this explicit, we recall that the general solution to the Dirac equation at any time is of the form

$$\psi(\vec{r}, t) = \frac{1}{(2\pi)^{3/2}} \int d^3 p \sqrt{\frac{m}{\omega_p}} \sum_\sigma \left[U(\vec{p}, \sigma) a(\vec{p}, \sigma) e^{-ip \cdot x} + V(\vec{p}, \sigma) b^*(\vec{p}, \sigma) e^{ip \cdot x} \right] ,$$

where $p^0 = \omega_p = \omega(\vec{p})$. Since the Fourier coefficients $a(p, \sigma)$ and $b^*(p, \sigma)$ are independent of time, they are determined by the initial condition (6.15) via the orthogonality relations

$$U^\dagger(\vec{p}, \sigma) U(\vec{p}, \sigma') = \frac{\omega_p}{m} \delta_{\sigma\sigma'}$$

$$V^\dagger(\vec{p},\sigma)V(\vec{p},\sigma') = \frac{\omega_p}{m}\delta_{\sigma\sigma'} \tag{6.16}$$

$$U^\dagger(\vec{p},\sigma)V(-\vec{p},\sigma') = 0.$$

One has

$$a(\vec{p},\sigma) = \frac{2m}{(2\pi)^{3/2}}\left(U^\dagger(\vec{p},\sigma)U\left(0,\frac{1}{2}\right)\right)\frac{1}{(\pi d^2)^{3/4}}\int d^3r\, e^{-\frac{r^2}{2d^2}}e^{-i\vec{p}\cdot\vec{r}},$$

$$b^*(\vec{p},\sigma) = \frac{2m}{(2\pi)^{3/2}}\left(V^\dagger(\vec{p},\sigma)U\left(0,\frac{1}{2}\right)\right)\frac{1}{(\pi d^2)^{3/4}}\int d^3r\, e^{-\frac{r^2}{2d^2}}e^{i\vec{p}\cdot\vec{r}}.$$

The remaining integral can be computed by the usual procedure of rewriting the exponent as a quadratic form and making a distortion of the integration contour in the complex plane, or by using the formulae

$$\int_{-\infty}^{\infty} dr^3\, f(|\vec{r}|)e^{-i\vec{p}\cdot\vec{r}} = -\frac{4\pi}{p}\frac{\partial}{\partial p}\int_0^\infty dr f(r)\cos pr$$

and

$$\int dr e^{-\frac{r^2}{2d^2}}\cos pr = \sqrt{\frac{\pi}{2}}e^{-\frac{p^2 d^2}{2}},$$

where $p = |\vec{p}|$. Either procedure gives

$$\int d^3r\, e^{-\frac{r^2}{2d^2}}e^{i\vec{p}\cdot\vec{r}} = (2\pi d^2)^{3/2}e^{-\frac{|\vec{p}|^2 d^2}{2}}.$$

Hence we have

$$a(\vec{p},\sigma) = 2m\left(\frac{d^2}{\pi}\right)^{3/4}e^{-\frac{|\vec{p}|^2 d^2}{2}}U^\dagger(\vec{p},\sigma)U\left(0,\frac{1}{2}\right)$$

$$b^*(\vec{p},\sigma) = 2m\left(\frac{d^2}{\pi}\right)^{3/4}e^{-\frac{|\vec{p}|^2 d^2}{2}}V^\dagger(\vec{p},\sigma)U\left(0,\frac{1}{2}\right).$$

Now, using (4.21) and (6.15) we have

$$U^\dagger(\vec{p},\sigma)U\left(\vec{0},\frac{1}{2}\right) = \sqrt{\frac{\omega+m}{2m}}\delta_{\sigma,\frac{1}{2}}$$

$$V^\dagger(\vec{p},\sigma)U\left(\vec{0},\frac{1}{2}\right) = (-1)^{\sigma+\frac{1}{2}}\sqrt{\frac{\omega+m}{2m}}\left(\chi^\dagger(-\sigma)\frac{\vec{p}\cdot\vec{\sigma}}{\omega+m}\chi\left(\frac{1}{2}\right)\right).$$

Hence

$$a(\vec{p},\sigma) = 2m\left(\frac{d^2}{\pi}\right)^{3/4}\sqrt{\frac{\omega+m}{2m}}e^{-\frac{|\vec{p}|^2 d^2}{2}}\delta_{\sigma,\frac{1}{2}},$$

$$b^*(\vec{p},\sigma) = (-1)^{\sigma+\frac{1}{2}}2m\left(\frac{d^2}{\pi}\right)^{3/4}\sqrt{\frac{\omega+m}{2m}}e^{-\frac{|\vec{p}|^2 d^2}{2}}\left(\chi^\dagger(-\sigma)\vec{\sigma}\cdot\hat{n}\chi\left(\frac{1}{2}\right)\right)\frac{|\vec{p}|}{\omega+m},$$

where

$$\hat{n} = \frac{\vec{p}}{|\vec{p}|}, \quad \omega = \omega_p .$$

Hence the Fourier coefficients associated with the negative energy solutions are of order $|\vec{p}|/(\omega + m)$ smaller than those associated with the positive energy solutions. They only become relevant at momenta comparable with the rest mass of the particle. In particular this means that a non-relativistic Schrödinger limit can only exist provided the uncertainty in the momentum is smaller than the rest mass of the particle. Since the uncertainty in the momentum for the wave packet in question is

$$\Delta p d \simeq \hbar$$

this means that for a non-relativistic limit to exist, we must demand that

$$d > \frac{\hbar}{mc} = \lambda_c.$$

This condition is of course satisfied for plane waves, which correspond to $d = \infty$. In that case a non-relativistic limit exists for sufficiently small momenta. In fact, a positive energy plane wave at time $t = 0$ will remain a positive energy plane wave at all times. On the other hand, if the particle is localized initially within a Compton wave length, negative energy components will contribute significantly. Such a localization can for instance be achieved by the application of sufficiently strong external fields. The result of applying such fields is beautifully exemplified by the *Klein paradox*, which we now discuss.

6.4 The Klein Paradox

As we have seen in the previous section, negative energy solutions are expected to play a role if we try to localize a particle within a Compton wave length by applying for instance a sufficiently strong field. As we shall see later on, the effect of such a field can be understood in terms of electron-positron pair creation. This picture, however, also prevails if we consider the scattering of an electron by a static Coulomb potential in the absence of a radiation field. In that case we can always choose a gauge in which the vector potential $\vec{A}$ vanishes. The Dirac Hamiltonian correspondingly takes the form

$$H = -i\vec{\alpha} \cdot \vec{\nabla} + \beta m + V(\vec{r})$$

with

$$V(\vec{r}) = e\phi(\vec{r}), \quad A^0 = \phi .$$

We look for stationary solutions of the corresponding Dirac equation,

$$i\partial_t \psi = H\psi \tag{6.17}$$

having the form

$$\psi(\vec{r}, t) = e^{-iEt} \begin{pmatrix} \psi_1(\vec{r}) \\ \psi_2(\vec{r}) \end{pmatrix} . \tag{6.18}$$

Working in the *Dirac* representation, substitution of (6.18) into (6.17) leads to the coupled set of equations

$$E\psi_1 = -i\vec{\sigma} \cdot \vec{\nabla}\psi_2 + (m + V)\psi_1$$
$$E\psi_2 = -i\vec{\sigma} \cdot \vec{\nabla}\psi_1 + (-m + V)\psi_2 .$$

From the second equation we have

$$\psi_2(\vec{r}) = \frac{-i}{E + m - V(\vec{r})}\vec{\sigma} \cdot \vec{\nabla}\psi_1(\vec{r}) . \tag{6.19}$$

Substitution of this expression into the first equation leads to an equation for the "large" component $\psi_1(\vec{r})$:

$$[E - V(\vec{r}) - m]\psi_1(\vec{r}) = (-i\vec{\sigma} \cdot \vec{\nabla})\frac{1}{E - V(\vec{r}) + m}(-i\vec{\sigma} \cdot \vec{\nabla})\psi_1(\vec{r}) . \tag{6.20}$$

We now specialize to the case of one spatial dimension. Suppose $V(\vec{r})$ to be given by the simple potential barrier in one spatial z-dimension depicted in the figure below,

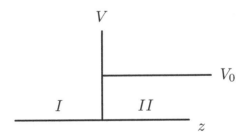

where the momentum of the plane waves takes the following values in regions I and II:

$$k^2 = \begin{cases} k_I^2 = E^2 - m^2 \\ k_{II}^2 = (E - V_0)^2 - m^2 \end{cases} . \tag{6.21}$$

For an incident spin 1/2 particle which is polarized in the z-direction we have from (6.20) and (6.19) for the incident and reflected waves in region I

$$\psi_{in}^{(I)}(z) = a_\uparrow e^{ik_I z}\begin{pmatrix} \chi(\uparrow) \\ \frac{k_I}{E+m}\chi(\uparrow) \end{pmatrix}$$

and

$$\psi_{refl}^{(I)}(z) = e^{-ik_I z}\left\{ b_\uparrow \begin{pmatrix} \chi(\uparrow) \\ \frac{-k_I}{E+m}\chi(\uparrow) \end{pmatrix} + b_\downarrow \begin{pmatrix} \chi(\downarrow) \\ \frac{k_I}{E+m}\chi(\downarrow) \end{pmatrix}\right\} .$$

Since we have no reflection in region II, the solution $\psi^{(II)}(x)$ has in this case the form

$$\psi^{(II)}(z) = e^{ik_{II} z}\left\{ d_\uparrow \begin{pmatrix} \chi(\uparrow) \\ \frac{k_{II}}{E+m-V_0}\chi(\uparrow) \end{pmatrix} + d_\downarrow \begin{pmatrix} \chi(\downarrow) \\ \frac{-k_{II}}{E+m-V_0}\chi(\downarrow) \end{pmatrix}\right\} .$$

Continuity at the potential barrier (current conservation) requires (we choose the potential barrier to lie at $z = 0$)

$$a_\uparrow + b_\uparrow = d_\uparrow \quad ; \quad b_\downarrow = d_\downarrow$$

$$(a_\uparrow - b_\uparrow) = \frac{k_{II}}{k_I} \frac{E + m}{E - V_0 + m} d_\uparrow \quad ; \quad b_\downarrow = -\frac{k_{II}}{k_I} \frac{E + m}{E + m - V_0} d_\downarrow .$$

It is thus convenient to define

$$r := \frac{k_{II}}{k_I} \frac{E + m}{E - V_0 + m} . \tag{6.22}$$

We then have to solve

$$2a_\uparrow = (1 + r)d_\uparrow , \quad 2b_\uparrow = (1 - r)d_\uparrow , \quad b_\downarrow = rd_\downarrow , \quad b_\downarrow = d_\downarrow.$$

Hence $b_\downarrow = d_\downarrow = 0$, i.e. there is no spin flip.

Now, according to (6.21) k_{II} is real for $E > V_0 + m$, $E < V_0 - m$. Hence free particle propagation in region II occurs for these values. Only the case $E < V_0 - m$ corresponding to free propagation in a classically *forbidden* region is of interest to us. The inequality states that for a potential strength of at least twice the mass of the particle there will be registered a current flow corresponding to the propagation of free particles.

The ratio of transmitted to incident current $J = \bar{\psi}\gamma^3\psi$ can be expressed in terms of this parameter r:

$$\frac{J_{tr}}{J_{in}} = \frac{4r}{(1 + r)^2}$$

$$\frac{J_{refl}}{J_{in}} = \frac{(1 - r)^2}{(1 + r)^2} = 1 - \frac{J_{tr}}{J_{in}} .$$

For the classically allowed energy range $E - V_0 > m$ we have $0 < r < 1$, and correspondingly the usually expected result

$$0 < \frac{J_{tr}}{J_{in}} < 1 .$$

For $E < V_0 - m$, on the other hand, we have $r < 0$. In this energy domain we thus find

$$\frac{J_{tr}}{J_{in}} < 0$$

corresponding to a "negative" transmission which represents an entirely new situation. At the same time we find

$$\frac{J_{refl}}{J_{in}} > 1$$

which indicates the existence of a *source* of particles in region II. Hence the following picture emerges: As soon as the potential V_0 exceeds twice the value of the electron rest mass, the current moving from right to left in region I exceeds the incident current of electrons. This indicates that pair creation has taken place in the background of the external Coulomb potential, as soon as it exceeded the minimum energy value required to produce a real electron-positron pair at rest.

6.5   Foldy–Wouthuysen Transformation

In Section 2 we have discussed the non-relativistic limit of the free Dirac equation. To order $O(1/m)$, the Hamiltonian for the positive energy solutions was given by (6.14) in this limit. We can repeat these considerations in the case of an electron coupled to an external electromagnetic field as described by the Dirac Hamiltonian

$$H = \vec{\alpha} \cdot (\vec{p} - e\vec{A}) + \beta m + e\phi , \tag{6.23}$$

where $\vec{p}$ stands for the momentum conjugate to the coordinate $\vec{r}$,

$$\vec{p} = \frac{1}{i}\vec{\nabla} \quad (p^i = -i\partial_i = i\partial^i)$$

and e stands for the (negative) electron charge. We shall consider the case of a *time independent* vector potential $(\phi(\vec{r}), \vec{A}(\vec{r}))$. As before, we work in the Dirac representation. The corresponding equations of motion then read

$$i\partial_0 \psi_1 = \vec{\sigma} \cdot \vec{\pi}\psi_2 + (e\phi + m)\psi_1 = E\psi_1$$

$$i\partial_0 \psi_2 = \vec{\sigma} \cdot \vec{\pi}\psi_1 + (e\phi - m)\psi_2 = E\psi_2 \tag{6.24}$$

where

$$\vec{\pi} = \vec{p} - e\vec{A}$$

and we looked for stationary positive and negative energy solutions. For $e\phi \ll m$ we have in the non-relativistic limit $E \approx m$ from the second equation,

$$\psi_2 = \frac{(\vec{\sigma} \cdot \vec{\pi})}{2m}\psi_1 + O\left(\frac{1}{m^2}\right) ,$$

or substituting into the first equation,

$$i\partial_0 \psi_1 = \left[\frac{(\vec{\sigma} \cdot \vec{\pi})^2}{2m} + (m + e\phi) + O\left(\frac{1}{m^2}\right)\right]\psi_1 = E\psi_1 .$$

With

$$(\sigma \cdot \pi)^2 = \pi^2 - e\vec{\sigma} \cdot \vec{B}$$

we conclude that

$$H = \left(m + \frac{(\vec{p} - e\vec{A})^2}{2m}\right) + e\phi - \frac{e}{2m}\vec{\sigma} \cdot \vec{B} + O\left(\frac{1}{m^2}\right) . \tag{6.25}$$

This is the *Pauli* Hamiltonian.

In order to unravel further interaction terms such as the spin-orbital interaction we must obtain an expansion to order $O(\frac{1}{m^2})$. A systematic method for obtaining an expansion to higher orders has been given by Foldy–Wouthuysen. It consists in seeking a unitary transformation,

$$\psi \to \psi' = U\psi, \quad U = e^{iS}, \quad S^\dagger = S , \tag{6.26}$$

such that in the new basis the Hamiltonian (6.23) becomes diagonal, implying the decoupling of *upper* and *lower* components. For a time-dependent electromagnetic field the generator S of this unitary transformation will be *time-* as well as space-dependent. Hence the Dirac wave function ψ' satisfies in general the equation

$$i\partial_t\psi' = H'\psi'$$

with the new Hamiltonian

$$H' = UHU^{-1} - Ui(\partial_t U^{-1}) . \tag{6.27}$$

It turns out that, except for the case of the free Dirac Hamiltonian, this diagonalization of H can only be achieved to every prescribed order of $1/m$. The systematic procedure for realizing this program has been given by Foldy–Wouthuysen (FW), and the unitary transformation (6.26) is correspondingly referred to as the *Foldy–Wouthuysen transformation*.

In order to characterize the general procedure, it is convenient to introduce the notion of *even* and *odd* operators, corresponding to *diagonal* and *off-diagonal* matrices in the *Dirac* representation. Correspondingly, we write the Dirac Hamiltonian (6.23) in the form

$$H = m\beta + O + \mathcal{E} ,$$

where O and $\mathcal{E}$ stand for the *odd* and *even* operators

$$O = \vec{\alpha} \cdot (\vec{p} - e\vec{A}), \quad \mathcal{E} = e\phi .$$

Our goal will be to bring H' into an analogous form

$$H' = m\beta + O' + \mathcal{E}'$$

with O' of order $0\left(\frac{1}{m}\right)^{r+1}$. As we now show this can be achieved by an iterative process.

The free Dirac Hamiltonian

Before we treat the interacting case, it is instructive to find the FW-transformation for a free Dirac Hamiltonian

$$H_0 = m\beta + O_0 + \mathcal{E}_0 \tag{6.28}$$

with[2]

$$O_0 = \vec{\alpha} \cdot \mathbb{P}, \quad \mathcal{E}_0 = 0.$$

We seek a unitary transformation

$$\psi \to \psi' = e^{iS_0}\psi$$

which decouples the upper and lower components of the Dirac wave function. Working in the Dirac basis, this means that we seek a transformation which eliminates

---

[2]Our notation emphasizes that $\mathbb{P}$ is an operator.

the *off-diagonal* matrices $\vec{\alpha}$ in favor of the *diagonal* ones, β and the *identity*. In the case in question, the generator S_0 will be independent of time, so that the transformed Hamiltonian will be given by

$$H_0' = e^{iS_0} H_0 e^{-iS_0} .$$

Since we wish to eliminate the *odd* term in (6.28), we may assume that the generator S_0 of the unitary transformation will be a function of $\vec{\alpha} \cdot \vec{\mathbb{P}}$. We correspondingly make the Ansatz

$$i S_0 = \Theta \beta \vec{\alpha} \cdot \vec{\mathbb{P}} , \qquad (6.29)$$

where Θ is allowed to depend on $\vec{\mathbb{P}}^2$. Since

$$[\beta, S_0]_+ = 0, \quad [\vec{\alpha} \cdot \vec{\mathbb{P}}, S_0]_+ = 0$$

we have

$$H_0' = H_0 e^{-2iS_0} .$$

Now, the exponential of S_0 is defined by its power series expansion

$$e^{-2iS_0} = \sum_{n=0}^{\infty} \frac{(-2iS_0)^n}{n!}$$

$$= \sum_{n=0}^{\infty} \frac{(-2iS_0)^{2n}}{(2n)!} + \sum_{n=0}^{\infty} \frac{(-2iS_0)^{2n+1}}{(2n+1)!} ,$$

or since

$$(-iS_0)^2 = \Theta^2 \beta(\vec{\alpha} \cdot \vec{\mathbb{P}})\beta(\vec{\alpha} \cdot \vec{\mathbb{P}}) = -\Theta^2(\vec{\alpha} \cdot \vec{\mathbb{P}})^2 = -\Theta^2 \vec{\mathbb{P}}^2 ,$$

we have

$$(-2iS_0)^{2n} = [(-2iS_0)^2]^n = (-1)^n (2\Theta|\vec{\mathbb{P}}|^2)^{2n}$$

and

$$(-2iS_0)^{2n+1} = (-2iS_0)(-1)^n (2\Theta|\vec{\mathbb{P}}|)^{2n}$$

$$= -\beta \frac{(\vec{\alpha} \cdot \vec{\mathbb{P}})}{|\vec{\mathbb{P}}|}(-1)^n (2\Theta|\vec{\mathbb{P}}|)^{2n+1}.$$

Hence

$$e^{-2iS_0} = \cos 2\Theta|\vec{\mathbb{P}}| - \beta \frac{\vec{\alpha} \cdot \vec{\mathbb{P}}}{|\vec{\mathbb{P}}|} \sin 2\Theta|\vec{\mathbb{P}}| .$$

We thus find

$$H_0' = (\vec{\alpha} \cdot \vec{\mathbb{P}} + m\beta) \left(\cos 2\Theta|\vec{\mathbb{P}}| - \beta \frac{\vec{\alpha} \cdot \vec{\mathbb{P}}}{|\vec{\mathbb{P}}|} \sin 2\Theta|\vec{\mathbb{P}}| \right) .$$

In order to eliminate the off-diagonal operator $\vec{\alpha} \cdot \vec{\mathbb{P}}$ we must choose Θ such that

$$\tan 2\Theta|\vec{\mathbb{P}}| = \frac{|\vec{\mathbb{P}}|}{m} . \qquad (6.30)$$

With this choice we have

$$H_0' = m\beta(\cos 2\Theta|\vec{\mathbb{P}}|)\left(1 + \frac{|\vec{\mathbb{P}}|}{m}\tan 2\Theta|\vec{\mathbb{P}}|\right)$$

$$= m\beta(\cos 2\Theta|\vec{\mathbb{P}}|)\left(1 + \left(\frac{\vec{\mathbb{P}}}{m}\right)^2\right).$$

Now,

$$\cos 2\Theta|\vec{\mathbb{P}}| = \frac{m}{\sqrt{m^2 + |\vec{\mathbb{P}}|^2}}$$

or we finally have

$$H_0' = \beta\sqrt{m^2 + |\vec{\mathbb{P}}|} .\tag{6.31}$$

Our unitary transformation has thus lead to the decomposition of the free Dirac Hamiltonian into the *direct sum* of two *non-local* Hamiltonians $\pm\sqrt{m^2 + |\vec{\mathbb{P}}|^2}$ with positive and negative energy spectrum.

From (6.30) we see that

$$\Theta = \frac{1}{2m} + O\left(\frac{1}{m^2}\right) ,$$

so that the modified Ansatz

$$iS_0 = \frac{\beta}{2m}\vec{\alpha} \cdot \vec{\mathbb{P}}\tag{6.32}$$

replacing (6.29) would lead to an H' with an *odd* operator of order $O(1/m^2)$. This observation will play a central role when we now come to discuss the general, interacting case.

Dirac Hamiltonian in presence of EM field

The general FW procedure now involves the following steps: Let $r + 1$ be the order to which we *wish* to diagonalize the Hamiltonian (6.27). The FW method then consists in the *iterative* computation of

$$H^{(l+1)} = e^{iS^{(l)}}H^{(l)}e^{-iS^{(l)}} - e^{iS^{(l)}}(i\partial_t e^{-iS^{(l)}}), \quad l = 0, 1, .., r ,$$

with

$$H^{(l)} = m\beta + O^{(l)} + \mathcal{E}^{(l)}$$

where $O^{(\ell)}$ and $\mathcal{E}^{(\ell)}$ are computed *up to* order $O(\frac{1}{m^{r+1}})$ at each iterative step,

$$O^{(l)} = \frac{1}{m^l}\left(a_0 + \frac{a_1}{m} + ... + \frac{a_{r-l}}{m^{r-l}}\right) + O\left(\frac{1}{m^{r+1}}\right) ,$$

with the Ansatz

$$iS^{(l)} = \frac{\beta}{2m}O^{(l)} , \quad iS^{(0)} = \frac{\beta}{2m}\alpha \cdot \vec{\mathbb{P}}$$

and the starting values

$$H^{(0)} = m\beta + \mathcal{E}^{(0)} + O^{(0)}$$
$$O^{(0)} = \vec{\alpha} \cdot \vec{\pi}, \quad \vec{\pi} = \vec{\mathbb{P}} - e\vec{A}$$
$$\mathcal{E}^{(0)} = e\phi \ .$$

Carrying out this iterative program up to corrections of $O\left(\frac{1}{m^3}\right)$, we arrive at[3]

$$H^{(3)} = \beta \left(m + \frac{(\vec{\mathbb{P}} - e\vec{A})^2}{2m} \right) + e\phi - \frac{e}{2m}\beta\vec{\Sigma} \cdot \vec{B} \tag{6.33}$$

$$- \left(\frac{ie}{8m^2}\vec{\Sigma} \cdot (\vec{\nabla} \times \vec{E}) - \frac{e}{4m^2}\vec{\Sigma} \cdot (\vec{E} \times \vec{\mathbb{P}}) \right) - \frac{e}{8m^2}div E + O\left(\frac{1}{m^3}\right)$$

where

$$\vec{\Sigma} = \begin{pmatrix} \vec{\sigma} & 0 \\ 0 & \vec{\sigma} \end{pmatrix} \ .$$

Interpretation of the result

To order $O(1/m)$ the Hamiltonian leads to the *Pauli equation* for the upper component of the transformed Dirac spinor: $\psi' = (\varphi, \chi)$

$$i\partial_t \varphi = \left[\frac{(\vec{\mathbb{P}} - e\vec{A})^2}{2m} - e\phi - \frac{e}{2m}\vec{\sigma} \cdot \vec{B} + m \right] \varphi \ .$$

The first term on the right-hand side of this equation if of course nothing but the first two leading terms in the expansion

$$\sqrt{(\vec{\mathbb{P}} - e\vec{A})^2 + m^2} = m \left(1 + \frac{(\vec{\mathbb{P}} - e\vec{A})^2}{2m^2} + O(m^{-4}) \right) \ .$$

The second term in (6.33) represents the electrostatic energy of a point-like charge. The third term describes the interaction of the particle spin $\vec{s} = \vec{\sigma}/2$ with the magnetic field, with an *anomalous* magnetic moment of gyromagnetic ratio $g = 2$:

$$\vec{\mu} = \frac{e}{m}\frac{\vec{\sigma}}{2} = g\frac{e}{2m}\vec{s} \ , \quad g = 2 \tag{6.34}$$

$$H_{Dip} = -\beta\vec{\mu} \cdot \frac{\vec{\sigma}}{2} \ .$$

This magnetic moment has been measured in the famous Stern–Gerlach experiment and represents one of the great successes of the Dirac equation. In reality this gyromagnetic ratio is not exactly equal to two due to so-called *radiative corrections* to be discussed at a later stage when we also quantize the electromagnetic field.

---

[3]C. Itzykson and J.-B. Zuber, *Quantum Field Theory* (McGraw-Hill Inc., 1980); J.D. Bjorken and Sidney D. Drell, *Relativistic Quantum Mechanics* (McGraw-Hill Inc., 1964).

Let us analyse the remaining terms in the Hamiltonian (6.33). For the fourth and the fifth terms we find

$$\frac{-ie}{8m^2}\vec{\Sigma}\cdot curl\vec{E} - \frac{e}{4m^2}\vec{\Sigma}\cdot(\vec{E}\times\mathbb{P}) = \frac{e}{8m^2}\epsilon_{ijk}\Sigma^i\{\mathbb{P}^j, E^k\}\ .$$

Note that this represents a *hermitian* operator. It acquires a familiar form if we restrict ourselves to a spherically symmetric potential

$$e\phi(\vec{r}) = V(r).$$

In that case $curl\ E = 0$, and

$$e\vec{\Sigma}\cdot(\vec{E}\times\mathbb{P}) = e\vec{\Sigma}\cdot(-\vec{\nabla}\phi\times\mathbb{P})$$

$$= -\frac{dV}{dr}\vec{\Sigma}\cdot(\vec{r}\times\mathbb{P}) = -\frac{1}{r}\frac{dV}{dr}\vec{\Sigma}\cdot\mathbb{L}$$

with

$$\vec{\mathbb{L}} = \vec{r}\times\mathbb{P}\ .$$

Hence the sum of the fifth and sixth terms in (6.33) just reduces in this case to the familiar *spin-orbit* interaction

$$H_{s.o} = -\frac{e}{2m^2}\frac{\vec{\Sigma}}{2}\cdot(\vec{E}\times\mathbb{P}) = \frac{1}{4m^2}\frac{1}{r}\frac{dV}{dr}\vec{\Sigma}\cdot\mathbb{L}\ . \tag{6.35}$$

This interaction term has a classical interpretation: Let us go into the rest frame of the electron where its spin is defined. In this reference system the electron sees a magnetic field B' whose strength is obtained by performing a Lorentz boost parametrized by $-v$:

$$\vec{B}' = -\gamma(v)\vec{v}\times\vec{E} \simeq \vec{E}\times\vec{v}\ .$$

Hence we can write the spin-orbit interaction also in the form ($\mathbb{P}\to\vec{p} = m\vec{v}$)

$$H_{s.o} = -\frac{e}{2m^2}\frac{\vec{\Sigma}}{2}\times(\vec{E}\times m\vec{v}) \approx -\frac{e}{2m}\vec{S}\cdot\vec{B}'\ .$$

This corresponds to the interaction of an electron-spin with a magnetic field $\vec{B}'$, with gyromagnetic ratio $g = 1$. This suggests writing (6.35) in the form ($g = 2$)

$$H_{s.o.} = (g-1)\frac{1}{2m^2}\frac{dV}{dr}\vec{S}\cdot\vec{\mathbb{L}}. \tag{6.36}$$

The appearance of the term

$$-\frac{1}{2m^2}\frac{dV}{dr}\vec{S}\cdot\vec{\mathbb{L}}$$

can be traced to the so-called *Thomas precession*[4] representing the precession of the electron spin with respect to the body-fixed coordinate system with angular velocity

$$\vec{\omega}_{th.} = -\frac{1}{2}\vec{v}\times\dot{\vec{v}}\ .$$

---

[4]Moeller, *The Theory of Relativity* (Oxford University Press, 1952).

This precession leads to a shift

$$\Delta H_{Th.} = \vec{S} \cdot \vec{\omega}_{Th}$$

in the energy. Now

$$m\dot{\vec{v}} = -\hat{r}\frac{dV}{dr} \ ,$$

or

$$\Delta H_{s.o} = \frac{1}{2m}\vec{S} \cdot (\vec{v} \times \hat{r})\frac{dV}{dr} = \frac{-1}{2m^2}\frac{dV}{dr}\vec{S} \cdot \vec{\mathbb{L}} \ .$$

This explains the correction to the gyromagnetic ratio in (6.36).

The last term in (6.33), the so-called *Darwin term*, has no classical interpretation, and can be roughly understood in terms of a quantum mechanical "Zitterbewegung" which we have not discussed.

Let us finally remark that the above analysis is somewhat misleading since the expectation value of the Hamiltonian (6.27) in the corresponding state ψ' is not the same as that of the original Hamiltonian H in the state ψ, since the FW transformation is in general time-dependent. Only in the case of a static electromagnetic field will the two expectation values agree.

Chapter 7

Second Quantization

In this chapter we take the first steps towards a quantum field theoretic formulation of the interaction of scalar bosons, fermions and photons. To this end we first consider the case of non-interacting particles. The essential step consists in replacing the fields of the Klein–Gordon, Dirac and Maxwell equations by *operator-valued fields* living in a *Fock space*. This process is referred to as *second Quantization*. The equations of motion for these operators can be derived from a lagrangian density via a variational principle. This will constitute the backbone of our general quantization procedure. It allows to rewrite the equations of motion for the operator-valued fields in the form of *Hamilton equations*, where the Hamiltonian is obtained from the Lagrange density via the usual Legendre transform. The 1-particle wave function is recovered in this second-quantized formulation by considering the appropriate matrix elements of the operators in Fock space.

7.1   Fock-space representation of fields

We have so far attempted to develop a 1-particle relativistic quantum mechanics, with a conventional probabilistic interpretation in terms of 1-particle wave functions satisfying a relativistically covariant differential equation. This had led us to the existence of also negative energy solutions, and a picture of e^+e^- pair production in the presence of an external Coulomb potential. This indicates that a consistent interpretation can only be achieved if we formulate relativistic quantum mechanics as a many-particle theory in which an arbitrary number of quanta can be created or destroyed, and suggests replacing the c-number fields $\phi(x), \psi(x)$ and $A^\mu(x)$ by *operator-valued fields* satisfying, *in the absence of interaction*, respectively:

(1) *The Klein–Gordon equation*

$$(\Box + m^2)\phi(x) = 0$$

with the solution

$$\phi(x) = \frac{1}{(2\pi)^{3/2}} \int \frac{d^3q}{\sqrt{2\omega_q}} [a(\vec{q})e^{-iq\cdot x} + a^\dagger(\vec{q})e^{iq\cdot x}] \qquad (7.1)$$

with $q^\mu = (\omega_q, \vec{q}), \omega_q = \sqrt{\vec{q}^2 + m^2}$.

(2) *the Dirac equation*

$$(i\partial\!\!\!/ - m)\psi(x) = 0$$

with the solution (4.9),

$$\psi_\alpha(x) = \frac{1}{(2\pi)^{3/2}} \int d^3p \sqrt{\frac{m}{\omega(\vec{p})}} \sum_\sigma [U(\vec{p}, \sigma)a(\vec{p}, \sigma)e^{-ip\cdot x} + V(\vec{p}, \sigma)b^\dagger(\vec{p}, \sigma)e^{ip\cdot x}] .$$

$$(7.2)$$

(3) *the Weyl equations* (4.52) and (4.53),

$$it \cdot \partial\varphi(x) = 0, \quad i\bar{t} \cdot \partial\bar{\varphi}(x) = 0$$

with the solutions (4.63),

$$\varphi_\sigma(x) = \frac{1}{(2\pi)^{3/2}} \int \frac{d^3p}{\sqrt{2|\vec{p}|}} D^{(\frac{1}{2})}[\mathcal{L}(\vec{p})]_{\sigma,-\frac{1}{2}} \left[a\left(\vec{p}, -\frac{1}{2}\right) e^{-ip\cdot x} + b^\dagger\left(\vec{p}, \frac{1}{2}\right) e^{ip\cdot x} \right],$$

$$\bar{\varphi}_{\bar{\sigma}}(x) = \frac{1}{(2\pi)^{3/2}} \int \frac{d^3p}{\sqrt{2|\vec{p}|}} D^{(\frac{1}{2})}[\mathcal{L}(\vec{p})]_{\bar{\sigma},\frac{1}{2}} \left[a\left(\vec{p}, \frac{1}{2}\right) e^{-ip\cdot x} + b^\dagger\left(\vec{p}, -\frac{1}{2}\right) e^{ip\cdot x} \right].$$

(4) *the Maxwell equation*

$$(\Box g_{\mu\nu} - \partial_\mu\partial_\nu)A^\nu(x) = 0$$

with the Coulomb-gauge solution $A_0 = 0$ and

$$\vec{A}(x) = \frac{1}{(2\pi)^{3/2}} \int \frac{d^3k}{\sqrt{2|\vec{k}|}} \sum_{\lambda=\pm1} \vec{\epsilon}(\vec{k}, \lambda)[a(\vec{k}, \lambda)e^{-ik\cdot x} + a^\dagger(\vec{k}, -\lambda)e^{ik\cdot x}] \qquad (7.3)$$

$$= \frac{1}{(2\pi)^{3/2}} \int \frac{d^3k}{\sqrt{2|\vec{k}|}} \sum_{\lambda=\pm1} [\vec{\epsilon}(\vec{k}, \lambda)a(\vec{k}, \lambda)e^{-ik\cdot x} + \vec{\epsilon}^*(\vec{k}, -\lambda)a^\dagger(\vec{k}, -\lambda)e^{ik\cdot x}] .$$

Lorentz transformation of field operators

The Fourier "coefficients" in the respective solutions are now operator-valued objects living in a *Fock space* to be defined below. Lorentz transformations are implemented in this Fock space by unitary operators $U[\Lambda]$, the operator-valued fields being required to transform as follows:

$$U[\Lambda]\phi(x)U^{-1}[\Lambda] = \phi(\Lambda x) \tag{7.4}$$

$$U[\Lambda]\psi_\alpha(x)U^{-1}[\Lambda] = \sum_\beta D^{(\frac{1}{2})}_{\alpha\beta}[\Lambda^{-1}]\psi_\beta(\Lambda x) \tag{7.5}$$

$$U[\Lambda]\varphi_\sigma(x)U^{-1}[\Lambda] = \sum_{\sigma'} D^{(\frac{1}{2},0)}_{\sigma\sigma'}[\Lambda^{-1}]\varphi_{\sigma'}(\Lambda x) \tag{7.6}$$

$$U[\Lambda]\tilde{\varphi}_{\bar{\sigma}}(x)U^{-1}[\Lambda] = \sum_{\tilde{\sigma}'} D^{(0,\frac{1}{2})}_{\bar{\sigma}\bar{\sigma}'}[\Lambda^{-1}]\tilde{\varphi}_{\bar{\sigma}'}(\Lambda x) \tag{7.7}$$

$$U[\Lambda]A^i(x)U^{-1}[\Lambda] = (\Lambda^{-1})^i_{\ j}A^j(\Lambda x) + \text{gauge}. \tag{7.8}$$

It will be one of our objectives in Chapter 9 to explicitly construct the operators $U[\Lambda]$ inducing such transformations. In contrast to the finite irreducible representations of the Lorentz group, these operators live in an infinite dimensional space, and are *unitary*.

Proposition

The Lorentz transformation properties (7.4) to (7.8) demand the following transformation properties for the Fourier coefficients:

(a) *Scalar field*

$$U[\Lambda]a(\vec{q})U^{-1}[\Lambda] = a'(\vec{q}) = \sqrt{\frac{\omega(\vec{\Lambda q})}{\omega(\vec{q})}}\,a(\vec{\Lambda q})\;. \tag{7.9}$$

(b) *Dirac and Weyl fields*

$$U[\Lambda]a(\vec{p},\sigma)U^{-1}[\Lambda] = \sqrt{\frac{\omega(\vec{\Lambda p})}{\omega(\vec{p})}}\sum_{\sigma'}D^{(\frac{1}{2})}[R_W^{-1}]_{\sigma\sigma'}a(\vec{\Lambda p},\sigma') \tag{7.10}$$

$$U[\Lambda]a^{\dagger}(\vec{p},\sigma)U^{-1}[\Lambda] = \sqrt{\frac{\omega(\vec{\Lambda p})}{\omega(\vec{p})}}\sum_{\sigma'}D^{(\frac{1}{2})*}[R_W^{-1}]_{\sigma\sigma'}a^{\dagger}(\vec{\Lambda p},\sigma') \tag{7.11}$$

$$= \sqrt{\frac{\omega(\vec{\Lambda p})}{\omega(\vec{p})}}\sum_{\sigma'}\{cD^{(\frac{1}{2})}[R_W^{-1}]c^{-1}\}_{\sigma\sigma'}a^{\dagger}(\vec{\Lambda p},\sigma') \tag{7.12}$$

where c has been defined in (4.13), and R_W^{-1} stands for the inverse of the Wigner rotation (2.45),

$$R_W^{-1} = L^{-1}(\vec{p})\Lambda^{-1}L(\vec{\Lambda p})\;.$$

(c) *Maxwell field*

$$U[\Lambda]a(\vec{k},\lambda)U^{-1}[\Lambda] = \sqrt{\frac{|\vec{\Lambda k}|}{|\vec{k}|}}e^{-i\lambda\Theta[\tilde{\Lambda}]}a(\vec{\Lambda k},\lambda) = \sqrt{\frac{|\vec{\Lambda k}|}{|\vec{k}|}}e^{i\lambda\Theta[\tilde{\Lambda}^{-1}]}a(\vec{\Lambda k},\lambda)$$

$$\tag{7.13}$$

$$U[\Lambda]a^{\dagger}(\vec{k},\lambda)U^{-1}[\Lambda] = \sqrt{\frac{|\vec{\Lambda k}|}{|\vec{k}|}}e^{i\lambda\Theta[\tilde{\Lambda}]}a^{\dagger}(\vec{\Lambda k},\lambda) = \sqrt{\frac{|\vec{\Lambda k}|}{|\vec{k}|}}e^{-i\lambda\Theta[\tilde{\Lambda}^{-1}]}a^{\dagger}(\vec{\Lambda k},\lambda)$$

where $\tilde{\Lambda}$ and $\Theta[\tilde{\Lambda}]$ have been defined in Section 6 of Chapter 2, and use has been made of $\Theta[\tilde{\Lambda}] = -\Theta[\tilde{\Lambda}^{-1}]$, following from (2.61). The reason for expressing the results in terms of the inverse of $\tilde{\Lambda}$ will become clear in the sequel and is vinculated to the appearance of Λ^{-1} in (7.8).

Proof of Proposition:

(**a**) The fact that (a) implies the Lorentz transformation (7.4) for the scalar field is obvious, if one makes use of the invariance property

$$\frac{d^3p}{\omega(\vec{p})} = \frac{d^3(\Lambda\vec{p})}{\omega(\Lambda\vec{p})} \ , \quad \frac{d^3k}{|\vec{k}|} = \frac{d^3(\Lambda\vec{k})}{|\Lambda\vec{k}|} \tag{7.14}$$

of the measure.

(**b**) Substitution of $U(\vec{p},\sigma)$ and $V(\vec{p},\sigma)$ given by (4.38) into the Fourier expansion for their fields, and making use of the usual group properties, results with (7.12) in

$$\psi'_\alpha(x) = \frac{1}{(2\pi)^{3/2}} \int d^3p \sqrt{\frac{m}{\omega(\vec{p})}} \sqrt{\frac{\omega(\Lambda\vec{p})}{\omega(p)}} \sum_{\sigma'} \left\{ \left(\begin{array}{c} D^{(\frac{1}{2})}[\Lambda^{-1}L(\Lambda\vec{p})]_{\bar{\alpha}\sigma'} \\ \bar{D}^{(\frac{1}{2})}[\Lambda^{-1}L(\Lambda\vec{p})]_{\underline{\alpha}\sigma'} \end{array} \right) a(\Lambda\vec{p},\sigma')e^{-ip\cdot x} \right.$$

$$+ \left. \left(\begin{array}{c} \{D^{(\frac{1}{2})}[\Lambda^{-1}L(\Lambda\vec{p})]c^{-1}\}_{\bar{\alpha}\sigma'} \\ -\{\bar{D}^{(\frac{1}{2})}[\Lambda^{-1}L(\Lambda\vec{p})]c^{-1}\}_{\underline{\alpha}\sigma'} \end{array} \right) b^\dagger(\Lambda\vec{p},\sigma')e^{ip\cdot x} \right\} . \tag{7.15}$$

Making again use of the group properties of the representation matrices and of the Lorentz invariance of the scalar product $p \cdot x$ and integration measure (7.14) one obtains (7.5). The corresponding proof in the Dirac representation just involves the transformation (4.41).

(**c**) Finally, the transformation law (7.8) is compatible with the Lorentz transformation properties of the gauge field in the Coulomb gauge. Substitution of (7.13) into the Lorentz-transformed vector potential in the Coulomb gauge yields

$$A'^i(x) = \frac{1}{(2\pi)^{3/2}} \int \frac{d^3q}{\sqrt{2|\vec{q}|}} \sqrt{\frac{|\Lambda\vec{q}|}{|\vec{q}|}} e^{-i\lambda\Theta[\tilde{\Lambda}]} \epsilon^i(\vec{q},\lambda)[a(\Lambda\vec{q},\lambda)e^{-iq\cdot x} + a^\dagger(\Lambda\vec{q},-\lambda)e^{iq\cdot x}] \tag{7.16}$$

where $\epsilon^\mu(\vec{q},\lambda)$ has been defined in (5.11), and has the property $\epsilon^0(\vec{q},\lambda) = 0$. We now observe that the polarization tensors constructed in Section 2 of Chapter 5 have the following remarkable property

Claim:

$$\left[(\Lambda^{-1})^i{}_j - (\Lambda^{-1})^0{}_j \frac{q^i}{|\vec{q}|} \right] \epsilon^j(\Lambda k, \lambda) = e^{i\lambda\Theta[\tilde{\Lambda}^{-1}]} \epsilon^i(q,\lambda) . \tag{7.17}$$

We now demonstrate this.

Proof:[1]
Since $A^\mu(x)$ represents a massless field, it transforms under the little group parametrized by the three independent parameters (2.57). For the *inverse* of the corresponding *finite* little group transformation (2.54) we have

$$\tilde{\Lambda}^{-1} = \begin{pmatrix} 1 + X^2/2 & X_1 & X_2 & -X^2/2 \\ X_1\cos\Theta + X_2\sin\Theta & \cos\Theta & \sin\Theta & X_1\cos\Theta - X_2\sin\Theta \\ -X_1\sin\Theta + X_2\cos\Theta & -\sin\Theta & \cos\Theta & X_1\sin\Theta - X_2\cos\Theta \\ X^2/2 & X_1 & X_2 & 1 - X^2/2 \end{pmatrix} , \tag{7.18}$$

---

[1] S. Weinberg, *Phys. Rev.* **135B** (1964) 1040.

where Θ is shorthand for $\Theta[\tilde{\Lambda}^{-1}]$, and

$$X^2 = X_1^2 + X_2^2 .$$

For an infinitesimal transformation the parameters Θ, X_1, X_2 reduce to $-\theta, -\chi_1, -\chi_2$, respectively, as defined in (2.57).

Now consider the polarization vectors (5.9),

$$(\epsilon_\pm^\mu) = \frac{1}{\sqrt{2}}(0, 1, \pm i, 0) . \tag{7.19}$$

Apply the Little-group transformation (7.18) to these vectors. We thus obtain

$$(\tilde{\Lambda}^{-1})^\mu{}_\nu \epsilon_\pm^\nu = \frac{1}{\sqrt{2}} \begin{pmatrix} X_1 \pm iX_2 \\ e^{\pm i\Theta} \\ \pm i e^{\pm i\Theta} \\ X_1 \pm iX_2 \end{pmatrix} = e^{\pm i\Theta} \epsilon_\pm^\mu + X_\pm \frac{\tilde{k}^\mu}{\kappa} \tag{7.20}$$

where $\tilde{k}^\mu = (\kappa, 0, 0, \kappa)$ is the Little-group *invariant* vector introduced in Section 2.6 of Chapter 2, and

$$X_\pm(\tilde{\Lambda}) = \frac{X_1 \pm iX_2}{\sqrt{2}} .$$

Remembering that $\tilde{\Lambda}$ can also be written in the form (2.54), we have for (7.20)

$$[\mathcal{L}(q)^{-1}\Lambda^{-1}]^\mu{}_\nu \epsilon_\pm^\nu(\Lambda q) = e^{\pm i\Theta} \epsilon_\pm^\mu + X_\pm(\tilde{\Lambda}) \frac{\tilde{k}^\mu}{\kappa} \tag{7.21}$$

where we have used (5.11). Multiplying now (7.21) from the left with $\mathcal{L}(q)$ we have

$$(\Lambda^{-1})^\mu{}_\nu \epsilon_\pm^\nu(\Lambda k) = e^{\pm i\Theta} \epsilon_\pm^\mu(q) + X_\pm(\tilde{\Lambda}) \frac{q^\mu}{\kappa}$$

where $q^\mu = \mathcal{L}^\mu{}_\nu(q)\tilde{k}$ (see (2.49)). Recalling that $\epsilon_\pm^0(k) = 0$, we may rewrite the last term in polarization tensor terms as follows:

$$X_\pm(\tilde{\Lambda})\frac{|\vec{q}|}{\kappa} = (\Lambda^{-1})^0{}_\nu \epsilon_\pm^\nu(\Lambda q) .$$

Collecting terms we arrive at (7.17).

Note that both sides vanish for $\mu = 0$, consistent with the Coulomb gauge where $A^0 = 0$. Hence, remembering that $\Theta[\tilde{\Lambda}^{-1}] = -\Theta[\tilde{\Lambda}]$, we obtain with (7.17),

$$U[\Lambda]A^i(x)U[\Lambda] = \frac{1}{(2\pi)^{3/2}} \int \frac{d^3(\vec{\Lambda q})}{\sqrt{2|\vec{\Lambda q}|}} \left[(\Lambda^{-1})^\mu{}_\nu - (\Lambda^{-1})^0{}_\nu \frac{q^\mu}{|\vec{q}|} \right]$$

$$\times \epsilon^\nu(\vec{\Lambda q}, \lambda)[a(\vec{\Lambda q}, \lambda)e^{-iq\cdot x} + a^\dagger(\vec{\Lambda q}, -\lambda)e^{iq\cdot x}] ,$$

or with the change of integration variable $\Lambda q \to q$,

$$A'^\mu(x) = (\Lambda^{-1})^\mu{}_\nu A^i(\Lambda x) + \partial^i f(\Lambda, x) . \tag{7.22}$$

The fact that the vector potential does not transform covariantly was to be expected, since the Coulomb gauge is a non-covariant gauge. The very important fact, however, is that the transformation law (7.22) is *gauge equivalent* to a co-variant transformation law, and thus guarantees the Lorentz covariance of *physical* quantities.

Corollary

The above transformation properties of the fields are in agreement with the trans-formation properties of 1-particle states as discussed in Chapter 2, if we require the *vacuum state* $|0\rangle$ to be invariant under Lorentz transformations

$$U[\Lambda]|0\rangle = |0\rangle \, , \qquad (7.23)$$

and define the 1-particle states by

$$|\vec{p}, \sigma\rangle = a^\dagger(\vec{p}, \sigma)|0\rangle \, , \qquad (7.24)$$

where σ now stands for helicity (in the case of massless particles) or z-component of the spin in the rest-frame of the particle (if the particle is massive). This constitutes the backbone of the second quantization. Indeed, it follows from (7.23), (7.24) and (7.11) that

$$U[\Lambda]|\vec{p}, \sigma > = (U[\Lambda]a^\dagger(\vec{p}, \sigma)U[\Lambda^{-1}])|0> = a'^\dagger(\vec{p}, \sigma)|0>$$

$$= \sqrt{\frac{\omega(\vec{\Lambda p})}{\omega(\vec{p})}} \sum_{\sigma'} |\vec{\Lambda p}, \sigma' > D^{(\frac{1}{2})}_{\sigma'\sigma}[R_W]$$

in agreement with (2.46). Similarly we have in the case of the Maxwell field,

$$U[\Lambda]|\vec{k}, \lambda > = \sqrt{\frac{|\vec{\Lambda k}|}{|\vec{k}|}} e^{i\lambda\Theta[\tilde{\Lambda}]}|\vec{\Lambda k}, \lambda > \, ,$$

in agreement with (2.62).

Equation (7.24) states that the operators $a^\dagger(\vec{p}, \sigma)$ create a *1-particle state* with momentum $\vec{p}$ and spin projection σ from the vacuum. Correspondingly we interpret $a(\vec{p}, \sigma)$ as operators *destroying* a 1-particle state of momentum $\vec{p}$ and spin projection σ. This means in particular that we must require

$$a(\vec{p}, \sigma)|0\rangle = 0 \, . \qquad (7.25)$$

We shall construct in the following section the many-particle states in terms of the annihilation and creation operators. Since these many particle states should obey *Bose–Fermi* statistics, we shall discover that the following commutation relations will have to be satisfied

$$[a(\vec{p}\,', \sigma'), a(\vec{p}, \sigma)] = 0, \quad [a(\vec{p}\,', \sigma'), a^\dagger(\vec{p}, \sigma)] = \delta_{\sigma\sigma'}\delta^3(\vec{p}\,' - \vec{p}) \qquad (7.26)$$

where $[A, B]$ stands for the generalized commutator

$$[A, B] = AB - (-1)^{\epsilon(A)\epsilon(B)} BA \qquad (7.27)$$

with $\epsilon(Q)$ representing the *Grassman signature* of the operators:

$$\epsilon(Q) = \begin{cases} 0 & (Q \text{ bosonic}) \\ 1 & (Q \text{ fermionic}) \end{cases} .$$

The above properties serve to define the so-called *Fock space*. As already mentioned, it will be one of our tasks to provide an explicit realization of the unitary operator $U[\Lambda]$ inducing the Lorentz transformation on the operator-valued fields in this Fock space.

Let us finally note that we recover the *1-particle wave functions* by identifying them with the matrix element of the corresponding Fock-space field between the vacuum and 1-particle state:

$$\langle 0|\phi(x)|\vec{q}\rangle = \frac{1}{(2\pi)^{3/2}\sqrt{2\omega_q}} e^{-iq\cdot x}$$

$$\langle 0|\psi_\alpha(x)|\vec{p}, \sigma, e\rangle = \frac{1}{(2\pi)^{3/2}} \sqrt{\frac{m}{\omega_p}} U_\alpha(\vec{p}, \sigma) e^{-ip\cdot x}$$

$$\langle \vec{p}, \sigma, -e|\psi_\alpha(x)|0\rangle = \frac{1}{(2\pi)^{3/2}} \sqrt{\frac{m}{\omega_p}} V_\alpha(\vec{p}, \sigma) e^{ip\cdot x}$$

$$\langle 0|A^\mu(x)|\vec{k}, \lambda\rangle = \frac{1}{(2\pi)^{3/2}} \frac{1}{\sqrt{2|\vec{k}|}} \epsilon^\mu(\vec{k}, \lambda) e^{-ik\cdot x} .$$

This establishes the link between the second-quantized and first-quantized formulations in the 1-particle sector. We now consider some operators acting on this Fock-space

Using (7.25) and the commutation relations (7.26) we obtain

$$\langle \vec{p}\,', \sigma'|\vec{p}, \sigma\rangle = \langle 0|a(\vec{p}\,', \sigma')a^\dagger(\vec{p}, \sigma)|0\rangle = \delta_{\sigma'\sigma}\delta^3(\vec{p}\,' - \vec{p}) ,$$

which corresponds to our non-relativistic normalization (2.42) of the 1-particle states. We define a general state of n identical free particles by

$$|\vec{p}_1\sigma_1, ..., \vec{p}_n\sigma_n\rangle = \frac{1}{\sqrt{n!}} a^\dagger(p_1, \sigma_1)...a^\dagger(\vec{p}_n, \sigma_n)|0\rangle . \tag{7.28}$$

The Grassman signature in the commutation relations (7.26) insures the symmetry (antisymmetry) of the state under the exchange of particle quantum numbers of bosons (fermions). The states (7.28) span an infinite dimensional space, called the *Fock* space. They are normalized as follows:

$$\langle \vec{p}_1'\sigma_1' \cdots \vec{p}_n'\sigma_n'|\vec{p}_1\sigma_1 \cdots \vec{p}_n\sigma_n\rangle = \frac{1}{n!} \sum_{P\{\alpha_1...\alpha_n\}} (-1)^{\epsilon(P)} \prod_{i=1}^n \delta_{\sigma_i'\sigma_{\alpha_i}} \delta^3(\vec{p}_i' - \vec{p}_{\alpha_i}) \tag{7.29}$$

where $\epsilon(P)$ is the "signature" of the permutation involved in the relabelling of the states: $(p_1, p_2, \cdots p_n) \to (p_{\alpha_1}, p_{\alpha_2}, \cdots p_{\alpha_n})$.

The states (7.28) are eigenstates of the *number operator* $\mathbb{N}_0$,

$$\mathbb{N}_0 = \sum_\sigma \int d^3p\, \mathcal{N}_0(\vec{p}, \sigma) , \quad \mathcal{N}_0(\vec{p}, \sigma) = a^\dagger(\vec{p}, \sigma)a(\vec{p}, \sigma), \quad \mathbb{N}_0^\dagger = \mathbb{N}_0 \tag{7.30}$$

counting the number of particles in such a state. Note the ordering of the operators in (7.30): the creation operator stands to the *left* of the destruction operator. Such an ordering is called *normal ordering*. Such normal ordering will play a fundamental role in our later considerations, when we seek a lower energy bound for the ground (vacuum) state. In particular $\mathbb{N}_0 |0\rangle = 0$.

It follows from (7.26) and (7.30) that we have

$$[\mathbb{N}_0, a(p, \sigma)] = -a(\vec{p}, \sigma)$$
$$[\mathbb{N}_0, a^\dagger(\vec{p}, \sigma)] = a^\dagger(\vec{p}, \sigma) .$$

We verify that the state (7.28) is an eigenstate of $\mathbb{N}_0$ with eigenvalue n:

$$\mathbb{N}_0 |\vec{p}_1 \sigma_1, ... \vec{p}_n \sigma_n\rangle = \frac{\mathbb{N}_0}{\sqrt{n!}} a^\dagger(\vec{p}_1, \sigma_1) ... a^\dagger(\vec{p}_n, \sigma_n) |0\rangle$$

$$= \frac{1}{\sqrt{n!}} \{ [\mathbb{N}_0, a^\dagger(\vec{p}_1, \sigma_1)] a^\dagger(\vec{p}_2, \sigma_2) ... a^\dagger(p_n, \sigma_n)$$

$$+ a^\dagger(\vec{p}_1, \sigma_1) \mathbb{N}_0 a^\dagger(\vec{p}_2, \sigma_2) ... a^\dagger(\vec{p}_n, \sigma_n) \} |0\rangle$$

$$= n |\vec{p}_1 \sigma_1, \cdots, \vec{p}_n \sigma_n\rangle .$$

We can at this stage define an operator whose eigenstates are just the Fock states (7.28) with eigenvalues given by the total energy of the state:

$$\mathbb{H}_0 = \sum_\sigma \int d^3 p \, \omega(\vec{p}) \mathbb{N}_0(\vec{p}, \sigma) .$$

$\mathbb{H}_0$ evidently plays the role of the *Hamiltonian* for the free-particle system. Remembering that the energy and momentum are the time and space components of a 4-vector, we are led to define the *momentum operator* in terms of the number operator as follows:

$$\vec{\mathbb{P}}_0 = \sum_\sigma \int d^3 p \, \vec{p} \, \mathbb{N}_0(\vec{p}, \sigma) .$$

The states (7.28) are eigenstates of this operator with the total momentum as eigenvalue:

$$\vec{\mathbb{P}}_0 |\vec{p}_1 \sigma_1, ... \vec{p}_n \sigma_n\rangle = \left(\sum \vec{p}_i \right) |\vec{p}_1 \sigma_1, ..., \vec{p}_n \sigma_n\rangle .$$

The fact that the states (7.28) can be simultaneously eigenstates of $\mathbb{H}_0$ and $\vec{\mathbb{P}}_0$ is a consequence of the fact that energy and momentum can be simultaneously measured. A general state $|\Psi\rangle$ of the free particle system need not be, however, an eigenstate of $\mathbb{H}_0, \vec{\mathbb{P}}_0$ or $\mathbb{N}_0$. It is represented in Fock space by

$$|\Psi\rangle = \frac{1}{\sqrt{n!}} \int d^3 p_1 ... d^3 p_n \sum_{\{\sigma\}} f(\vec{p}_1 \sigma_1, ..., \vec{p}_n \sigma_n) a^\dagger(\vec{p}_1, \sigma_1) ... a^\dagger(\vec{p}_n, \sigma_n) |0\rangle. \quad (7.31)$$

This is called the *Tamm–Dankhoff representation* of a general state. Note that we are working here implicitly in the *Heisenberg picture* where the states are time-independent and the operators carry all the time dependence. The coefficient functions in (7.31) evidently must have the symmetry properties of the corresponding

n-particle states. For states normalized to unity

$$\langle \Psi | \Psi \rangle = 1$$

we must have

$$\sum_{\{\sigma\}} \int d^3p_1 ... d^3p_n |f(\vec{p}_1\sigma_1, ..., \vec{p}_n\sigma_n)|^2 = 1 .$$

The *Tamm–Dankhoff* representation (7.31) of a general state is also applicable in the case of an interacting system. However, unlike the case of a free particle system, we can, in general, not find a basis in which the Hamiltonian and the number-operator are simultaneously diagonal. This is a consequence of the fact that, in general, the Hamiltonian and number operators do not commute:

$$[\mathbb{H}, \mathbb{N}] \neq 0 \quad (interacting \; system).$$

There are nevertheless examples for which the particle number is conserved. In this case the infinite dimensional Fock space separates into an infinite number of sectors labelled by the eigenvalues of the number operator. Such models represent one class of theories which can be completely solved to the extent that eigenvalue problems in non-relativistic quantum mechanics can be regarded as solvable. In particular, the concept of a Schrödinger n-particle wave function can be introduced, and eigenstates of the position operator can be defined (which in general case turns out to be highly problematic). These models have a high pedagogical value, but are otherwise only of marginal interest to the particle physicist.[2]

7.2   Commutation relations

Using the commutation relations (7.26) for the annihilation and creation operators, it is straightforward to compute the corresponding commutation relations for the operator-valued fields.

(i) Scalar field

From (7.1) and (7.26) we immediately obtain,

$$[\phi(x), \phi(y)] = i\Delta(x - y; m^2) \tag{7.32}$$

where $\Delta(x - y; m^2)$ is the *Lorentz invariant commutator function*

$$i\Delta(x - y; m^2) = \frac{1}{(2\pi)^3} \int \frac{d^3q}{2\omega_q} [e^{-iq\cdot(x-y)} - e^{+iq\cdot(x-y)}]_{q^0=\omega_q} . \tag{7.33}$$

Making use of

$$\epsilon(q^0)\delta(q^2 - m^2) = \frac{\delta(q^0 - \omega_q) - \delta(q^0 + \omega_q)}{2\omega_q} ,$$

---

[2]T.D. Lee, *Phys. Rev.* **95** (1954) 1329; S.S. Schweber, *An Introduction to Relativistic Quantum Field Theory* (Row, Peterson and Company, 1961).

we may rewrite the commutator function in the compact form

$$i\Delta(x - y; m^2) = \frac{1}{(2\pi)^3} \int d^4q\, \epsilon(q^0)\delta(q^2 - m^2)e^{-iq\cdot(x-y)} . \qquad (7.34)$$

Setting $x^0 = y^0$, the integrand becomes an odd function of q^0 showing that

$$\Delta(z; m^2) \,|_{z^0=0} = 0 . \qquad (7.35)$$

In fact, we have more generally

$$\Delta(z; m^2) = 0, \quad z^2 < 0 .$$

This follows from (7.35) and the invariance of the forward and backward lightcones under proper Lorentz transformations, implying

$$\epsilon(q'^0) = \epsilon(q^0) .$$

We therefore conclude that

$$\Delta(\Lambda z; m^2) = \Delta(z; m^2) .$$

On the other hand, taking the derivative of expression (7.33) with respect to x^0 and setting again $x^0 = y^0$, we find

$$\left[\frac{\partial}{\partial x^0} \Delta(x - y; m^2) \right]_{x^0=y^0} = -\delta^3(x - y) , \qquad (7.36)$$

implying for the equal time commutator of the field $\phi(x)$,

$$[\phi(x), \partial_0\phi(y)]_{ET} = i\delta^3(x - y) , \qquad (7.37)$$

where ET stands for "equal time". It is useful to know that the commutator function (7.32) may in fact be computed in closed form, the result being

$$\Delta(z; m^2) = -\frac{1}{2\pi}\epsilon(z^0)\delta(z^2) + \frac{m}{4\pi\sqrt{z^2}}\epsilon(z^0)\theta(z^2)J_1(m\sqrt{z^2})$$

where J_n stands for the Bessel function of the first kind.[3]

(ii) Dirac field

From (7.2) and (7.26) we obtain (recall definition (7.27))

$$[\psi_\alpha(x), \bar\psi_\beta(y)] = \int \frac{d^3p}{(2\pi)^3} \frac{m}{\omega_p} \left\{ \sum_\sigma U_\alpha(\vec{p}, \sigma)\bar{U}_\beta(\vec{p}, \sigma)e^{-ip\cdot(x-y)} \right.$$
$$\left. + \sum_\sigma V_\alpha(\vec{p}, \sigma)\bar{V}_\beta(\vec{p}, \sigma)e^{ip\cdot(x-y)} \right\} . \qquad (7.38)$$

---

[3]See Magnus, Oberhettinger and Soni, *"Formulas and Theorems for the Special Functions of Mathematical Physics"*.

Recalling from (4.43) that

$$\sum_\sigma U(\vec{p},\sigma)\bar{U}(\vec{p},\sigma) = \Lambda_+(p) = \frac{m+\not{p}}{2m} \ ,$$

$$-\sum_\sigma V(\vec{p},\sigma)\bar{V}(\vec{p},\sigma) = \Lambda_-(p) = \frac{m-\not{p}}{2m} \ ,$$

we find this time

$$[\psi_\alpha(x),\bar{\psi}_\beta(y)] = \int \frac{d^3p}{(2\pi)^3}\frac{1}{2\omega_p}\left\{(m+\not{p})_{\alpha\beta}e^{-ip\cdot(x-y)} - (m-\not{p})_{\alpha\beta}e^{ip\cdot(x-y)}\right\} \ ,$$

which we can write in the form

$$[\psi_\alpha(x),\bar{\psi}_\beta(y)] = iS_{\alpha\beta}(x-y;m) \ ,$$

with

$$S(x-y;m) = (m+i\not{\partial})\Delta(x-y;m^2) \ . \tag{7.39}$$

Setting again $x^0 = y^0$ and using (7.35) as well as (7.36), we conclude that

$$[\psi_\alpha(x),\psi_\beta^\dagger(y)]_{ET} = \delta_{\alpha\beta}\delta^3(\vec{x}-\vec{y}) \ . \tag{7.40}$$

(iii) The radiation field in the Coulomb gauge

From (7.3) and (7.26) we obtain

$$[A^i(x),A^j(y)] = \frac{1}{(2\pi)^3}\int \frac{d^3k}{2|\vec{k}|}\sum_{\lambda=\pm1}\epsilon^i(\vec{k},\lambda)\epsilon^{j*}(\vec{k},\lambda)\left(e^{-ik\cdot(x-y)} - e^{ik\cdot(x-y)}\right) \ . \tag{7.41}$$

Now, because of the completeness relation,

$$\sum_{\lambda=\pm1}\epsilon^i(\vec{k},\lambda)\epsilon^{j*}(\vec{k},\lambda) + \frac{k^ik^j}{\vec{k}^2} = \delta^{ij} \ , \tag{7.42}$$

we may write (7.41) in the form

$$[A^i(x),A^j(y)] = \frac{1}{(2\pi)^3}\int \frac{d^3k}{2|\vec{k}|}\left(\delta^{ij} - \frac{k^ik^j}{\vec{k}^2}\right)\left(e^{-ik\cdot(x-y)} - e^{ik\cdot(x-y)}\right)$$

or alternatively

$$[A^i(x),A^j(y)] = \left(\delta^{ij} - \frac{\partial^i\partial^j}{\vec{\nabla}^2}\right)iD(x-y) =: iD^{ij}(x-y) \ ,$$

where

$$iD(x-y) = \frac{1}{(2\pi)^3}\int d^4k\,\epsilon(k^0)\delta(k^2)e^{-ik\cdot(x-y)} \ .$$

Setting $x^0 = y^0$ and using (7.35) we find

$$[A^i(x), A^j(y)]_{ET} = 0 .$$

On the other hand, recalling the definition of the electric field,

$$E^i = F^{i0} = \partial^i A^0 - \partial^0 A^i$$

and using (7.36), we obtain $(m = 0, A_0 = 0)$

$$[E^i(x), A^j(y)]_{ET} = i\left(\delta^{ij} - \frac{\partial^i \partial^j}{\vec{\nabla}^2}\right) \delta^3(\vec{x} - \vec{y}) . \tag{7.43}$$

Finally, it is evident from (7.34) that the scalar invariant function satisfies the *Klein–Gordon* equation just as the corresponding Fock space scalar field:

$$(\Box + m^2)\Delta(z; m) = 0 .$$

Using

$$(m + i\partial\!\!\!/)(m - i\partial\!\!\!/) = m^2 + \Box$$

it thus immediately follows that the Dirac-commutator function (7.39) satisfies the Dirac equation

$$(i\partial\!\!\!/ - m)S(x - y; m) = 0 ,$$

just as its corresponding Fock-space Dirac field. In the same way one finds for the commutator function of the radiation field in the Coulomb gauge,

$$\Box D^{ij}(z) = 0, \quad \partial_i D^{ij}(z) = 0 .$$

The solution of these equations is of course not unique; the equal-time commutator (7.37) plays however the role of initial conditions in the *Cauchy initial value problem* determining uniquely the form of the solutions, as given above.

7.3   P, C, T from equations of motion

We now want to discuss the transformation properties of the Dirac equation under parity (P), charge conjugation (C) and time reversal (T). In order to obtain a representation for the charge conjugation operation, we shall need to introduce an electromagnetic interaction. We do this again via the *minimal coupling prescription*

$$i\partial_\mu \to i\partial_\mu - eA_\mu$$

already introduced in (3.10). The Dirac equation in the presence of an external electromagnetic field thus reads

$$[\gamma^\mu(i\partial_\mu - eA_\mu) - m]\psi = 0 . \tag{7.44}$$

This equation is evidently invariant under the simultaneous substitutions

$$A_\mu(x) \to A_\mu(x) + \partial_\mu \Lambda(x), \quad \psi(x) \to e^{-ie\Lambda(x)}\psi(x) , \tag{7.45}$$

where e denotes the charge of the electron. In different terms: The solution of Eq. (7.44) is a functional of the gauge field with the property

$$\psi(x; A_\mu + \partial_\mu \Lambda) = e^{-ie\Lambda(x)} \psi(x; A_\mu).$$

We next examine the transformations under *space reflection, charge conjugation* and *time reversal.*

P, C, T

Physically the operations P, C and T correspond to the transformations

$$
\begin{aligned}
P &: (t, \vec{r}) \to (t, -\vec{r}), \quad \vec{J}(x) \to -\vec{J}(x_P), \quad \vec{p} \to -\vec{p} \\
T &: (t, \vec{r}) \to (-t, \vec{r}), \quad \vec{J}(x) \to -\vec{J}(x_T), \quad \vec{p} \to -\vec{p} \\
C &: e \to -e
\end{aligned}
$$

where $\vec{J}(x)$ is the electromagnetic current.

Parity P

We demand covariance of the Dirac equation under the parity transformation by requiring that in the coordinate system obtained by space reflection of the coordinate axes, Eq. (7.44) should read

$$[i\gamma^\mu \partial'_\mu - e\gamma^\mu A'_\mu(x') - m]\psi'(x') = 0, \quad x' = (t, -\vec{r}),$$

with the transformation law

$$A'_0(x') = A_0(x), \quad \vec{A}'(x') = -\vec{A}(\vec{x})$$

for the vector potential. We thus see that we must have (in analogy to (2.5))

$$\psi'(x_P) = \gamma^0 \psi(x) .$$

We conclude that the parity transformation in $L_-^\uparrow$ is realized as follows:

$$\mathcal{D}[\Lambda_P] = \gamma^0 . \tag{7.46}$$

We thus have

$$\mathcal{D}[\Lambda_P^{-1}] \begin{pmatrix} \gamma^0 \\ \vec{\gamma} \end{pmatrix} \mathcal{D}[\Lambda_P] = \begin{pmatrix} \gamma^0 \\ -\vec{\gamma} \end{pmatrix}, \quad \mathcal{D}[\Lambda_P^{-1}] \gamma_5 \mathcal{D}[\Lambda_P] = -\gamma_5.$$

We infer from here the following properties of bilinear densities:

$$
\begin{aligned}
&\bar{\psi}(x)\psi(x) \text{ (scalar (S))} \\
&\bar{\psi}(x)\gamma^5 \psi(x) \text{ (pseudoscalar (PS))} \\
&\bar{\psi}(x)\gamma^\mu \psi(x) \text{ (vector (V))} \\
&\bar{\psi}(x)\gamma^5 \gamma^\mu \psi(x) \text{ (axialvector (A))} .
\end{aligned}
$$

The parity operator $P = \gamma^0$ acts as follows on the Dirac spinors:

$$\gamma^0 U(\vec{p}, \sigma) = U(-\vec{p}, \sigma), \quad \gamma^0 V(\vec{p}, \sigma) = -V(-\vec{p}, \sigma) .$$

In the Weyl representation this corresponds to the substitution,

$$P : D[L(\vec{p})] \leftrightarrow \bar{D}[L(\vec{p})] ,$$

which makes explicit the need for the Dirac wave function to transform with respect to both D and $\bar{D}$ representations, if parity is a symmetry of the Dirac equation. The spin is not affected, as expected.

Charge conjugation

The operation of charge conjugation reverses the charge of a particle. Correspondingly it follows from Eq. (7.44) that the charge conjugate wave function should satisfy this equation with $e \to -e$:

$$[\gamma^\mu(i\partial_\mu + eA_\mu) - m]\psi_c(x) = 0 .$$

Taking the complex conjugate of this equation results in

$$[\gamma^{\mu*}(-i\partial_\mu + eA_\mu) - m]\psi_c^*(x) = 0 .$$

We now recall from (4.14) that the matrix C has the property

$$C\gamma^\mu C^{-1} = -\gamma^{\mu*} , \quad C^2 = 1 . \tag{7.47}$$

Using this property and comparing this result with (7.44) we conclude that

$$\psi(x) = C\psi_c^*(x) . \tag{7.48}$$

The fact that the complex conjugate of the wave function appears on the right does not mean that the operator implementing the transformation of charge conjugation is an anti-unitary operator. This will become clear in the following section when we treat the above symmetry transformations from the operator point of view.

Time reversal

Consider two observers O and O' in their respective inertial systems S and S'. System S' is obtained from system S by simply inverting the time axis. Both observers describe the same physical process, referred to their respective axes. Thus, for observer O' currents run in the opposite direction, while the charge density will continue to have the same sign:

$$\vec{J}'(x') = -\vec{J}(x), \quad J'^0(x') = J^0(x), \quad x' = (-t, \vec{r}) .$$

We conclude therefore that the vector potential transforms as follows under time reversal

$$A'^0(x') = A^0(x), \quad \vec{A}'(x') = -\vec{A}(x) .$$

Hence covariance of the equation of motion under time reversal demands

$$(-i\gamma^0\partial_0 + i\gamma^i\partial_i - e\gamma^0 A^0(x) - e\vec{\gamma}\cdot\vec{A}(x) - m)\psi_T(-x^0,\vec{x}) = 0 \ .$$

Taking the complex conjugate of this equation, we have

$$(i\gamma^{0*}\partial_0 - i\gamma^{i*}\partial_i - e\gamma^{0*}A^0(x) - e\vec{\gamma}^*\cdot\vec{A}(x) - m)\psi_T^*(-x^0,\vec{x}) = 0$$

we make use this time of the property (4.17) in order to cast this equation into the form

$$(i\slashed{\partial} - e\slashed{A}(x) - m)C\psi_T^*(x_T) = 0 \ .$$

Comparing with (7.44) we conclude that

$$\psi(x) = C\psi_T^*(x_T) \ . \tag{7.49}$$

7.4   P, C, T in second quantization

Let us examine now how the discrete symmetries P, C and T are realized in Fock space. We assume that they can be unitarily implemented by the operators U_P, U_C and U_T on the states as follows[4]

$$U_P|\vec{p},\sigma\rangle = \eta_P^*|-\vec{p},\sigma\rangle \ ,$$

$$U_C|\vec{p},\sigma,e\rangle = \eta_C^*|\vec{p},\sigma;-e\rangle \ ,$$

$$U_T|\vec{p},\sigma\rangle = \eta_T^*\sum_{\sigma'}c_{\sigma\sigma'}|-\vec{p},\sigma'\rangle$$

with corresponding phases $\bar{\eta}_i$ for the antiparticle states. The matrix $c = -i\sigma_2$ accounts for the spin-flip under time-reversal. η_P, η_C and η_T are phase factors representing degrees of freedom. The above transformation laws are physically natural and imply that the *destruction* operators $a(k,\sigma)$ and $b(k,\sigma)$ of particles and antiparticles are, respectively:

$$U_P a(\vec{p},\sigma)U_P^{-1} = \eta_P a(-\vec{p},\sigma) \ , \quad U_P b(\vec{p},\sigma)U_P^{-1} = \bar{\eta}_P b(-\vec{p},\sigma) \ ,$$
$$U_C a(\vec{p},\sigma)U_C^{-1} = \eta_C b(\vec{p},\sigma), \quad U_C b(\vec{p},\sigma)U_C^{-1} = \bar{\eta}_C a(\vec{p},\sigma) \ ,$$
$$U_T a(\vec{p},\sigma)U_T^{-1} = \eta_T\sum_{\sigma'}c_{\sigma\sigma'}a(-\vec{p},\sigma') \ , \quad U_T b(\vec{p},\sigma)U_T^{-1} = \bar{\eta}_T\sum_{\sigma'}c_{\sigma\sigma'}b(-\vec{p},\sigma') \ .$$

When computing the corresponding transformation law of the operator-valued fields, we use the fact that the operators U_P and U_C only act on the Fock-space operators and commute with c-numbers. This does not apply to U_T, which is represented by an *anti-unitary* operator, which has the property of complex conjugation when commuted through a c-number. Making use of the Dirac spinors (4.38), and of $D[L(\vec{p})]^* = CD[L(-\vec{p})]C^{-1}$, one finds the following transformation laws for the

---

[4]S. Weinberg, *Phys. Rev.* **133** (1964) B1318; For a general discussion of phases see G. Feinberg and S. Weinberg, *Nuovo Cimento* **14** (1959) 571.

operator fields (now labeled by "hat"), provided we choose $\bar{\eta}_T^* = \eta_T, \bar{\eta}_C^* = \eta_C$ and $\eta_p^* = -\eta_P$:

$$U_P \hat{\psi}_\alpha(x) U_P^{-1} = \eta_P \gamma_{\alpha\beta}^0 \hat{\psi}_\beta(x_P) \tag{7.50}$$

$$U_C \hat{\psi}_\alpha(x) U_C^{-1} = \eta_C C_{\alpha\beta} \hat{\psi}_\beta^\dagger(x), \tag{7.51}$$

$$U_T \hat{\psi}_\alpha(x) U_T^{-1} = \eta_T C_{\alpha\beta} \hat{\psi}_\beta(x_T). \tag{7.52}$$

These transformation laws seem to differ from the transformation laws for the Dirac field *wave functions* found above. We now demonstrate how this apparent discrepancy is resolved.

Charge conjugation

The following steps lead from the transformation law (7.51) for the Fock-space field $\hat{\psi}_\alpha(x)$ to the transformation law for the 1-particle wave function:

$$\begin{aligned}
\psi_\alpha(x) :&= \langle 0|\hat{\psi}_\alpha(x)|\vec{p}, \sigma; e\rangle \\
&= \langle 0|U_C^{-1} U_C \hat{\psi}_\alpha(x) U_C^{-1} U_C|\vec{p}, \sigma; e\rangle \\
&= \langle 0|C_{\alpha\beta}\hat{\psi}_\beta^\dagger(x)|\vec{p}, \sigma; -e\rangle \\
&= C_{\alpha\beta}\langle -e; \vec{p}, \sigma|\hat{\psi}_\beta(x)|0\rangle^* = C_{\alpha\beta}\psi_\beta^*(x)_c
\end{aligned}$$

in agreement with (7.48).

Time reversal

Antiunitary operators A have the fundamental property

$$\langle \alpha|A^\dagger|\beta\rangle = \langle A\alpha|\beta\rangle^*. \tag{7.53}$$

Hence the adjoint of a unitary operator is defined in a different way. Using property (7.53), the following steps lead from the transformation law (7.52) for the Fock-space field to the transformation law for the corresponding 1-particle wave function:

$$\begin{aligned}
\psi_\alpha(x) &= \langle 0|\hat{\psi}_\alpha(x)|\vec{p}, \sigma\rangle = \langle 0|(U_T^{-1} U_T \hat{\psi}_\alpha(x) U_T^{-1} U_T)|\vec{p}, \sigma\rangle \\
&= \langle 0|U_T^{-1}\rangle(U_T \hat{\psi}_\alpha(x) U_T^{-1})(U_T|\vec{p}, \sigma\rangle) \\
&= C_{\alpha\beta}\langle 0|\hat{\psi}_\beta(x_T)|\vec{p}, \sigma\rangle_T^*
\end{aligned}$$

or

$$\psi_\alpha(x) = C_{\alpha\beta}\psi_\beta^{T*}(x_T)$$

in agreement with (7.49).

We thus see that the transformation laws for the operator-valued field and corresponding wave functions have a very natural interpretation for the 1-particle states.

Chapter 8

Canonical Quantization

Given some classical field equations, we show in this chapter how to systematically arrive at the corresponding quantum theory satisfying the correspondence principle. The approach is essentially the same as in non-relativistic quantum mechanics and presupposes that the equations of motion can be derived from a Lagrange density via a variational principle. The quantization is then realized in phase space by going over to the Hamiltonian formulation via the usual Legendre transformation. Although there have been attempts of quantization without invoking the existence of a lagrangian, the *canonical* approach has the great advantage of providing a systematic construction of conserved "currents", and hence of the generators of symmetry transformations. It also provides a powerful method for discovering and implementing the *constraints* of the theory in question. Since most interesting theories in particle physics correspond to constrained Hamiltonian systems, we shall have occasion to exemplify the power of the canonical approach for constrained Hamiltonian systems as developed by Dirac.

8.1 Lagrangian formulation and Euler–Lagrange equations

Quantum Field Theory can be regarded as the extension of quantum mechanics with relativistic kinematics to a relativistic quantum mechanics with an infinite number of degrees of freedom, where this number is in general not conserved in the course of an interaction. This extension can be formalized by expanding the fields of QFT in a complete discrete set of orthonormal basis functions as follows[1]

$$\phi(\vec{r}, t) = \sum_n q_n(t)\varphi_n(\vec{r}) \tag{8.1}$$

with $\phi(x)$ standing for a *generic* field, and with $\{\varphi_n(\vec{r})\}$ satisfying

$$\int d^3r\, \varphi_n^*(\vec{r})\varphi_m(\vec{r}) = \delta_{nm}, \quad \sum_n \varphi_n(\vec{r})\varphi_n^*(\vec{r}') = \delta^3(\vec{r} - \vec{r}') . \tag{8.2}$$

---

[1] For pedagogical reasons we take this set to be discrete.

We have the inversion formula

$$q_n(t) = \int d^3r \, \phi(\vec{r}, t) \varphi_n^*(\vec{r}) \,. \tag{8.3}$$

In the following we consider real fields $\phi(x)$ for simplicity, unless explicitly stated otherwise.

With respect to the basis $\{\varphi_n(x)\}$, the $q_n(t), n = 0, ..., \infty$ represent the new degrees of freedom. Let $L = L(\{q_n(t), \dot{q}_n(t)\})$ be the lagrangian for the corresponding classical system and let

$$S = S[\{q_n\}] = \int_{-\infty}^{\infty} dt L(\{q_n(t), \dot{q}_n(t)\})$$

be the corresponding action. The time integration is taken to extend from $t = -\infty$ to $t = \infty$. The extrema of this action happen to play a central role in the semiclassical description of the corresponding quantum system, as given by the WKB method when applied to a quantum mechanical system with an infinite number of degrees of freedom. Requiring as usual that the variation of the $q_n(t)$ vanishes at the endpoints, these extrema correspond to solutions of the *Euler–Lagrange equations*,

$$\frac{\delta S}{\delta q_n} = 0 \implies \frac{d}{dt} \frac{\partial L}{\partial \dot{q}_n} - \frac{\partial L}{\partial q_n} = 0 \,. \tag{8.4}$$

Let us define a Lagrangian density

$$\mathcal{L}(x) =: \mathcal{L}(\phi(x), \partial_\mu \phi(x)) \tag{8.5}$$

via the relation

$$\int d^3r \mathcal{L}\left(\sum_n q_n(t)\varphi_n(\vec{r}), \, \partial_\mu \sum_n q_n(t)\varphi_n(\vec{r})\right) = L(\{q_n(t), \dot{q}_n(t)\}) \,. \tag{8.6}$$

Correspondingly, we have for the associated action and equations of motion,

$$S = \int d^4x \mathcal{L}(x), \qquad \frac{\delta S[\phi]}{\delta \phi(x)} = 0 \,. \tag{8.7}$$

Note that in QFT the integration in time extends from minus infinity to plus infinity. Note further that relativistic covariance tells us that the *time* derivative can only occur in combination with the *spatial gradient* as shown in (8.6). We have implicitly assumed the Lagrange density to depend at most on the first derivative of the fields. This will turn out to be important for the formulation of the *initial value problem* as well as for the *renormalizability* of the theory.

From (8.6) we obtain

$$\frac{\partial L}{\partial q_n(t)} = \int d^3r \left(\frac{\partial \mathcal{L}}{\partial \phi(\vec{r}, t)} \frac{\partial \phi(\vec{r}, t)}{\partial q_n(t)} + \frac{\partial \mathcal{L}}{\partial \partial_i \phi(\vec{r}, t)} \frac{\partial \partial_i \phi(\vec{r}, t)}{\partial q_n(t)}\right)$$

$$= \int d^3r \left(\frac{\partial \mathcal{L}}{\partial \phi(\vec{r}, t)} \varphi_n(\vec{r}) + \frac{\partial \mathcal{L}}{\partial \partial_i \phi(\vec{r}, t)} \partial_i \varphi_n(\vec{r})\right)$$

$$= \int d^3r \left(\frac{\partial \mathcal{L}}{\partial \phi(\vec{r}, t)} - \partial_i \frac{\partial \mathcal{L}}{\partial \partial_i \phi(\vec{r}, t)}\right) \varphi_n(\vec{r}) + \int d^3r \, \partial_i \left[\left(\frac{\partial \mathcal{L}}{\partial \partial_i \phi(\vec{r}, t)}\right) \varphi_n(\vec{r})\right]$$

and

$$\frac{d}{dt}\frac{\partial L}{\partial \dot{q}_n} = \frac{d}{dt}\int d^3r \frac{\partial \mathcal{L}}{\partial \partial_0\phi(\vec{r},t)}\frac{\partial \partial_0\phi(\vec{r},t)}{\partial \dot{q}_n(t)} = \frac{d}{dt}\int d^3r \frac{\partial \mathcal{L}}{\partial \partial_0\phi(\vec{r},t)}\varphi_n(\vec{r}) \ .$$

Since the $\varphi_n(x)$ form a complete set of orthonormal basis functions, we conclude that the Lagrange equations (8.4) are equivalent to requiring

$$\partial_\mu \frac{\partial \mathcal{L}}{\partial \partial_\mu\phi(x)} - \frac{\partial \mathcal{L}}{\partial \phi(x)} = 0 \ . \tag{8.8}$$

This set of equations for the generic fields $\phi(x)$ are the field equations of *classical* field theory associated with the Lagrangian density (8.5). As one easily verifies, they are not affected by the following redefinition of the Lagrange density

$$\mathcal{L} \to \mathcal{L}' = \mathcal{L}(\phi(x), \partial_\mu\phi(x)) + \partial_\nu F^\nu(\phi(x)) \ ,$$

where the arbitrary function $F(\phi(x))$ only depends on the fields, and not on their derivatives. Indeed, we have

$$\mathcal{L}' = \mathcal{L} + \left(\frac{\partial F^\mu}{\partial \phi(x)}\right)\partial_\mu\phi(x)$$

$$\partial_\mu \frac{\partial \mathcal{L}'}{\partial \partial_\mu\phi(x)} = \partial_\mu \frac{\partial \mathcal{L}}{\partial \partial_\mu\phi(x)} + \frac{\partial^2 F^\mu}{\partial \phi(x)^2}\partial_\mu\phi(x)$$

$$\frac{\partial \mathcal{L}'}{\partial \phi(x)} = \frac{\partial \mathcal{L}}{\partial \phi(x)} + \frac{\partial^2 F^\mu}{\partial \phi(x)^2}\partial_\mu\phi(x) \ ,$$

so that

$$\partial_\mu \frac{\partial \mathcal{L}'}{\partial \partial_\mu\phi(x)} - \frac{\partial \mathcal{L}'}{\partial \phi(x)} = \partial_\mu \frac{\partial \mathcal{L}}{\partial \partial_\mu\phi(x)} - \frac{\partial \mathcal{L}}{\partial \phi(x)} \ .$$

This invariance of the Euler–Lagrange equations under the transformation is a consequence of the fact that the added term only contributes a surface term to the corresponding action and hence does not contribute to the variation (8.7) due to our choice of boundary conditions.

The Lagrange densities associated with the field equations for the free (real and complex) scalar field, the Dirac field and the Maxwell field are easily seen to be (treat ϕ, ϕ^*, ψ and $\bar{\psi}$ as independent variables)

$$\mathcal{L}_S^{(0)} = \frac{1}{2}\partial_\mu\phi\partial^\mu\phi - \frac{m^2}{2}\phi^2 \tag{8.9}$$

$$\mathcal{L}_S^{(0)} = \partial_\mu\phi^*\partial^\mu\phi - m^2\phi^*\phi \tag{8.10}$$

$$\mathcal{L}_F^{(0)} = \bar{\psi}(i\slashed{\partial} - m)\psi \tag{8.11}$$

$$\mathcal{L}_{em}^{(0)} = -\frac{1}{4}F^{\mu\nu}F_{\mu\nu} \ . \tag{8.12}$$

Note that the respective lagrangians depend quadratically on the fields, which is a sufficient condition for the corresponding QFT's to be explicitly solvable. Indeed, such a field dependence implies linear Euler–Lagrange equations, which admit explicit solutions, as we have seen.

We may introduce interactions by adding higher order polynomials to the respective lagrangians. A particularly important role in the formal development of field theory plays the lagrangian

$$\mathcal{L} = \frac{1}{2}\left(\partial_\mu\phi\partial^\mu\phi - m^2\phi^2\right) - \frac{\lambda}{4!}\phi^4 \tag{8.13}$$

for an interacting real scalar field, and

$$\mathcal{L} = \partial_\mu\phi^*\partial^\mu\phi - m^2\phi^*\phi - \frac{\lambda}{2!}(\phi^*\phi)^2$$

for an interacting complex scalar field, where a quartic self-interaction for the scalar field was added to (8.9) and (8.10) respectively. One refers to the corresponding QFT as the ϕ^4-theory. The corresponding Euler–Lagrange equations for the complex scalar field read

$$\frac{\delta S[\phi]}{\delta\phi^*(x)} = 0 \Rightarrow (\Box + m^2)\phi(x) = J(x)$$

with $J(x)$ the source

$$J(x) = -\lambda|\phi(x)|^2\phi(x) \ .$$

It is the nonlinearity of these equations which renders the corresponding field theory no longer exactly solvable.

We introduce the interaction with the electromagnetic field via the usual principle of *minimal substitution*:

$$i\partial_\mu \to i\partial_\mu - qA_\mu \ ,$$

where q is the charge to which the electromagnetic field is coupled. (For an electron $q = e$.) We thus obtain[2]

$$\mathcal{L}_{SED} = -\frac{1}{4}F^{\mu\nu}F_{\mu\nu} + (D_\mu\phi)^*(D^\mu\phi) - m^2\phi^*\phi$$

$$= \mathcal{L}_{em}^{(0)} + \mathcal{L}_S^{(0)} - qA^\mu(\phi^*i \overleftrightarrow{\partial}_\mu \phi) + q^2A_\mu A^\mu\phi^*\phi \tag{8.14}$$

$$\mathcal{L}_{QED} = -\frac{1}{4}F^{\mu\nu}F_{\mu\nu} + \bar\psi(i\slashed{D} - m)\psi$$

$$= \mathcal{L}_{em}^{(0)} + \mathcal{L}_F^{(0)} - e\bar\psi\slashed{A}\psi \ , \tag{8.15}$$

with

$$iD_\mu = i\partial_\mu - qA_\mu \ , \quad D_\mu^\dagger = -D_\mu$$

and

$$\phi^*(x) \overleftrightarrow{\partial}_\mu \phi(x) \equiv \phi^*(x)\partial_\mu\phi(x) - (\partial_\mu\phi^*(x))\phi(x).$$

Note that the simultaneous substitutions (gauge transformation)

$$A_\mu(x) \to A'_\mu(x) = A_\mu(x) + \frac{1}{q}\partial_\mu\Lambda(x) \ ,$$

$$\phi \to \phi'(x) = e^{-i\Lambda(x)}\phi(x) \tag{8.16}$$

$$\psi(x) \to \psi'(x) = e^{-i\Lambda(x)}\psi(x)$$

---

[2]In order to obtain in this way a gauge invariant lagrangian for the scalar field $\phi(x)$, it must be a complex field.

leave these equations form-invariant, since

$$iD_\mu\phi(x) \to (i\partial_\mu - qA'_\mu)\phi'(x) = e^{i\Lambda(x)}iD_\mu\phi(x) ,$$

and correspondingly

$$(iD_\mu\phi(x))^* \to e^{-i\Lambda(x)}(iD_\mu\phi(x))^* .$$

Noting that

$$\partial^\mu \left(\frac{\partial\mathcal{L}^{(0)}_{em}}{\partial\partial^\mu A^\nu} \right) = -\partial^\mu F_{\mu\nu} ,$$

we obtain from (8.14) the Euler–Lagrange equations for

Scalar Electrodynamics

$$\partial_\mu F^{\mu\nu} = j^\nu$$
$$(D_\mu D^\mu + m^2)\phi = 0 ,$$

where

$$j^\mu(x) = q\phi^*(x)i \stackrel{\leftrightarrow}{\partial^\mu} \phi(x) - 2q^2 A^\mu\phi^*\phi ,$$

and from (8.15) the Euler–Lagrange equations for

Electrodynamics[3]

$$\partial_\mu F^{\mu\nu} = J^\nu$$
$$(i\slashed{D} - m)\psi = 0 ,$$

where

$$J^\nu = e\bar{\psi}\gamma^\nu\psi .$$

The quantization of these theories will play a central role in the forthcoming chapters.

It turns out that the lagrangian describing the electromagnetic interaction of matter represents *constrained systems*, requiring particular care in their quantization, while the lagrangian (8.13) of the ϕ^4 theory describes an ordinary unconstrained system. We shall therefore begin by considering the quantization of the lagrangian of the ϕ^4-theory.

---

[3] $E^k = F^{k0}, \quad \partial_k E^k = \partial_k F^{k0} = 4\pi\rho = q\bar{\psi}\gamma^0\psi.$
In the Lorentz gauge:
$$\Box A^\nu = q\bar{\psi}\gamma^\nu\psi.$$

8.2 Canonical quantization: unconstrained systems

Starting from the lagrangian $L(\{q_n(t), \dot{q}_n(t)\})$ defined in (8.6), the transition to the Hamiltonian is achieved as usual by a Legendre transformation:

$$H(\{q_n, p_n\}) = \sum_n p_n \dot{q}_n - L(\{q_n, \dot{q}_n\}) \tag{8.17}$$

with

$$p_n(t) = \frac{\partial L_n(t)}{\partial \dot{q}_n}$$

the momentum canonically conjugate to $q_n(t)$, satisfying the Poisson brackets

$$\{q_n(t), p_n(t)\} = \delta_{nm} \; . \tag{8.18}$$

The corresponding Hamilton equations read

$$\dot{q}_n(t) = \frac{\partial H}{\partial p_n(t)} \; , \quad \dot{p}_n(t) = -\frac{\partial H}{\partial q_n(t)} \; . \tag{8.19}$$

Recalling (8.3), we further define the density $\pi(\vec{r}, t)$ via

$$p_n(t) = \frac{\partial L}{\partial \dot{q}_n(t)} = \int d^3 r \, \pi(\vec{r}, t) \varphi_n^*(\vec{r}) \tag{8.20}$$

where

$$\pi(\vec{r}, t) = \frac{\partial \mathcal{L}}{\partial \partial_0 \phi(\vec{r}, t)} \; . \tag{8.21}$$

The field $\pi(\vec{r}, t)$ is called the momentum canonically conjugate to the field $\phi(\vec{r}, t)$. It can be written also in the form of a variational derivative as follows:

$$\pi(x) = \frac{\delta L}{\delta \partial_0 \phi(x)} \; ,$$

with the additional "rule" (we now set $\vec{r} = \vec{x}$)

$$\frac{\delta}{\delta \phi(\vec{y}, t)} \phi(\vec{x}, t) = \delta^3(\vec{x} - \vec{y}) \; . \tag{8.22}$$

Introducing expressions (8.20) into (8.17) and using (8.2), we equivalently have

$$H = \sum_n p_n(t) \dot{q}_n(t) - L$$

$$= \sum_n \int d^3 x \int d^3 y \, \pi(\vec{x}, t) \partial_0 \phi(\vec{y}, t) \varphi_n(\vec{x}) \varphi_n(\vec{y}) - L$$

$$= \int d^3 x \mathcal{H}(x) = H[\phi, \pi] \; ,$$

where $\mathcal{H}(x)$ is the Legendre transform of $\mathcal{L}(x)$,

$$\mathcal{H}(x) = \pi(x) \partial_0 \phi(x) - \mathcal{L}(x)$$

and is referred to as the *Hamiltonian density* associated with the Lagrange density $\mathcal{L}(x)$. Since

$$\{q_n(t), p_m(t)\} = \int d^3x \int d^3y \{\phi(\vec{x}, t), \pi(\vec{y}, t)\} \varphi_n(\vec{x}) \varphi_m(\vec{y}) \qquad (8.23)$$

we see that the *Poisson brackets* (8.23) in turn imply the Poisson brackets

$$\{\phi(x), \phi(y)\} = \{\pi(x), \pi(y)\} = 0, \quad \{\phi(x), \pi(y)\} = \delta^3(\vec{x} - \vec{y}) \qquad (8.24)$$

for the fields, with $\pi(x)$ given by (8.21). These are just the Poisson brackets corresponding to the equal time commutation relations we explicitly computed in Chapter 7. The Hamiltonian equations (8.19) correspondingly take the form

$$\partial_0 \phi(x) = \frac{\delta H}{\delta \pi(x)}, \quad \partial_0 \pi(x) = -\frac{\delta H}{\delta \phi(x)}, \qquad (8.25)$$

and are equivalent to the Euler–Lagrange equations of motion (8.8). When written in the form (8.25), the equations of motion are the natural extension of the corresponding equation of motion in Quantum Mechanics. Notice that the time has played everywhere a special role. Indeed, we could have taken over directly the results from Quantum Mechanics, if we had chosen to label the fields and conjugate momenta as follows, $\phi(x) = \phi_{\vec{x}}(t), \pi(x) = \pi_{\vec{x}}(t)$, with the index $\vec{x}$ labelling the coordinates and conjugate momenta of a quantum mechanical system with an continuous infinite set of degrees of freedom. Note that from the quantum mechanical point of view the degrees of freedom of a *free* field theory would nevertheless be coupled through the gradiant term $(\nabla\phi)^2$.

Let us define the Poisson brackets in the continuum case as follows,[4]

$$\{A, B\} = \int d^3z \left(\frac{\delta A}{\delta \pi(\vec{z}, t)} \frac{\delta B}{\delta \pi(\vec{z}, t)} - (-)^{\epsilon(A)\epsilon(B)} \frac{\delta B}{\delta \pi(\vec{z}, t)} \frac{\delta A}{\delta \pi(\vec{z}, t)} \right),$$

where A and B are any functionals of the phase-space variables, and where the variational derivatives are computed using the rule (8.22). In terms of these Poisson brackets the Hamilton equations (8.25) can be written in the form

$$\partial_0 \phi(x) = \{\phi(x), H\}, \quad \partial_0 \pi(x) = \{\pi(x), H\}. \qquad (8.26)$$

So far the discussion has been on classical level. The transition to the quantum theory is now achieved by letting the fields be *operator-valued* and making the substitution

$$i\{A(\vec{x}, t), B(\vec{y}, t)\} \longrightarrow [A(x), B(y)]_{ET}, \qquad (8.27)$$

where $[A, B]_{ET}$ denotes the generalized *commutator* of operator-valued quantities A and B at equal times defined in (7.27). This transition is, as we shall see, not as simple as it appears, due to the fact that the Euler–Langrange equations involve products of fields at the same point. As one can infer already from the commutation relations

$$[\phi(x), \pi(y)]_{ET} = i\delta^3(\vec{x} - \vec{y}) \qquad (8.28)$$

---

[4]Note that Poisson brackets are always defined at equal times.

such operator products are singular and will thus require a careful definition leading to the highly non-trivial machinery of *renormalization theory*.

In order to be able to make such a transition, it is of course essential that both, the Poisson brackets and the commutators formally satisfy the same algebraic relations. Indeed, the following properties follow from the definition of the Poisson brackets:

Properties of Poisson brackets shared with commutators.

(i) *Antisymmetry:*
$$\{f, g\} = -\{g, f\}$$

(ii) *Linearity:*
$$\{f_1 + f_2, g\} = \{f_1, g\} + \{f_2, g\}$$

(iii) *Product-law:*
$$\{f_1 f_2, g\} = f_1\{f_2, g\} + (-1)^{\epsilon(f_2)\epsilon(g)}\{f_1, g\}f_2$$

(iv) *Jacobi-Identity:*
$$\{f, \{g, h\}\} + \{h, \{f, g\}\} + \{g, \{h, f\}\} = 0. \tag{8.29}$$

These properties hold for the Poisson bracket of phase-space functions f, g, as well as the commutator of operators $\hat{f}, \hat{g}$.

Example: free scalar field theory

We consider as prototype the theory described by the lagrangian *density*

$$\mathcal{L} = \frac{1}{2}\partial^\mu \phi \partial_\mu \phi - \frac{1}{2}m^2\phi^2, \quad \phi \quad real.$$

We expand $\phi(x)$ as in (8.1), with $\varphi_n(\vec{r})$ the eigenfunctions of $-i\vec{\nabla}$, which we imagine as calculated in a box of finite volume V subject to, say, periodic boundary conditions in order to render them denumerable:

$$\varphi_n(\vec{r}) = \frac{1}{\sqrt{V}}e^{i\vec{q}_n \cdot \vec{r}}.$$

Let us resume the index n and the vector index in a single index i. In terms of the coordinates $q_i(t)$, the corresponding lagrangian then reads

$$L = \sum_i \frac{1}{2}(\dot{q}_i^2 - \omega_i^2 q_i^2), \quad \omega_i^2 = q_i^2 + m^2,$$

where we have made use of the orthonormality of the momentum eigenfunctions. The momenta conjugate to the coordinate $q_i(t)$, defined as usual by $p_i = \frac{\partial L}{\partial \dot{q}_i}$, yield the expected result $p_i = \dot{q}_i$. The *Poisson bracket* of the coordinates and their conjugate momenta are required to have the canonical form

$$\{q_k(t), q_l(t)\} = \{p_k(t), p_l(t)\} = 0, \quad \{q_k(t), p_l(t)\} = \delta_{kl}.$$

The transition to the Hamiltonian is achieved as usual by the Legendre transformation (8.17). For the case in question the Hamiltonian reads:

$$H = \sum_i \frac{1}{2}(p_i^2 + \omega_i^2 q_i^2) \; .$$

The corresponding Hamilton equations

$$\dot{q}_k = \frac{\partial H}{\partial p_k}, \quad \dot{p}_k = -\frac{\partial H}{\partial q_k} \tag{8.30}$$

are equivalent to the Euler–Lagrange equations (8.4), as is well known.

For the theory in question, $p_n = \dot{q}_n$, so that we may eliminate the "velocities" $\dot{q}_k(t)$ in favor of the canonically conjugate momenta. This possibility is characteristic of an *unconstrained* or *non-singular* system. The latter terminology stems from the observation that the possibility of solving Eq. (8.30) for the velocities $\dot{q}_k(t)$ in terms of the coordinates and conjugate momenta is a consequence of the invertibility of the matrix (Hessian)

$$W_{kl} = \frac{\partial^2 L}{\partial \dot{q}_k \partial \dot{q}_l} \; . \tag{8.31}$$

The crucial role which the matrix W plays is seen by noting that the Euler–Lagrange equations (8.4) can also be written in the form

$$\sum_k W_{kl}(q, \dot{q})\ddot{q}_l = \frac{\partial L}{\partial q_k} - \frac{\partial^2 L}{\partial \dot{q}_k \partial q_l}\dot{q}_l \; .$$

This shows that, if $\det W = 0$, the dynamics of the system is not uniquely determined by the specification of the coordinates and the velocities at some initial time. In that case one says that the lagrangian is *singular*.

So far we have considered the simple case of quantizing non-singular lagrangians. It turns out that most of the theories of interest do not belong to this class of systems. This is in particular the case for *gauge theories*, such as electrodynamics of scalar bosons and Dirac fermions. A systematic method for quantizing such systems has been provided by Dirac, and will be discussed in the following section.

8.3   Canonical quantization: constrained systems

A comprehensive discussion of the quantization of constrained systems would in itself constitute a semester course and will therefore not be attempted here.[5] We shall therefore limit ourselves to present the *algorithm* due to Dirac for quantizing such systems.[6]

---

[5]See Marc Henneaux and Claudio Teitelboim, *Quantization of Gauge Systems* (Princeton University Press, 1992); Heinz J. Rothe and Klaus D. Rothe, *Classical and Quantum Dynamics of Constrained Hamiltonian Systems* (World Scientific, 2010).

[6]P.A.M. Dirac, *Lectures on Quantum Mechanics* (Belfer Graduate School of Science, Yeshiva University, New York, 1964).

It will again be convenient to discuss the problem in terms of the infinite set of discrete "coordinates" $q_i(t)$. The transition to the continuum formulation following the procedure of the preceding section is then straightforward. Our starting point is thus the lagrangian $L(q(t), q(t))$ and the corresponding canonically conjugate momenta

$$p_i = \frac{\partial L(q, \dot{q})}{\partial \dot{q}_i} \qquad i = 1, ..., n , \tag{8.32}$$

where we supposed for the moment that the number of degrees of freedom is finite. We now consider the *singular* case where the Hessian (8.31) is of rank $R_w < n$. In that case we can solve Eq. (8.32) as a function of the n coordinates, R_w momenta p_b and the remaining $n - R_w$ velocities $\dot{q}_\beta$, with $\beta = 1, ..., n - R_w$:

$$\dot{q}_a = f_a(\{q_i\}, \{p_b\}, \{\dot{q}_\beta\}) \quad b = 1, ..., R_w . \tag{8.33}$$

Substituting (8.33) into (8.32) we formally obtain

$$p_i = g_i(q, \{p_b\}, \{\dot{q}_\beta\}) .$$

It is clear that for $i = b = 1, ..., R_w$ these equations reduce to an *identity*. On the other hand, for $i = \beta \in [R_w + 1, n]$, we have

$$p_\beta = g_\beta(q, \{p_b\}), \quad \beta = R_w + 1, ..., n , \tag{8.34}$$

where $g_\beta(q, \{p_b\})$ cannot depend on the velocities $\{\dot{q}_\beta\}$, since we could otherwise eliminate further velocities, contrary to our hypothesis.

Equation (8.34) represents $m_P = n - R_w$ *constraints*, following from the *definition* of the momenta p_β; they are thus called *primary* constraints. They imply that the dynamics in phase space evolves on the subspace Γ_P (subscript "P" for "primary") defined by

$$\Gamma_P : \phi_\beta = 0, \quad \beta = R_w + 1, ..., n$$

with

$$\phi_\beta(q, p) := p_\beta - g_\beta(q, \{p_b\}) .$$

Dirac algorithm

It is convenient to have the label β on ϕ_β and v_β run from 1 to m_P. Following Dirac, the dynamics of the system is then governed by the equations of motion

$$\dot{p}_i = \{p_i, H_T\}_{\Gamma_\Omega}$$

and constraints,

$$\Gamma_\Omega : \{\phi_\beta(q, p) = 0 , \varphi_\rho(q, p) = 0\} , \quad \beta = 1, \cdots, m_P$$

where H_T is the *total* Hamiltonian

$$H_T = H_c + \sum_{\beta=1}^{m_P} v_\beta \phi_\beta , \tag{8.35}$$

with $m_P = n - R_w$ *arbitrary parameters* v_β, H_c the *canonical* Hamiltonian obtained by Legrendre transformation of the lagrangian,

$$H_c = \sum_{i=1}^{n} p_i \dot{q}_i - L(q, \dot{q}) ,$$

and $\varphi_\rho = 0$ further *secondary* constraints computed iteratively as follows: Require the persistence in time of the *primary* constraints, that is

$$\dot{\phi}_\beta = \{\phi_\beta, H_T\}_{\Gamma_P = 0} .$$

If the rank r of the matrix

$$||\{\phi_\alpha, \phi_\beta\}||, \quad \alpha, \beta = 1, ..., n - R_w = m_P$$

is $r = n - R_w$, then these equations can be solved for the $n - R_w$ "velocities" v_β, and we are finished: there are no further (secondary) constraints. If $r < n - R_w = m_P$, then only r of these arbitrary parameters will be determined, and there will be $m_S = m_P - r$ further constraints which are called secondary. We denote these by φ_ρ. Requiring now persistence in time of these secondary constraints,

$$\dot{\varphi}_\rho = \{\varphi_\rho, H_T\}_{\Gamma_P} = 0, \quad \phi_\beta = 0, \quad \beta = 1, ...m_P$$
$$\varphi_\rho = 0, \quad \rho = 1, ...m_S$$

we may either discover further secondary constraints or the process terminates in the sense that the equations are *trivially* satisfied, and so on. The formalization of this procedure is cumbersome and will not be given here, since for the theories of interest in these lectures the iterative process actually ends at this point, as we shall demonstrate.

Once we have unravelled in this way all the constraints, we collect them into a single "vector", whose number of components we suppose to be given by M:

$$\{\Omega_A\} = (\phi_1...\phi_{m_P}, \varphi_1...\varphi_{m_S}), \quad m_P + m_S = M .$$

We now look at the *rank* of the matrix

$$||\{\Omega_A, \Omega_B\}||, \quad A, B = 1, ..., M.$$

Let the rank of this matrix be R. If $R < M$, we can form M-R linear combinations $\Omega_{A_1}^{(1)}$ of these constraints with the properties,

$$\{\Omega_{A_1}^{(1)}, \Omega_B\} \approx 0 , \forall B, \quad A_1 = 1, ...M^{(1)}, \quad M^{(1)} = M - R \tag{8.36}$$

and the remaining constraints with the properties

$$\{\Omega_{A_2}^{(2)}, \Omega_{B_2}^{(2)}\} \not\approx 0 , \quad \det||\{\Omega_{A_2}^{(2)}, \Omega_{B_2}^{(2)}\}|| \not\approx 0 . \tag{8.37}$$

The sign of *weak inequality* means here that the relations should hold on the subspace of phase-space defined by the constraints $\Omega_A = 0$. In particular, one can show that

$$\{\Omega_{A_1}^{(1)} \Omega_{B_1}^{(1)}\} = \sum_{C_1}^{M^{(1)}} f_{A_1 B_1 C_1} \Omega_{C_1}^{(1)} .$$

Constraints which have weakly vanishing Poisson brackets with *all* (primary and secondary) constraints as in (8.36) are called *first class* (denoted here by a superscript "1"); constraints with the property (8.37) are called *second class* (denoted here by a superscript "2"). For the respective equations of motion we have accordingly

$$\{\Omega^{(1)}_{A_1}, H_c\} \approx 0 \,, A_1 = 1 \,, \cdots \,, M^{(1)}$$

$$\{\Omega^{(2)}_{A_2}, H_c\} + \sum_{\beta} \{\Omega^{(2)}_{A_2}, \phi^{(2)}_{\beta}\} v^{(2)}_{\beta} \approx 0 \,, \quad A_2 = 1 \,, \cdots \,, M^{(2)}. \tag{8.38}$$

From (8.38) we see that we can solve for as many of the arbitrary "velocities" as there are second-class constraints:

$$v^{(2)}_{A_2} \approx - \sum_{B_2=1}^{M^{(2)}} \mathcal{R}_{A_2 B_2} \{\Omega^{(2)}_{B_2}, H_c\} \,,$$

where

$$\mathcal{R} = Q^{-1}, \quad Q = ||\{\Omega^{(2)}_{A_2}, \Omega^{(2)}_{B_2}\}|| \,.$$

Note from (8.38) that $M^{(2)} \le m_P$. The remaining parameters $v^{(1)}_{\alpha}$ remain undetermined!

The freedom in the choice of the parameters $v^{(1)}_{\beta}$ associated with *first class primary* constraints reflects the existence of a *local* (gauge) symmetry of the Hamiltonian system under the infinitesimal transformation

$$\delta_G H = \{H, G\} \,, \tag{8.39}$$

with $G(t)$ the *generator* of the symmetry transformation, as given by

$$G(t) = \sum_{A_1=1}^{M^{(1)}} \epsilon_{A_1}(t) \Omega^{(1)}_{A_1}(q, p) \tag{8.40}$$

where $\epsilon_{A_1}(t)$ are infinitesimal parameters. Note that in (8.39) a *strong* equality is meant, i.e. the infinitesimal transformation represents a symmetry transformation of the Hamiltonian on full phase space Γ, and is thus not limited to the subspace of phase space on which the actual motion takes place. Nevertheless, the transformation generated by (8.40) only leaves the equations of motion invariant on a subspace of the parameters $\epsilon_{A_1}(t)$, as we now illustrate for the case of quantum electrodynamics.

8.4 QED as a constrained system

Consider the Lagrange density

$$\mathcal{L} = -\frac{1}{4} F_{\mu\nu} F^{\mu\nu} + \bar{\psi}(i\partial\!\!\!/ - e A\!\!\!/ - m)\psi \tag{8.41}$$

with

$$F^{\mu\nu} = \partial^\mu A^\nu - \partial^\nu A^\mu$$

and

$$L = \int d^3x \mathcal{L} \ .$$

We have for the momenta conjugate to $A^\mu(x)$ and $\psi_\alpha(x)$ and their respective Poisson brackets

$$\pi_0(x) = \frac{\delta L}{\delta \partial_0 A^0(x)} = 0 \ ,$$

$$\pi_i(x) = \frac{\delta L}{\delta \partial_0 A^i(x)} = -F^{i0}(x) = -E^i(x)$$

$$\{E^i(x), A^j(y)\} = \delta^3(\vec{x} - \vec{y})\delta^{ij} \tag{8.42}$$

and

$$\pi_{\psi_\alpha}(x) = \frac{\delta^{(r)} L}{\delta \partial_0 \psi_\alpha(x)} = i\psi_\alpha^\dagger(x) \ , \quad \{\psi_\alpha(x), \pi_{\psi_\beta}(y)\} = \delta^3(\vec{x} - \vec{y}) \ ,$$

where the superscript "r" denotes "right derivative". Notice that the substitution $i\{\ \} \to [\ ]$ for $\{E^i(x), A^j(y)\}$ would be inconsistent with the Coulomb gauge. Indeed, unlike the case of the scalar field, we see that we have two *primary* constraints:

$$\phi_1 = \pi_0 \approx 0 \ , \tag{8.43}$$

$$\phi_2 = \pi_{\psi_\alpha} - i\psi_\alpha^\dagger \approx 0 \ . \tag{8.44}$$

The first constraint is a result of the absence of the velocity $\dot{A}^0$ in the lagrangian (8.41). The second constraint is the consequence of the lagrangian depending only linearly on the "velocity" $\dot{\psi}_\alpha$.

The *canonical* Hamiltonian obtained as the Legendre transform of the lagrangian (8.41) is given by

$$H_c = \int d^3x [\pi_\mu \partial_0 A^\mu + \pi_\psi \partial_0 \psi - \mathcal{L}] \ . \tag{8.45}$$

The primary constraint (8.44) is implemented by the Poisson bracket

$$\{\psi_\alpha(x), i\psi_\beta^\dagger(y)\} = \delta^3(\vec{x} - \vec{y}) \ . \tag{8.46}$$

With this Poisson bracket we may strongly set $\pi_{\psi_\alpha} = i\psi_\alpha^\dagger$ in (8.45), obtaining

$$H_c = \int d^3x [\frac{1}{2}\vec{\pi}^2 + \frac{1}{4}F^{ij}F_{ij} + \bar{\psi}(i\vec{\gamma}\cdot\vec{\partial} - e\vec{\gamma}\cdot\vec{A} + m)\psi + A^0\mathcal{G}] \ , \tag{8.47}$$

where $\mathcal{G}$ is the "Gauss operator"

$$\mathcal{G}(x) = \partial_j \pi_j(x) + e\bar{\psi}(x)\gamma^0 \psi(x) \ .$$

Note that with the usual definitions of the electric and magnetic fields

$$E^i = F^{i0} = -\pi_i \ , \quad B^i = -\frac{1}{2}\epsilon_{ijk}F^{jk}$$

the Hamiltonian (8.47) can also be written in the form

$$H_c = \int d^3x \left[\frac{1}{2}(\vec{E}^2 + \vec{B}^2) + \bar{\psi}(i\vec{\gamma}\cdot\vec{\partial} - e\vec{\gamma}\cdot\vec{A} + m)\psi + A^0 \mathcal{G} \right] .$$

The *total* Hamiltonian is thus given by (note that we are now left with only one primary constraint)

$$H_T = H_c + \int d^3x\, v\pi_0 .$$

The requirement that the primary constraint $\pi_0 \approx 0$ be satisfied at all times now leads to *Gauss' law* as the *secondary* constraint:

$$\{\pi_0(x), H_T\} = -\mathcal{G} \approx 0 . \tag{8.48}$$

Demanding that Gauss' law (8.48) be satisfied at all times requires

$$\{\mathcal{G}, H_T\}_{\substack{\pi_0=0\\\mathcal{G}=0}} = 0 .$$

Since

$$\{\mathcal{G}, H_T\} = -\int d^3x\, \partial_i\partial_j F^{ij} = 0$$

there are no further constraints and the Dirac algorithm terminates at this point. The Hamilton equations thus read

$$\{A^\mu(x), H_T\}_{\Gamma_\Omega} = \partial_0 A^\mu(x)$$
$$\{\psi_\alpha(x), H_T\}_{\Gamma_\Omega} = \partial_0\psi(x)$$
$$\Gamma_\Omega := \{\pi_0 = 0,\ \mathcal{G} = 0\} .$$

Explicitly we have

$$\partial_0 A^0 = v$$
$$\partial_0 A^i = \pi_i + \partial^i A^0$$
$$\partial_0 \pi_0 = 0$$
$$\partial_0 \pi_j = \partial^i F_{ij} + e\bar{\psi}\gamma^j\psi$$
$$\partial_0 \psi = \gamma^0\vec{\gamma}\cdot\vec{\partial}\psi - i\gamma^0(m + e\!\!\!/A)\psi$$

and

$$\Gamma_\Omega : \left\{ \begin{array}{c} \phi = \pi_0 = 0 \\ \varphi = \mathcal{G} = \partial_j\pi_j + e\bar{\psi}\gamma^0\psi = 0 \end{array} \right\} .$$

By eliminating all canonical momenta from these equations, one easily checks that one is led to the Euler–Lagrange equations

$$\partial_\mu F^{\mu\nu} = e\bar{\psi}\gamma^\nu\psi ,$$
$$(i\!\!\!/\partial - e\!\!\!/A - m)\psi = 0 .$$

We next examine how gauge transformations are realized.

Gauge transformations

Grouping the primary constraint and secondary constraint in the "vector"

$$(\Omega_A) = (\phi, \varphi) = (\pi_0, \mathcal{G}) , \qquad (8.49)$$

we find

$$\{\Omega_A, \Omega_B\} = 0 ,$$

that is, all constraints are *first class*. The primary and secondary constraints thus generate a local symmetry corresponding to the infinitesimal transformations

$$\delta f = \{f, G\}$$

with the generator given by

$$G = \int d^3x (\epsilon_1(x)\pi_0(x) + \epsilon_2(x)\mathcal{G}(x)) . \qquad (8.50)$$

Explicitly one computes

$$\delta A^0(x) = \epsilon_1(x), \quad \delta A^i(x) = \partial^i \epsilon_2(x) ,$$
$$\delta \psi(x) = -ie\epsilon_2(x) .$$

It however turns out that these transformations do not represent a symmetry of the Hamilton *equations* constructed from the *total* Hamiltonian, unless certain relations between the parameters in (8.50) are satisfied. We do not dwell on deriving these relations in general, and refer the reader to the literature.[7] For the case in question it is easy to guess what these relations should read. Indeed, the Hamilton equations are equivalent to the Euler–Lagrange equations, which in turn are known to be invariant under the infinitesimal gauge transformations (see (8.16))

$$\delta A^\mu = \partial^\mu \Lambda, \ \delta \psi = -ie\Lambda\psi . \qquad (8.51)$$

This shows that we must choose

$$\epsilon_1(x) = \partial_0 \Lambda , \quad \epsilon_2 = \Lambda(x) . \qquad (8.52)$$

The invariance of the lagrangian (8.41) under the local transformations (8.51) means that the dynamics corresponding to different choices of the parameters $\epsilon_a(x)$ subject to the restriction (8.52) should correspond to the same physics.

Dirac brackets

In order to make the transition to the quantum theory, we must implement the constraints *strongly*, or else implement the constraints on the *states* by working with only gauge-invariant states, of which the ground state is a particular example. The latter method turns out to be fashionable in the case of the so-called *covariant*

---

[7] Marc Henneaux and Claudio Teitelboim, *Quantization of Gauge Systems* (Princeton University Press, 1992); H.J. Rothe and K.D. Rothe, *Classical and Quantum Dynamics of Constrained Hamiltonian Systems* (World Scientific, 2010).

gauges, while the first method is usually preferred in the so-called *non-covariant* gauges such as the *Coulomb* gauge.

In order to implement the Coulomb gauge strongly, we must impose *subsidiary conditions* which serve to fix all remaining arbitrary parameters. This is called *gauge fixing*. In order to implement the Coulomb gauge strongly, these subsidiary conditions read[8]

$$\chi_1 = A^0 \approx 0 \ ,$$
$$\chi_2 = \vec{\nabla} \cdot \vec{A} \approx 0 \ . \tag{8.53}$$

Adjoining these conditions to the first-class constraints (8.49) to form the "vector"

$$(\Phi_K) = (\Omega_{A_1}^{(1)}, \chi_{B_1}) \ ,$$

we define the matrix

$$Q : Q_{KL} = \{\Phi_K, \Phi_L\} \ . \tag{8.54}$$

Following Dirac, we construct the so-called *Dirac brackets of two phase space quantities* A *and* B in terms of this matrix as follows:

$$\{A, B\}_\mathcal{D} = \{A, B\} - \sum_{K,L} \{A, \Phi_K\} \mathcal{R}_{KL} \{\Phi_L, B\} \ ,$$

where $\mathcal{R}$ denotes the inverse of the matrix (8.54). One can easily check that these brackets *vanish identically* if either A or B is replaced by one of the constraints or subsidiary conditions. This allows one to replace phase-space functions of the classical description by operators in the quantum description, with the corresponding replacement

$$i\{A, B\}_\mathcal{D} \longrightarrow [\hat{A}, \hat{B}]. \tag{8.55}$$

Now, making use of the explicit form of the matrix (8.54),

$$Q(x, y) = \begin{pmatrix} 0 & 0 & -1_{xy} & 0 \\ 0 & 0 & 0 & -\Delta_{xy} \\ 1_{xy} & 0 & 0 & 0 \\ 0 & \Delta_{xy} & 0 & 0 \end{pmatrix} ,$$

where

$$1_{xy} \equiv \delta^3(\vec{x} - \vec{y}), \quad \Delta_{xy} \equiv \vec{\nabla}_x^2 \delta^3(\vec{x} - \vec{y}) \ ,$$

one finds for the inverse of Q

$$\mathcal{R}(x, y) = \begin{pmatrix} 0 & 0 & 1_{xy} & 0 \\ 0 & 0 & 0 & \frac{1}{4\pi|\vec{x}-\vec{y}|} \\ -1_{xy} & 0 & 0 & 0 \\ 0 & -\frac{1}{4\pi|\vec{x}-\vec{y}|} & 0 & 0 \end{pmatrix} .$$

---

[8]Note that (with the exception of the free, non-interacting case) these conditions are in general inconsistent on lagrangian level, while on the hamiltonian level this is not so.

Correspondingly one computes

$$i\{A^i(x), \pi_j(y)\}_D = i\left(\delta^{ij} - \frac{\partial^i \partial^j}{\vec{\nabla}^2}\right)\delta^3(\vec{x} - \vec{y}) ,$$

in agreement with the subsidiary condition (8.53). With the substitution (8.55), these are the equal time commutation relations for the gauge field with the electric field $E^j = -\pi_j$ in the Coulomb gauge.

Chapter 9

Global Symmetries and Conservation Laws

We have seen in Sections 3 and 4 of Chapter 8 how to construct, following Dirac, the generator of *local* symmetries in the case of hamiltonian systems with so-called first class constraints (gauge theories). They are characterized by the fact that the infinitesimal parameters of the symmetry transformations depend on the space-time coordinates and may thus take different values at different space-time points.

There exist another important class of symmetries called *continuous global* symmetries, characterized by the fact that the infinitesimal parameters of the symmetry transformations do not depend on the space-time coordinates, but nevertheless take continuous values. Hence the denomination is *global* and *continuous*. Examples of such symmetries are the space-time translations and Lorentz transformations of the Poincaré group, the conformal transformations, as well as the internal symmetry transformations associated with the *isospin* group $SU(2)$ of the strong interactions, the *color* group $SU(3)$, and *flavor* group $SU(N)$.

In this chapter we show how to construct systematically the generators of continuous global symmetries in a lagrangian field theory following Noether's construction. The possibility of doing this is another important reason for insisting in a formulation of field theory within a *lagrangian* framework. We first formulate *Noether's theorem*[1] for a general lagrangian field theory involving up to first-order derivatives in the fields, and then specialize to the case of internal symmetries and Poincaré invariance.

9.1 Noether's Theorem

Consider the Lagrange density

$$\mathcal{L}(x) = \mathcal{L}(\{\phi_\alpha(x)\}, \{\partial_\mu \phi_\alpha(x)\})$$

---

[1] E. Noether, Nachricht. Ges. Wiss. Goettingen (1918) 171.

depending on n (in general complex) fields $\phi_\alpha(x), a = 1, ..., n$, and their first derivatives $\partial_\mu \phi_\alpha(x)$, where $\phi_\alpha(x)$ stands generically for a bosonic or fermionic field. We then have

Noether's Theorem

For every transformation depending *continuously* on N *global* parameters ϵ^a, $a = 1, ..., N$, which leaves the action

$$S[\phi] = \int d^4 x \mathcal{L}(x) \tag{9.1}$$

invariant, there exist N conserved "currents"

$$\partial_\mu J^{\mu,a} = 0 \ , \quad a = 1, ... N \ .$$

The "charges"

$$Q^a = \int d^3 x J^{0,a}(x)$$

associated with these currents are the *generators* of the symmetry transformation in question.

Proof:

Consider an infinitesimal transformation of the fields

$$\phi_\alpha(x) \to \phi'_\alpha(x) = \phi_\alpha(x) + \delta\phi_\alpha(x),$$
$$\phi^*_\alpha(x) \to \phi'^*_\alpha(x) = \phi^*_\alpha(x) + \delta\phi^*_\alpha(x) \ , \tag{9.2}$$

where the fields and their complex conjugate are regarded as independent.

So far the transformation need not be a symmetry of the action. The Lagrange density correspondingly transforms as follows

$$\mathcal{L}(x) \longrightarrow \mathcal{L}'(x) = \mathcal{L}(x) + \delta\mathcal{L}(x)$$
$$\delta\mathcal{L}(x) = \mathcal{L}(\phi + \delta\phi, \phi^* + \delta\phi^*, \partial_\mu\phi + \partial_\mu\delta\phi, \partial_\mu\phi^* + \partial_\mu\delta\phi^*)$$
$$- \mathcal{L}(\phi, \phi^*, \partial_\mu\phi, \partial_\mu\phi^*) \ ,$$

i.e.

$$\delta\mathcal{L} = \sum_\alpha \left(\frac{\partial\mathcal{L}}{\partial\phi_\alpha} \delta\phi_\alpha + \frac{\partial\mathcal{L}}{\partial\partial_\mu\phi_\alpha} \partial_\mu(\delta\phi_\alpha) \right) + \text{h.c.}$$
$$= \sum_\alpha \left(\frac{\partial\mathcal{L}}{\partial\phi_\alpha} - \partial_\mu \frac{\partial\mathcal{L}}{\partial\partial_\mu\phi_\alpha} \right) \delta\phi_\alpha + \sum_\alpha \partial_\mu \left(\frac{\partial\mathcal{L}}{\partial\partial_\mu\phi_\alpha} \delta\phi_\alpha \right) + \text{h.c.} \tag{9.3}$$

where h.c. stands for "hermitian conjugate". So far no restriction has been imposed on the transformation (9.2). If we now assume that this transformation is of such a form as to leave the action (9.1) *identically* invariant, then it must have the property that $\delta\mathcal{L}$ at most contributes a surface term to the action, i.e.

$$\delta_\epsilon \mathcal{L}(x) = \partial_\mu \mathcal{F}^\mu(x; \epsilon) \quad (\text{off} - \text{shell}) \ , \tag{9.4}$$

where ϵ stands for the N parameters ϵ_a, $a = 1, \cdots, N$ parametrizing the symmetry transformation. Note that (9.4) must be also true for arbitrary, *off-shell* field configurations, i.e. those which do not satisfy the Euler–Lagrange equations.

For field configurations which satisfy the Euler–Lagrange equations

$$\frac{\partial \mathcal{L}}{\partial \phi_\alpha} - \partial_\mu \frac{\partial \mathcal{L}}{\partial \partial_\mu \phi_\alpha} = 0 \ , \qquad \frac{\partial \mathcal{L}}{\partial \phi_\alpha^*} - \partial_\mu \frac{\partial \mathcal{L}}{\partial \partial_\mu \phi_\alpha^*} = 0$$

it follows from (9.3) that $\delta \mathcal{L}$ is always a four-divergence of a 4-vector field:

$$\delta \mathcal{L} = \sum_\alpha \partial_\mu \left(\frac{\partial \mathcal{L}}{\partial \partial_\mu \phi_\alpha} \delta \phi_\alpha \right) + \text{h.c.} \quad \text{(on-shell)} \ .$$

We therefore conclude that if (9.2) represents a *symmetry* transformation of the action, then

$$\text{on-shell}: \ \partial_\mu \mathcal{F}^\mu(x; \epsilon) - \sum_\alpha \left\{ \partial_\mu \left(\frac{\partial \mathcal{L}}{\partial \partial_\mu \phi_\alpha} \delta \phi_\alpha \right) + \text{h.c.} \right\} = 0 \ .$$

Define the vector-field

$$K^\mu := \sum_\alpha \left(\frac{\partial \mathcal{L}}{\partial \partial_\mu \phi_\alpha} \delta \phi_\alpha + \text{h.c.} \right) - \mathcal{F}^\mu \ . \tag{9.5}$$

Then K^μ is conserved for fields ϕ_α satisfying the Euler–Lagrange equations,

$$\partial_\mu K^\mu = 0 \quad \text{(on-shell)} \ . \tag{9.6}$$

For linear transformations, $\delta \phi_\alpha$ can be written in the form

$$\delta \phi_\alpha = \epsilon_a D_{\alpha\beta}^a \phi_\beta \tag{9.7}$$

with $\{\epsilon_a\}$ N *space-time independent* real parameters and $D_{\alpha\beta}(\epsilon)$ an operator. In that case (9.5) reduces to

$$K^\mu = \sum_a \epsilon_a \left(\sum_{\alpha,\beta} \frac{\partial \mathcal{L}}{\partial \partial_\mu \phi_\alpha} D_{\alpha\beta}^a \phi_\beta + \text{h.c.} - \mathcal{F}^{\mu,a} \right) \ , \quad \mathcal{F}^\mu(x; \epsilon) = \epsilon_a \mathcal{F}^{\mu,a}(x) \ . \tag{9.8}$$

Since K^μ depends *continuously* on the parameters ϵ_a, we obtain from (9.6) N conservation laws:

$$\partial_\mu J^{\mu,a} = 0, \quad J^{\mu,a} = \sum_{\alpha,\beta} \left(\frac{\partial \mathcal{L}}{\partial \partial_\mu \phi_\alpha} D_{\alpha\beta}^a \phi_\beta + \text{h.c.} \right) - \mathcal{F}^{\mu,a} \ . \tag{9.9}$$

With the time components $J^{0,a}$ of the currents $J^{\mu,a}$ we associate the charges

$$Q^a = \int d^3x \left(\sum_{\alpha,\beta} (\pi_\alpha D_{\alpha\beta}^a \phi_\beta + \text{h.c.}) - \mathcal{F}^{0,a} \right) \ . \tag{9.10}$$

These charges are time independent "on-shell" as a result of the conservation laws (9.6). Making use of the canonical commutation relations as well as the time-independence of the charges, we shall see that the charges (9.10) are the generators of the symmetry transformation (9.7) in the sense that[2]

$$\delta\phi_\alpha(x) = i\epsilon_a [Q^a, \phi_\alpha(x)] , \tag{9.11}$$

$$\delta\phi_\alpha^*(x) = i\epsilon_a [Q^a, \phi_\alpha^*(x)] . \tag{9.12}$$

The corresponding unitary operator generating the transformation

$$U\phi_\alpha(x)U^{-1} = \phi_\alpha'(x)$$

is

$$U[\epsilon] = e^{i\epsilon_a Q^a} .$$

We now discuss some important examples.

9.2 Internal symmetries

So far the label α on the fields ϕ_α ($\alpha = 1, \cdots, n$) has merely served to label different fields. Suppose now that these fields belong to an n-dimensional representation of a continuous, *internal* symmetry group parametrized by A parameters, such as $SU(N)$ ($A = N^2 - 1$). This means that under a transformation of the group, labelled by the parameters ϵ^a, $a = 1, ...A$,

$$\phi_\alpha(x) \to \phi_\alpha'(x) = D_{\alpha\beta}(\epsilon)\phi_\beta(x) \qquad \alpha = 1, ..., n ,$$

where $D(\epsilon)$ are the $n \times n$ representation matrices of the group in question. For a unitary group they may be written in the form

$$D(\epsilon) = e^{-i\sum_{a=1}^A \epsilon_a \lambda^a} , \qquad \lambda_{\alpha\beta}^{a*} = \lambda_{\beta\alpha}^a ,$$

where λ^a, $a = 1, ...A$ are the A generators of the group in the representation in question. For an infinitesimal transformation we thus have

$$\phi_\alpha'(x) \simeq \phi_\alpha(x) - i \sum_a \epsilon_a \lambda_{\alpha\beta}^a \phi_\beta(x) . \tag{9.13}$$

A Lagrange density respecting an *internal* symmetry is generally left strictly invariant by such a transformation, since no derivatives with respect to space-time are involved, that is

$$\delta_\epsilon \mathcal{L} = 0 \to \mathcal{F}^\mu = 0 .$$

Hence the conserved quantities (9.8) are given in this case by

$$K^\mu = \sum_{\alpha=1}^n \left(\frac{\partial \mathcal{L}}{\partial \partial_\mu \phi_\alpha} \delta_\epsilon \phi_\alpha + \text{h.c.} \right) . \tag{9.14}$$

---

[2]We take our system to be unconstrained.

Now, from (9.13) we have for an infinitesimal transformation

$$\delta\phi_\alpha = -i\sum_a \epsilon^a \lambda^a_{\alpha\beta}\phi_\beta, \quad \delta\phi^*_\alpha = i\sum_a \epsilon^a \phi^*_\beta \lambda^a_{\beta\alpha} . \tag{9.15}$$

Hence we obtain from (9.14) the A conserved currents

$$J^{\mu,a} = -i\sum_{\alpha,\beta}\left(\frac{\partial\mathcal{L}}{\partial\partial_\mu\phi_\alpha}\lambda^a_{\alpha\beta}\phi_\beta - \phi^*_\beta\lambda^a_{\beta\alpha}\frac{\partial\mathcal{L}}{\partial\partial_\mu\phi^*_\alpha}\right) .$$

It is important to realize that the *complex* fields and their *complex conjugate* fields are treated here as independent degrees of freedom.

With the A conserved currents are associated A *operator* valued conserved charges, which are the generators of the symmetry transformation (9.15):

$$Q^a = -i\int_{y^0=t} d^3y[\pi_\alpha(y)\lambda^a_{\alpha\beta}\phi_\beta(y) - \phi^\dagger_\beta(y)\lambda^a_{\beta\alpha}\pi^\dagger_\alpha(y)] . \tag{9.16}$$

Indeed, one has ($U = e^{i\epsilon^a Q^a}$)

$$U[\epsilon]\phi_\alpha(x)U^{-1}[\epsilon] \simeq 1 + i\epsilon^a[Q^a, \phi_\alpha(x)] ,$$

or using the commutation relations (8.28) we have

$$i[Q^a, \phi_\gamma(x)] = \int d^3y[\pi_\alpha(y), \phi_\gamma(x)]\lambda^a_{\alpha\beta}\phi_\beta(y) = -i\lambda^a_{\gamma\beta}\phi_\beta(x) .$$

One thus recovers the transformation law (9.15).

It remains to verify that the generators (9.16) satisfy the Lie-algebra of the group in question,

$$[Q^a, Q^b] = if^{ab}_c Q^c \tag{9.17}$$

with f^{ab}_c being the *structure coefficients*. Let us check this for the case of the three-dimensional representation of the orthogonal group $O(3)$ whose three generators $\lambda^i, i = 1, 2, 3$ satisfy the Lie algebra

$$[\lambda^i, \lambda^j] = i\epsilon_{ijk}\lambda^k, \quad \lambda^i_{jk} = -i\epsilon_{ijk} ,$$

with ϵ_{ijk} the totally antisymmetric Levi–Cevita tensor in three dimensions. In this case we can write the generators (9.16) in the compact form

$$\vec{Q} = -\int d^3y(\vec{\pi}\times\vec{\phi})(\vec{y},t) .$$

Making use of the canonical commutation relations

$$[\phi_i(x), \pi_j(y)]_{ET} = i\delta_{ij}\delta^3(\vec{x}-\vec{y})$$

and of the property

$$\sum_{\ell=1}^{3}\epsilon_{ij\ell}\epsilon_{i'j'\ell} = (\delta_{ii'}\delta_{jj'} - \delta_{ij'}\delta_{ji'})$$

of the Levi–Cevita tensor, one obtains

$$[Q^i, Q^j] = -i\epsilon_{ijk} \int d^3y (\pi(\vec{y}, t) \times \phi(\vec{y}, t))_k = i\epsilon_{ijk} Q^k \ , \tag{9.18}$$

where we have made use of the time independence of Q^i as guaranteed by Noether's theorem. We identify (9.18) with the Lie algebra of the rotation group with structure constants ϵ_{ijk}.

9.3   Translational invariance

Consider the (infinitesimal) translation of the space-time coordinates

$$x^\mu \to x'^\mu = x^\mu + \epsilon^\mu, \ \mathcal{L}'(x) = \mathcal{L}(x + \epsilon) \ .$$

This transformation is characterized by the four infinitesimal parameters ϵ^μ, $\mu = 0, 1, 2, 3$, that is, the "group" index is just the Lorentz index. The invariance of the four-dimensional volume element d^4x under this transformation evidently implies the invariance of the action (9.1) under the infinitesimal transformation. We have

$$\mathcal{L}'(x) = \mathcal{L}(x) + \epsilon^\mu \frac{\partial \mathcal{L}}{\partial x^\mu} + O(\epsilon^2) \ .$$

In accordance with (9.4) we thus find

$$\delta\mathcal{L}(x) = \partial_\mu \mathcal{F}^\mu(x; \epsilon) \tag{9.19}$$

with

$$\mathcal{F}^\mu(x; \epsilon) = \epsilon_\nu \mathcal{F}^{\mu,\nu}(x), \ \mathcal{F}^{\mu,\nu}(x) = g^{\mu\nu} \mathcal{L}(x) \ .$$

On the other hand we have $\phi'_\alpha(x) = \phi_\alpha(x + \epsilon)$, i.e.

$$\phi'_\alpha(x) - \phi_\alpha(x) = \delta\phi_\alpha(x) = \epsilon^\mu \frac{\partial \phi_\alpha(x)}{\partial x^\mu} \ , \tag{9.20}$$

so that

$$\delta\phi_\alpha(x) = \epsilon_\mu D^\mu_{\alpha\beta} \phi_\beta, \ D^\mu_{\alpha\beta} = \delta_{\alpha\beta} \partial^\mu \ .$$

Hence we have from (9.9)

$$J^{\mu,\nu} = \sum_\alpha \left(\frac{\partial \mathcal{L}}{\partial \partial_\mu \phi_\alpha} \partial^\nu \phi_\alpha + \text{h.c.} \right) - g^{\mu\nu} \mathcal{L} \ . \tag{9.21}$$

By construction we have the conservation law

$$\partial_\mu J^{\mu,\nu} = 0 \ . \tag{9.22}$$

The tensor $J^{\mu,\nu}$ is called the *energy-momentum* tensor.[3] The reason for this terminology is that the corresponding quantities

$$Q^\nu = \int d^3x J^{0,\nu} \equiv P^\nu \tag{9.23}$$

---

[3] The energy-momentum tensor is usually denoted by $\Theta^{\mu\nu}$.

have the meaning of the *energy-momentum 4-vector* associated with the fields $\phi_\alpha(x)$. Indeed, we recognize in $J^{0,0}(x)$ and $J^{0,i}(x)$ the hamiltonian and momentum density, respectively, associated with the Lagrange density $\mathcal{L}(x)$:

$$J^{0,0} = \sum_\alpha (\pi_\alpha \partial^0 \phi_\alpha + \text{h.c.}) - \mathcal{L} \equiv \mathcal{H} , \tag{9.24}$$

$$J^{0,i} = \sum_\alpha (\pi_\alpha \partial^i \phi_\alpha + \text{h.c.}) \equiv \mathcal{P}^i . \tag{9.25}$$

Integration over the space-coordinates gives the generators of time and space coordinates, respectively:

$$Q^0 = P^0 = \int d^3x \mathcal{H}(x) , \quad Q^i = P^i = \int d^3x \mathcal{P}^i(x) .$$

Dropping a surface term, we see using (9.22), that the corresponding charges are conserved:

$$\frac{d}{dt} Q^\nu = \int d^3x \partial_0 J^{0,\nu} = \int d^3x \partial_\mu J^{\mu,\nu} = 0 .$$

Note that we have been able to write the generators Q^ν entirely in terms of canonical coordinates. According to Noether's theorem, the 4-vector P^μ should be the generator of space-time translations of the fields. Indeed, from

$$i\partial_0 \phi_\alpha(x) = [\phi_\alpha(x), H] , \quad i\partial_0 \pi_\alpha(x) = [\pi_\alpha(x), H]$$

and the commutation relations

$$[\phi_\alpha(x), P^i] = \int d^3y \sum_\beta [\phi_\alpha(\vec{x}, t), \pi_\beta(\vec{y}, t)] \partial^i \phi_\beta(\vec{y}, t) = i\partial^i \phi_\alpha(x)$$

$$[\pi_\alpha(x), P^i] = -\int d^3y \sum_\beta \partial^i \pi_\beta(\vec{y}, t) [\pi_\alpha(\vec{x}, t), \phi_\beta(\vec{y}, t)] = i\partial^i \pi_\alpha(x) ,$$

we conclude that

$$[\phi_\alpha(x), P^\mu] = i\partial^\mu \phi_\alpha(x), \quad [\pi_\alpha(x), P^\mu] = i\partial^\mu \pi_\alpha(x) .$$

To order $O(\epsilon)$,

$$e^{i\epsilon_\mu P^\mu} \phi_\alpha(x) e^{-i\epsilon_\mu P^\mu} = \phi_\alpha(x) + i\epsilon_\mu [P^\mu, \phi_\alpha(x)] = \phi_\alpha(x) + \epsilon^\mu \partial_\mu \phi_\alpha(x),$$

which is just the infinitesimal form for a translation. Note that all these results are independent of a specific form of the translational invariant $\mathcal{L}$.

9.4   Lorentz transformations

Consider the Lorentz transformation

$$x^\mu \to x'^\mu = \Lambda^\mu{}_\nu x^\nu ,$$ (9.26)

parametrized by the *six* finite parameters $\Lambda^\mu{}_\nu$. The invariance of the action (9.1) under such a transformation follows from the invariance of the four-dimensional volume element d^4x under this transformation, and the assumption that the Lagrange density $\mathcal{L}$ transforms like a Lorentz scalar:

$$\mathcal{L}(x) \to \mathcal{L}'(x) = \mathcal{L}(\Lambda x) .$$

For an infinitesimal Lorentz transformation

$$\Lambda^\mu{}_\nu = g^\mu{}_\nu + \omega^\mu{}_\nu$$ (9.27)

we correspondingly have

$$\delta\mathcal{L} = \mathcal{L}'(x) - \mathcal{L}(x) \simeq \mathcal{L}((1+\omega)x) - \mathcal{L}(x)$$
$$= -\frac{1}{2}\omega_{\mu\nu}(x^\mu \partial^\nu - x^\nu \partial^\mu)\mathcal{L}(x) .$$

It is easy to see that the right-hand side has the expected form (9.4). Indeed, noting that

$$\partial_\sigma(x^\mu g^{\sigma\nu} - x^\nu g^{\sigma\mu})\mathcal{L} = (x^\mu \partial^\nu - x^\nu \partial^\mu)\mathcal{L} ,$$ (9.28)

we see that in this case

$$\mathcal{F}^\sigma = -\frac{1}{2}\omega_{\mu\nu}(x^\mu g^{\sigma\nu} - x^\nu g^{\sigma\mu})\mathcal{L} .$$

Hence we conclude that K^σ in (9.8) is given by

$$K^\sigma = \sum_\alpha \frac{\partial\mathcal{L}}{\partial\partial_\sigma\phi_\alpha}\delta\phi_\alpha + \frac{1}{2}\omega_{\mu\nu}(x^\mu g^{\sigma\nu} - x^\nu g^{\sigma\mu})\mathcal{L} .$$ (9.29)

Let us now suppose the fields $\phi_\alpha(x)$ to belong to an *irreducible representation* (A, B) of the Lorentz group. In this case, we have under a Lorentz transformation (compare with (4.22) and (4.23))

$$\phi'_\alpha(x) := U[\Lambda]\phi_\alpha(x)U[\Lambda^{-1}] = D^{(A,B)}_{\alpha\beta}[\Lambda^{-1}]\phi_\beta(\Lambda x) .$$

According to our discussion in Chapter 2, the representation matrices $D^{(A,B)}$ corresponding to the Lorentz transformation (9.27) have the form

$$D^{(A,B)}[\Lambda] = e^{\frac{i}{2}\omega_{\mu\nu}S^{\mu\nu}} ,$$

with

$$S_{\mu\nu} := S^{(A,B)}_{\mu\nu}$$

the six (algebraic) generators of the transformation. For an infinitesimal Lorentz transformation, we thus have

$$D^{(A,B)}[\Lambda^{-1}] \simeq 1 - \frac{i}{2}\omega_{\mu\nu}S^{\mu\nu} \ ,$$

$$\phi'_\alpha(x) \simeq \left(1 - \frac{i}{2}\omega_{\mu\nu}S^{\mu\nu}\right)_{\alpha\beta} \phi_\beta((1+\omega)x)$$

$$\simeq \phi_\alpha(x) - \frac{i}{2}\omega_{\mu\nu}(S^{\mu\nu})_{\alpha\beta}\phi_\beta(x) - \frac{1}{2}\omega_{\mu\nu}(x^\mu\partial^\nu - x^\nu\partial^\mu)\phi_\alpha(x) \ .$$

We conclude that

$$\delta\phi_\alpha(x) = \frac{1}{2}\omega_{\mu\nu}(M^{\mu\nu})_{\alpha\beta}\phi_\beta(x) \ , \tag{9.30}$$

where

$$(M^{\mu\nu})_{\alpha\beta} = -(x^\mu\partial^\nu - x^\nu\partial^\mu)\delta_{\alpha\beta} - i(S^{\mu\nu})_{\alpha\beta} \ .$$

We may thus rewrite (9.29) as follows

$$K^\sigma = \frac{1}{2}\omega_{\mu\nu}J^{\sigma,\mu\nu} \tag{9.31}$$

$$J^{\sigma,\mu\nu} = \sum_\alpha \left(\frac{\partial\mathcal{L}}{\partial\partial_\sigma\phi_\alpha}(M^{\mu\nu})_{\alpha\beta}\phi_\beta + \text{h.c.}\right) + (x^\mu g^{\sigma\nu} - x^\nu g^{\sigma\mu})\mathcal{L}. \tag{9.32}$$

We now observe that $J^{\sigma,\mu\nu}$ can be written in terms of the energy-momentum tensor (9.21) as follows:

$$J^{\sigma,\mu\nu} = (x^\mu J^{\sigma,\nu} - x^\nu J^{\sigma,\mu}) + S^{\sigma,\mu\nu} \tag{9.33}$$

with

$$S^{\sigma,\mu\nu} = i\sum_{\alpha,\beta} \frac{\partial\mathcal{L}}{\partial\partial_\sigma\phi_\alpha}(S^{\mu\nu})_{\alpha\beta}\phi_\beta \ .$$

Hence, as in the case of translations, the generators $Q^{\mu,\nu}$ can be expressed entirely in terms of canonical coordinates using (9.24) and (9.25):

$$Q^{0i} = \int d^3x \left(x^0\mathcal{P}^i - x^i\mathcal{H} + S^{0,0i}\right) \ ,$$

$$Q^{ij} = \int d^3x \left(x^i\mathcal{P}^j - x^j\mathcal{P}^i + S^{0,ij}\right) \ .$$

From (9.32) we construct the corresponding Noether charges

$$Q^{\mu\nu} = \int d^3x\, J^{0,\mu\nu} \tag{9.34}$$

which are conserved:

$$\frac{d}{dt}Q^{\mu\nu} = \int d^3x\, \partial_\sigma J^{\sigma,\mu\nu} = 0 \ .$$

These charges generate the infinitesimal transformation (9.30) on the fields. Indeed, with

$$U[\Lambda] = e^{\frac{i}{2}\omega_{\mu\nu}Q^{\mu\nu}} \simeq 1 + \frac{i}{2}\omega_{\mu\nu}Q^{\mu\nu} \ ,$$

we have

$$U[\Lambda]\phi_\alpha(x)U[\Lambda^{-1}] = \phi'_\alpha(x) \approx \phi_\alpha(x) + \frac{i}{2}\omega_{\mu\nu}[Q^{\mu\nu},\phi_\alpha(x)] \ .$$

From (9.34), (9.24) and (9.25) we have

$$Q^{0i} = \int d^3x(x^0\mathcal{P}^i - x^i\mathcal{H}) - i\sum_\alpha \int d^3x \pi_\alpha S^{0i}_{\alpha\beta}\phi_\beta(x) \tag{9.35}$$

and

$$Q^{ij} = \int d^3x(x^i\mathcal{P}^j - x^j\mathcal{P}^i) - i\sum_\alpha \int d^3x \pi_\alpha S^{ij}_{\alpha\beta}\phi_\beta(x) \ . \tag{9.36}$$

It follows that

$$i[Q^{\mu\nu},\phi_\alpha(x)] = \frac{1}{2}(M^{\mu\nu})_{\alpha\beta}\phi_\beta(x)$$

implying the transformation law (9.30) for the fields.

It remains to show that together with the generators of space-time translations constructed in Section 9.3, the generators (9.34) of the Lorentz transformations satisfy the Poincaré algebra

$$[P_\mu, P_\nu] = 0$$
$$[Q_{\mu\nu}, P_\lambda] = -i(g_{\mu\lambda}P_\nu - g_{\nu\lambda}P_\mu)$$
$$[Q_{\mu\nu}, Q_{\lambda\rho}] = i(g_{\mu\lambda}Q_{\nu\rho} + g_{\nu\rho}Q_{\mu\lambda} - g_{\mu\rho}Q_{\nu\lambda} - g_{\nu\lambda}Q_{\mu\rho}) \ .$$

This involves the use of (9.35) and (9.36). We leave this exercise to the interested reader.

Chapter 10

The Scattering Matrix

The experimental physicist concerns himself with the scattering of particles on a target by recording the rate at which different particles are registered in his counter and by measuring the mass, momenta, spin and polarization of these particles. From his measurements he can indirectly infer the existence of unstable particles (resonances) which have been produced during the interaction. All this information is contained in the so-called *differential scattering cross-section*. In terms of this quantity, he can then compare his results with the theoretical predictions.

The possibility of defining such a scattering cross-section presupposes that the particles no longer interact with the target when they hit the counters. In theoretical language, this means that we must be able to define *asymptotic* states behaving like free particles. For the field theorist this requirement is embodied in the so-called *asymptotic condition*. In a field theoretical context this requires the existence of asymptotic fields essentially given by Fock-space operators. This turns out to be a highly non-trivial requirement. It is particularly problematic in the case of gauge theories, where the fields are subject to long-range interactions such as the Coulomb interaction. Thus an electron is really never free, since it always carries with it an "electromagnetic tail". This creates difficulties in defining a cross-section, which can, however, be tamed, as Block and Nordsiek[1] have shown. In the case of non-abelian gauge theories it gives rise to very interesting phenomena such as *quark confinement* and *color neutrality* of physical states.

We begin in this chapter by deriving a general expression for the differential cross-section in terms of the so-called scattering matrix elements. We then turn to the LSZ formalism[2] expressing these matrix elements in terms of the fundamental fields of the field theory in question. This will require the computation of the vacuum expectation value of *time-ordered* products of fields. These expectation values can in general be computed only as power series expansions in the parameters (coupling constants) characterizing the strength of the interaction. In the following chapter we shall derive a general formula allowing us to calculate systematically the scattering matrix to a given order in the coupling constants.

---

[1] F. Bloch and A. Nordsiek, *Phys. Rev.*, **52** (1937) 54.
[2] H. Lehmann, K. Symanzik and W. Zimmermann, *Nuovo Cimento*, **1** (1955) 205.

10.1   The S-matrix and T-matrix

Let us consider the scattering of a particle in an external potential. We shall exclude in our discussion bound states, which would have in general to be included.

Let $|\Psi_\alpha^{(\pm)}(E)\rangle$ be eigenstates of $H = H_0 + H_{int}$ with energy E in the Heisenberg picture,

$$H|\Psi_\alpha^{(\pm)}(E)\rangle = E|\Psi_\alpha^{(\pm)}(E)\rangle \,, \tag{10.1}$$

where α stands for the remaining quantum numbers and $(\pm)$ reflects the asymptotic behaviour at $t = \pm\infty$. Let $|\Phi_\alpha(E)\rangle$ further be the corresponding eigenstates of H_0, again with eigenvalue E:

$$H_0|\Phi_\alpha(E)\rangle = E|\Phi_\alpha(E)\rangle \,.$$

For scattering states we take H and H_0 to have the same spectrum. The label α stands for all quantum numbers labelling the state, except for the total energy. From (10.1) we then have formally the "Lippmann–Schwinger equation"[3]

$$|\Psi_\alpha^{(\pm)}(E)\rangle = |\Phi_\alpha(E)\rangle + \frac{1}{E - H_0 \pm i\epsilon}V|\Psi_\alpha^{(\pm)}(E)\rangle \tag{10.2}$$

where we have set $H_{int} = V$.

Interpretation of $|\Psi^{(\pm)}(E) >$

Using the completeness of free particle states (no bound states!)

$$\sum_\beta \int dE' |\Phi_\beta(E')\rangle\langle\Phi_\beta(E')| = 1 \,,$$

we have the integral equation

$$|\Psi_\alpha^{(\pm)}(E)\rangle = |\Phi_\alpha(E)\rangle + \sum_\beta \int dE' \, |\Phi_\beta(E')\rangle\frac{T_{\beta\alpha}^{(\pm)}(E', E)}{E - E' \pm i\epsilon} \tag{10.3}$$

where

$$T_{\beta\alpha}^{(\pm)}(E', E) := \langle\Phi_\beta(E')|V|\Psi_\alpha^{(\pm)}(E)\rangle \,.$$

This equation defines the elements of the so-called "T-matrix", which will play a central role in the computation of the differential cross-section. In order to see that $|\Psi^{(-)}\rangle$ and $|\Psi^{(+)}\rangle$ correspond to in- and out-going scattering states, we must build wave packets in time. We therefore return to the Schrödinger picture, where

$$|\Psi_{E,\alpha}^{(\pm)}(t) >= |\Psi_\alpha^{(\pm)}(E) > e^{-iEt} \,,$$

---

[3]B.A. Lippmann and J. Schwinger, *Phys. Rev.* **79** (1956) 469; M.L. Goldberger and K.M. Watson, *Collision Theory* (John Wiley and Sons, New York, 1964).

and similarly for $|\Phi_{E,\alpha}(t)>$. Smearing these states out in energy with a distribution function $g_\alpha(E)$ sharply concentrated about the energy $E = E_\alpha$ we obtain

$$|\Psi_{g_\alpha}^{(\pm)}(t)\rangle = \int dE |\Psi_\alpha^{(\pm)}(E)\rangle g_\alpha(E) e^{-iEt} \tag{10.4}$$

$$|\Phi_{g_\alpha}(t)\rangle = \int dE |\Phi_\alpha(E)\rangle g_\alpha(E) e^{-iEt} .$$

Exchanging E and E' integrations, we have upon substituting (10.3) into (10.4),

$$|\Psi_{g_\alpha}^{(\pm)}(t)\rangle = |\Phi_{g_\alpha}(t)\rangle + \sum_\beta \int dE' |\Phi_\beta(E')\rangle I_{\beta\alpha}^{(\pm)}(t; E') , \tag{10.5}$$

where

$$I_{\beta\alpha}^{(\pm)}(t; E') = \int dE \frac{T_{\beta\alpha}^{(\pm)}(E', E)}{E - E' \pm i\epsilon} g_\alpha(E) e^{-iEt} . \tag{10.6}$$

Consider $|\Psi_\alpha^{(+)}(t)\rangle$. As we let $t \to -\infty$ we may close the E integration contour at infinity in the upper half plane.[4] The singularities of $T_{\gamma\alpha}^{(+)}(E', E)$ in E are known to lie on the unphysical, second Riemann sheet. Hence the resulting closed contour contains no singularities, and we obtain

$$I_{\beta\alpha}^{(+)}(t, E') \xrightarrow{t \to -\infty} 0 ,$$

i.e.

$$|\Psi_{g_\alpha}^{(+)}(t)\rangle \xrightarrow{t \to -\infty} |\Phi_{g_\alpha}(t)\rangle .$$

In the same way we find

$$|\Psi_{g_\alpha}^{(-)}(t)\rangle \xrightarrow{t \to +\infty} |\Phi_{g_\alpha}(t)\rangle .$$

On the other hand, for $t \to \infty$ we can close in (10.6) the contour in the lower half complex E-plane. Using the Cauchy residue theorem, we thus find

$$I_{\beta\alpha}^{(+)}(t; E') \to -2\pi i T_{\beta\alpha}^{(+)}(E', E') g_\alpha(E') e^{-iE't} .$$

Hence we have from (10.5) for $t \to \infty$,

$$|\Psi_{g_\alpha}^{(+)}(t)\rangle \to |\Phi_{g_\alpha}(t)\rangle - 2\pi i \sum_\beta \int dE' |\Phi_\beta(E')\rangle T_{\beta\alpha}^{(+)}(E', E') g_\alpha(E') e^{-iE't} . \tag{10.7}$$

The introduction of the smearing function has merely served to expose the time evolution of a wave packet in time, and has allowed us to close the E-contour at infinity. In the limit of sharp energy, i.e. $g_\alpha(E') = \delta(E' - E)$, the expression (10.7) reduces to

$$|\Psi_{E,\alpha}^{(+)}(t)\rangle \to |\Phi_{E,\alpha}(t)\rangle - 2\pi i \sum_\beta |\Phi_\beta(E)\rangle T_{\beta\alpha}(E) e^{-iEt} , \tag{10.8}$$

---

[4]We take $g_\alpha(E)$ to have a finite support, so that we may extend the integration to the whole real axis.

where

$$\mathcal{T}_{\beta\alpha}(E) \equiv \mathcal{T}_{\beta\alpha}^{(+)}(E, E) = <\Phi_\beta(E)|V|\Psi_\alpha^{(+)}(E)> .$$

This result would have followed also directly from (10.3) by formally closing the E' contour of integration at infinity.

The picture is thus as follows: The state $|\Psi_g^{(+)}(t)\rangle$ represents an incoming free wave packet described by the state $|\Phi_g(t)\rangle$ and an outgoing "spherical" wave, whereas $|\Psi_g^{(-)}(t)\rangle$ represents an incoming "spherical wave" and an outgoing free wave packet described by the state $|\Phi_g(t)\rangle$.

The QM wave function and scattering amplitude

Projecting (10.2) on the eigenstates of the position operator, we have for the "outgoing" wave function in the Heisenberg picture and a *local* potential,

$$\langle\vec{r}|\Psi_{\vec{p}}^{(+)}\rangle = \langle\vec{r}|\vec{p}\rangle + \int d^3p' \frac{\langle\vec{r}|\vec{p}'\rangle\langle\vec{p}'|V|\Psi_{\vec{p}}^{(+)}\rangle}{\frac{\vec{p}^2}{2m} - \frac{\vec{p}'^2}{2m} + i\epsilon}$$

$$= \langle\vec{r}|\vec{p}\rangle + \int d^3r' \int d^3r'' \int d^3p' \frac{\langle\vec{r}|\vec{p}'\rangle\langle\vec{p}'|\vec{r}'\rangle\langle\vec{r}'|V|\vec{r}''\rangle\langle\vec{r}''|\psi_{\vec{p}}^{(+)}\rangle}{\frac{\vec{p}^2}{2m} - \frac{\vec{p}'^2}{2m}}$$

$$= \frac{e^{i\vec{p}\cdot\vec{r}}}{(2\pi)^{3/2}} + \int d^3r' \int \frac{d^3p'}{(2\pi)^3} \frac{e^{i\vec{p}'\cdot(\vec{r}-\vec{r}')}}{\vec{p}^2 - \vec{p}'^2} 2mV(\vec{r}')\psi_{\vec{p}}^{(+)}(\vec{r}')$$

$$= \frac{e^{i\vec{p}\cdot\vec{r}}}{(2\pi)^{3/2}} + \int d^3r G^{(+)}(\vec{r} - \vec{r}')2mV(\vec{r}')\psi_{\vec{p}}^{(+)}(\vec{r}') ,$$

where

$$G^{(+)}(\vec{r} - \vec{r}') = \int \frac{d^3p'}{(2\pi)^3} \frac{e^{i\vec{p}'\cdot(\vec{r}-\vec{r}')}}{\vec{p}^2 - \vec{p}'^2} = -\frac{1}{4\pi} \frac{e^{ip|\vec{r}-\vec{r}'|}}{|\vec{r} - \vec{r}'|} \qquad (10.9)$$

with $p = |\vec{p}|$, and we have used

$$<\vec{p}'|V|\vec{p}> = V(\vec{p})\delta^3(\vec{p}' - \vec{p}) .$$

Note that the momentum label $\vec{p}$ now stands at the same time for the momentum and energy of the initial state in the far past. Furthermore, we have identified $\sum_\beta \int dE'$ in (10.3) with $\int d^3p'$. Now, for $\vec{r} \to \infty$,

$$|\vec{r} - \vec{r}'| \to r - \frac{\vec{r} \cdot \vec{r}'}{r}$$

or

$$p|\vec{r} - \vec{r}'| \to pr - p\hat{r} \cdot \vec{r}' ,$$

so that

$$<\vec{r}|\Psi_{\vec{p}}^{(+)}\rangle \xrightarrow[r\to\infty]{} \frac{e^{i\vec{p}\cdot\vec{r}}}{(2\pi)^{3/2}} + \frac{e^{ipr}}{r}f_{\vec{p}}(\theta, \phi) \qquad (10.10)$$

with

$$f_{\vec{p}}(\theta, \phi) = -\frac{1}{4\pi} \int d^3r' e^{-i|\vec{p}|\hat{r}\cdot\vec{r}'} 2mV(\vec{r}')\psi_{\vec{p}}^{(+)}(\vec{r}')$$

the "scattering amplitude", where θ and ϕ are the longitudinal and asymuthal angles measured with respect to the incident direction $\vec{r}$.

The S-matrix

The S-matrix element S_{fi} is defined as the projection of the outgoing spherical wave $|\Psi_i^{(+)}(t)\rangle$ on the plane-wave packet $|\Phi_f(t)\rangle$, as $t \to \infty$:

$$S_{fi} = \left\{ \langle \Phi_f(t) | \Psi_i^{(+)}(t) \rangle \right\}_{t\to\infty} . \tag{10.11}$$

Taking the projection of $|\Psi_i^{(+)}(t) >$ in (10.8) on $|\Phi_f(E) > e^{-iEt}$ we obtain

$$S_{fi} = \left\{ \langle \Phi_f(t) | \Psi_i^{(+)}(t) \rangle \right\}_{t\to\infty} = \delta_{fi} - 2\pi i \delta_{\alpha\beta} \delta(E_f - E_i) \mathcal{T}(E)_{fi} . \tag{10.12}$$

In and Out states

Since the scalar product in the definition (10.11) of the S-matrix is time independent, we can also write

$$S_{fi} = \langle \Psi_f^{(-)}(0) | \Psi_i^{(+)}(0) \rangle .$$

Note that at time zero the Schrödinger and Heisenberg pictures coincide.

Following convention we call $|\Psi_i^{(+)}(0) >$ and $|\Psi_f^{(-)}(0) > in$ states and *out* states, respectively. The designation corresponds to the fact, that these states just correspond to the incoming and outgoing free particle states labelled by "i" and "f" respectively, after these states have gone through a time evolution from $t = -\infty$ ($t = +\infty$) to $t = 0$.

The completeness of the *in* and *out*-states implies that there exists an isomorphism between the free particle *in* and *out* states, given by

$$< f, out| =< f, in|S \tag{10.13}$$

or

$$< f, out|i, in >=< f, in|S|i, in >= S_{fi} . \tag{10.14}$$

The orthogonality of the *in* and *out* states evidently implies the *unitary* of the S-matrix:

$$\sum_n S_{fn} S_{ni} = \delta_{fi} , \quad or \quad SS^\dagger = 1 . \tag{10.15}$$

In order to establish the connection between these operators and the quantum fields, we observe that in the Heisenberg picture the operators are dependent on time, while the states are independent of time. Since in the Schrödinger picture the *in* and *out* states correspond to incoming and outgoing plane waves, respectively, we expect now in the Heisenberg picture, that this role is taken up by the field-operators. This is materialized in the Lehmann–Symanzik–Zimmermann (LSZ) asymptotic condition

$$\phi(x) \xrightarrow[t\to\mp\infty]{} \sqrt{Z}\phi_{in/out}(x) , \tag{10.16}$$

where

$$(\Box + m^2)\phi_{in/out}(x) = 0 \,,$$

with the solution

$$\phi_{in/out}(x) = \frac{1}{(2\pi)^{\frac{3}{2}}} \int \frac{d^3p}{\sqrt{2\omega_p}} \left[a_{in/out}(\vec{p})e^{-ip\cdot x} + a^\dagger_{in/out}(\vec{p})e^{ip\cdot x} \right] \,.$$

The reason for the (renormalization) constant Z is that we want the *in* and *out* fields to be canonically normalized, while the field $\phi(x)$, which includes the effect of all interactions, will in general no longer satisfy the canonical commutation relations demanded on the semiclassical level.

Correspondingly we define creation and destruction operators to act as follows on the bare vacuum state $|0>$:

$$|\vec{p}, in> = a^\dagger_{in}(\vec{p})|0>$$
$$|\vec{p}, out> = a^\dagger_{out}(\vec{p})|0> \,,$$

with the property

$$a_{in}(\vec{p})|0>= 0\,, \quad a_{out}(\vec{p})|0>= 0$$

and the commutation relation for either case,

$$[a(\vec{p}), a^\dagger(\vec{p}')] = \delta^3(\vec{p} - \vec{p}') \,,$$

in agreement with (7.26).

The LSZ asymptotic condition will play a central role in order to establish the connection between the S-matrix elements and the quantum fields of the Quantum Field Theory in question. But before doing this, let us examine how the differential cross-section is related to the S-matrix elements.

10.2   Differential cross-section

Experiments usually involve only <u>two</u> particles in the initial state, the "target", and the "projectile" to be scattered on the target. The two-particle state prepared by the experimentalist corresponds to a wave packet

$$|i, in> = \int d^3q_a d^3q_b \tilde{f}_{p_a}(\vec{q}_a)\tilde{f}_{p_b}(\vec{q}_b)|\vec{q}_a, \vec{q}_b, in> \,,$$

where

$$|q_a, q_b, in> = |q_a, in>|q_b, in>$$

and $\vec{p}$ corresponds to the momentum around which $\tilde{f}_p(\vec{q})$ is concentrated. Because of $<i, in|i, in>= 1$, we have

$$\int d^3q |\tilde{f}_p(\vec{q})|^2 = 1.$$

If we turn off the interaction, we must have for the scattering-matrix $S = 1$. Since the experimentalist is only interested in particles that are scattered, we write

$$S = 1 + iT \ , \tag{10.17}$$

where the (T-matrix) T only contains the scattered part. Translational invariance then allows us to write

$$< f, in|T|\vec{p}_a, \vec{p}_b, in > = (2\pi)^4 \delta^4(P_f - p_a - p_b) < f, in|T|\vec{p}_a, \vec{p}_b, in > \ , \tag{10.18}$$

where the final state is taken to be a free n-particle state of the total 4-momentum $P_f^\mu = p_1^\mu + \cdots + p_n^\mu$.

Differential cross-section for $2 \to n$-scattering

The number of transitions of two initial free "wave packets" to a final state $|f>$ is given by[5]

$$dW_{fi} \equiv | < f|T|\tilde{f}_{p_a}, \tilde{f}_{p_b} > |^2 [df] \tag{10.19}$$

$$= \int d^3 q'_a d^3 q'_b d^3 q_a d^3 q_b \tilde{f}_{p_a}(\vec{q}_a) \tilde{f}_{p_b}(\vec{q}_b) \tilde{f}^*_{p_a}(\vec{q}'_a) \tilde{f}^*_{p_b}(\vec{q}'_b)$$

$$\times [(2\pi)^4]^2 \delta^4(P_f - q_a - q_b)\delta^4(P_f - q'_a - q'_b) < f|T|\vec{q}_a, \vec{q}_b > < f|T|\vec{q}'_a, \vec{q}'_b >^* [df]$$

with

$$[df] = \prod_{i=1}^{n} d^3 p_i \ .$$

The form of this volume element follows from the completeness relation

$$\sum_n \int d^3 p_1 ... d^3 p_n |\vec{p}_1 ... \vec{p}_n > < \vec{p}_1 ... \vec{p}_n| = 1 \ .$$

We assume $\tilde{f}_p(q)$ to be strongly peaked around the momentum $\vec{p}$, and substitute in (10.19) the following

$$< f|T|\vec{q}_a, \vec{q}_b > \longrightarrow < f|T|\vec{p}_a, \vec{p}_b >$$

$$< f|T|\vec{q}'_a, \vec{q}'_b > \longrightarrow < f|T|\vec{p}_a, \vec{p}_b >$$

$$\delta^4(P_f - q_a - q_b)\delta^4(P_f - q'_a - q'_b) \to \delta^4(q'_a + q'_b - q_a - q_b)\delta^4(P_f - p_a - p_b)$$

$$= \delta^4(P_f - p_a - p_b) \int \frac{d^4 x}{(2\pi)^4} e^{-i(q_a + q_b - q'_a - q'_b) \cdot x}$$

so that

$$dW_{fi} = (2\pi)^4 \delta^4(P_f - p_a - p_b)| < f|T|\vec{p}_a, \vec{p}_b > |^2 \mathcal{F}(\vec{p}_a, \vec{p}_b)[df] \tag{10.20}$$

---

[5]In this section we omit the label "in" for the states.

with

$$\mathcal{F}(\vec{p}_a, \vec{p}_b) = \int d^4x |f_{p_a}(x)|^2 |f_{p_b}(x)|^2 , \qquad (10.21)$$

where

$$f_{p_i}(x) = \int d^3q \, \tilde{f}_{p_i}(\vec{q}) e^{-iq \cdot x}, \quad p_i^2 = m_i^2 .$$

The measurement dW_{fi} should be independent of the inertial reference system. The volume element $[df]$ is however not Lorentz invariant. Correspondingly, the T-matrix element is not Lorentz invariant, either. We can however make the Lorentz covariance manifest if we recognize that $< \vec{p}_1'...\vec{p}_n'|\mathcal{T}|\vec{p}_a, \vec{p}_b >$ always has the structure

$$| < \vec{p}_1'...\vec{p}_n'|\mathcal{T}|\vec{p}_a, \vec{p}_b > |^2 = \frac{1}{(2\pi)^6 2\omega_a 2\omega_b} \prod_{i=1}^{n} \frac{1}{(2\pi)^3 2\omega_i'} | < \vec{p}_1' \cdots \vec{p}_n'|\mathcal{M}|\vec{p}_a, \vec{p}_b > |^2 ,$$

$$(10.22)$$

where now $< \vec{p}_1'...\vec{p}_n'|\mathcal{M}|\vec{p}_a, \vec{p}_b >$ is a Lorentz covariant quantity. This quantity one refers to as "Stapp M-function".[6] The appearance of the $1/2\omega$-factors is a consequence of the non-covariant normalization of our states in Fock-space.

In order to interpret (10.20), we note to begin with that we take our 1-particle states

$$|\tilde{f}_p > = \int d^3q \tilde{f}_p(\vec{q})|\vec{q} >$$

to be normalized such that

$$< \tilde{f}_{p_i}|\tilde{f}_{p_i} > = n_i ,$$

with n_i = the number of particles of type "i", or

$$\int d^3q |\tilde{f}_{p_i}(\vec{q})|^2 = n_i .$$

Hence

$$\int_{V \to \infty} d^3x |f_i(\vec{x}, t)|^2 = \int d^3x \int d^3q \int d^3q' \, \tilde{f}_{p_i}^*(\vec{q}) \tilde{f}_{p_i}(\vec{q}') e^{-i(q-q') \cdot x}$$

$$= (2\pi)^3 \int d^3q |\tilde{f}_{p_i}(\vec{q})|^2 = (2\pi)^3 n_i ,$$

so that obviously

$$\rho_i(\vec{x}, t) = \frac{1}{(2\pi)^3} |f_i(\vec{x}, t)|^2$$

is the density of particles of type "i" at time $x^0 = t$. Hence we may write for $\mathcal{F}(\vec{p}_a, \vec{p}_b)$ in (10.21),

$$\mathcal{F}(\vec{p}_a, \vec{p}_b) = (2\pi)^6 \int_{-\Delta T/2}^{\Delta T/2} dt \int d^3x \rho_a(x) \rho_b(x) ,$$

---

[6]H. Stapp, *Phys. Rev.* **125** (1962) 2139. Notice that the replacement of T by M amounts to the change of our non-covariant normalization of states $< \vec{p}'|\vec{p} > = \delta(\vec{p}' - \vec{p})$ to $< \vec{p}'|\vec{p} > = (2\pi)^3 2\omega(p)\delta(\vec{p}' - \vec{p})$.

where the limit $\Delta T \to \infty$ is to be taken at the end of the calculation. The time-dependence of $\rho(x)$ takes account of the fact, that a wave packet will expand with time, unless we are dealing with a plane wave. Since the latter case represents our limiting situation, we are allowed to regard our integrand above to be constant in space and time. The number of transitions $i \to [f, f + df]$ is correspondingly given from (10.20) by

$$dW_{fi} = (2\pi)^{10}\delta^4(P_f - \vec{p}_a - \vec{p}_b)| < f|\mathcal{T}|\vec{p}_a, \vec{p}_b >|^2 (\Delta T \rho_a)(V\rho_b)[df] .$$

We now go to the lab system of particle "b". We evidently have

$$V\rho_b = n_b .$$

For the interpretation of $\rho_a \Delta T$ we observe that $\Delta n_a = \rho_a \Delta T v_{ab}$ is the number of particles "a", that cross a unit surface normal to the relative velocity v_{ab} in a time ΔT. We define

$$d\sigma_{fi} = \frac{\text{\# of transitions } i \to [f, f + df] \text{ in time } \Delta T}{(\text{\# of target} - \text{particles } b) \times \Delta n_a} \quad (10.23)$$

$$= \frac{(2\pi)^{10}\delta^4(\sum p_i' - p_a - p_b)| < p_1'...p_n'|\mathcal{T}|p_a, p_b >|^2}{n_b(\rho_a \Delta T v_{ab})} n_b(\rho_a \Delta T)[df]$$

$$= (2\pi)^4\delta^4\left(\sum p_i' - p_a - p_b\right) \frac{| < p_1'...p_n'|\mathcal{M}|\vec{p}_a, \vec{p}_b >|^2}{2\omega_a 2\omega_b v_{ab}} \prod_i \frac{d^3 p_i}{(2\pi)^3 2\omega_i}$$

where we have used (10.22).

Now, in the Lab system $\vec{p}_b = 0$, i.e.

$$v_{ab} = \frac{|\vec{p}_a|_{lab}}{\omega_a}$$

or

$$(2\omega_a 2\omega_b v_{ab})_{\vec{p}_b=0} = 4m_b|\vec{p}_a|_{Lab} = 4m_b(\omega_a^2 - m_a^2)_{lab}^{1/2}$$

$$= 4[(p_a \cdot p_b)^2 - m_a^2 m_b^2]_{lab}^{1/2} ,$$

so that in a general system

$$2\omega_a 2\omega_b v_{ab} = 2\sqrt{(s - m_a^2 - m_b^2)^2 - 4m_a^2 m_b^2}$$

with

$$s = (p_a + p_b)^2 , \quad s - m_a^2 - m_b^2 = 2p_a \cdot p_b . \quad (10.24)$$

We thus see that $2\omega_a 2\omega_b v_{ab}$ is also a Lorentz invariant quantity.

Putting things together, we finally obtain

$$d\sigma_{a+b\to n} = \frac{(2\pi)^4\delta^4(\Sigma p_i - p_a - p_b)| < \vec{p}_1...\vec{p}_n|\mathcal{M}|\vec{p}_a, \vec{p}_b >|^2}{2\sqrt{(s - m_a^2 - m_b^2)^2 - 4m_a^2 m_b^2}} \prod_{i=1}^n \frac{d^3 p_i}{(2\pi)^3 2\omega_i} . \quad (10.25)$$

Unitarity and Optical Theorem

Unitarity of the S-matrix (10.15) expresses the fundamental principle of probability conservation. From (10.17) we have

$$(T - T^\dagger) = T^\dagger T$$

or

$$2iImT = T^\dagger T \ .$$

With

$$< f|T|i >= (2\pi)^4\delta(P_f - P_i)T_{fi}$$

and the completeness relation

$$\sum_n \int \prod_k d^3p_k |\vec{p}_1, \cdots \vec{p}_n >< \vec{p}_1 \cdots \vec{p}_n| = 1 \ .$$

We thus have

$$(\mathcal{T}_{fi} - \mathcal{T}_{fi}^*) = i\sum_n (2\pi)^4\delta^4(P_n - P_i)\mathcal{T}_{nf}^*\mathcal{T}_{ni} \tag{10.26}$$

where the sum includes the integration over the intermediate momenta.

Let us consider 2-particle elastic forward scattering, $|f >= |i >$. The left-hand side then reduces to $2iIm\mathcal{T}_{ii}$. In the basis of the covariant normalization of the states $< \vec{p}\,'|\vec{p} >= (2\pi)^3 2\omega(p)\delta(\vec{p}\,' - \vec{p})$ this reads

$$2Im\mathcal{M}_{ii} = i\sum_n (2\pi)^4\delta^4(P_n - P_i)\mathcal{M}_{ni}^*\mathcal{M}_{ni} \prod_{k=1}^n \frac{d^3p_k}{(2\pi)^3 2\omega_k}$$

with $\mathcal{M}$ the Stapp matrix.

On the other hand, we have from (10.25) for the total cross-section

$$\sigma_{tot} = \sum_n \frac{(2\pi)^4\delta^4(P_n - p_a - p_b)| < \vec{p}_1...\vec{p}_n|\mathcal{M}|\vec{p}_a, \vec{p}_b > |^2}{2\sqrt{(s - m_a^2 - m_b^2)^2 - 4m_a^2m_b^2}} \prod_{k=1}^n \frac{d^3p_k}{(2\pi)^3 2\omega_k} \ .$$

Comparing with (10.25) we see that[7]

$$Im\mathcal{M}_{ii} = \sqrt{\lambda(s, m_a^2, m_b^2)}\sigma_{tot}.$$

This is the *Optical Theorem*.

---

[7]The square root appearing in the denominator arises in different contexts and is denoted in the literature by $\sqrt{\lambda(s, m_a^2, m_b^2)}$ with $\lambda(x_1, x_2, x_3) = (x_1^2 + x_2^2 + x_3^2) - 2x_1x_2 - 2x_2x_3 - 2x_3x_1$.

10.3   LSZ reduction formula

In this section we derive a fundamental formula for the scattering matrix which shall allow us to compute the S-matrix elements in terms of fundamental quantities — the so-called *n-point functions* — of the field theory in question. The result was first derived by Lehmann, Symanzik and Zimmermann,[8] and hence is referred to as the *LSZ-reduction formula*. For simplicity we shall consider the derivation for the case of a scalar field. The corresponding results for the Dirac and electromagnetic field are obtained in an analogous way, and we limit ourselves to listing the results for these cases at the end of this section. We shall illustrate the procedure for the first two essential steps. The remaining reduction process is then self-evident.

First step in the LSZ-reduction process

The elements of the S-matrix are defined by (10.14). In order to illustrate the method we shall take all the particles participating in the scattering process to be of one kind (no anti-particles). The final result will then be stated at the end for the general case where antiparticles are also involved. The first step in the LSZ reduction process consists in rewriting the transition amplitude (10.14) as

$$\langle \vec{p}_1', \vec{p}_2' \cdots out | \vec{p}_1, \vec{p}_2 \cdots in \rangle = \langle \vec{p}_1', \vec{p}_2' \cdots out | a_{in}^\dagger(\vec{p}_1) | \vec{p}_2 \cdots in \rangle \ , \tag{10.27}$$

where $a_{in}^\dagger(\vec{p}_1)$ is the creation operator in Fock-space for a 1-particle *in*-state of momentum $\vec{p}_1$. We then express $a_{in}^\dagger(\vec{p})$ in terms of the *in*-field having the Fourier decomposition,

$$\varphi_{in}(x) = \int d^3q [f_{\vec{q}}(x) a_{in}(\vec{q}) + f_{\vec{q}}^*(x) a_{in}^\dagger(\vec{q})] \ , \tag{10.28}$$

where $f_{\vec{q}}(x)$ is the plane wave

$$f_{\vec{q}}(x) = \frac{1}{\sqrt{(2\pi)^3 2\omega_q}} \, e^{-iq\cdot x} \ , \quad q^0 = \omega(\vec{q}) \ .$$

Using the orthogonality properties

$$i \int d^3x f_{\vec{q}'}^*(x) \overset{\leftrightarrow}{\partial_0} f_{\vec{q}}(x) = \delta^3(\vec{q}' - \vec{q})$$

$$i \int d^3x f_{\vec{q}'}(x) \overset{\leftrightarrow}{\partial_0} f_{\vec{q}}(x) = i \int d^3x f_{\vec{q}'}^*(x) \overset{\leftrightarrow}{\partial_0} f_{\vec{q}}^*(x) = 0 \tag{10.29}$$

with

$$f \overset{\leftrightarrow}{\partial_0} g \equiv f \partial_0 g - \partial_0 f \cdot g$$

one easily checks that (10.28) may be inverted to give

$$a_{in}^\dagger(\vec{p}) = - \int_t d^3x f_{\vec{p}}(x) i \overset{\leftrightarrow}{\partial_0} \varphi_{in}^\dagger(x)$$

$$= -Z_\varphi^{-1/2} \int_{t=-\infty} d^3x f_{\vec{p}}(x) \, i\overset{\leftrightarrow}{\partial_0} \varphi^\dagger(x) \tag{10.30}$$

---

[8] H. Lehmann, K. Symanzik and W. Zimmermann, *Nuovo Cimento* **1** (1955) 205.

where we have used the asymptotic condition (10.16).

We now use the identity

$$\int_{-\infty}^{\infty} dt \partial_t \int d^3x F(\vec{x}, t) = \int d^3x F(\vec{x}, t) \Big|_{t=-\infty}^{t=+\infty} \tag{10.31}$$

in order to rewrite (10.30) in the form

$$a_{in}^{\dagger}(\vec{p}) = -Z_{\varphi}^{-1/2} \int_{t=+\infty} d^4x f_{\vec{p}}(x) \, i\overset{\leftrightarrow}{\partial_0} \varphi^{\dagger}(x) + Z_{\varphi}^{-1/2} \int d^4x \partial_0(f_{\vec{p}}(x) i \overset{\leftrightarrow}{\partial_0} \varphi^{\dagger}(x)) \ ,$$

or

$$a_{in}^{\dagger}(\vec{p}) = a_{out}^{\dagger}(\vec{p}) + Z_{\varphi}^{-1/2} \int d^4x \partial_0(f_p(x) i \overset{\leftrightarrow}{\partial_0} \varphi^{\dagger}(x)) \ . \tag{10.32}$$

Now, from the definition of $\overset{\leftrightarrow}{\partial}$ it follows that

$$\partial_0(f \overset{\leftrightarrow}{\partial_0} g) = f \partial_0^2 g - \partial_0^2 f \cdot g \ . \tag{10.33}$$

Hence, using the equation of motion

$$\partial_0^2 f = (\vec{\nabla}^2 - m^2) f \ , \tag{10.34}$$

we may write (10.32) after a partial integration with respect to the spatial coordinates (a partial integration with respect to time leads to surface term contributions!) in the form

$$a_{in}^{\dagger}(\vec{p}) = a_{out}^{\dagger}(\vec{p}) + iZ_{\varphi}^{-1/2} \int d^4x \varphi^{\dagger}(x)(\overset{\leftarrow}{\Box} + m^2) f_{\vec{p}}(x) \ . \tag{10.35}$$

Note that it also follows from here, that

$$a_{out}(\vec{p}) = a_{in}(\vec{p}) - iZ_{\varphi}^{-1/2} \int d^4x f_{\vec{p}}^*(x)(\overset{\rightarrow}{\Box} + m^2) \varphi(x) \ . \tag{10.36}$$

Substituting (10.35) into (10.27), we are thus left with

$$\langle \vec{p}_1' \dots out | \vec{p}_1 \dots in \rangle = \langle \vec{p}_1' \cdots out | a_{out}^{\dagger}(\vec{p}_1) | \vec{p}_2 \dots in \rangle \tag{10.37}$$

$$+ iZ_{\varphi}^{-1/2} \int d^4x \langle \vec{p}_1' \dots out | \varphi^{\dagger}(x) | \vec{p}_2 \cdots in \rangle (\overset{\leftarrow}{\Box} + m^2) f_{\vec{p}_1}(x) \ .$$

Now

$$\langle \vec{p}_1' \dots out | a_{out}^{\dagger}(\vec{p}_1) = \sum_{k=1} \delta^3(\vec{p}_1 - \vec{p}_k')\langle \vec{p}_1' \dots \hat{\vec{p}}_k' \dots out | \ ,$$

where the "hat" on the momentum $\hat{p}_k$ indicates that the particle with this momentum is to be omitted in the state $\langle \vec{p}_1', \dots, \vec{p}_n'|$. The first term in (10.37) contributes a "disconnected" piece. We thus have

$$\langle \vec{p}_1' \dots out | \vec{p}_1 \dots in \rangle = iZ_{\varphi}^{-\frac{1}{2}} \int d^4x \langle \vec{p}_1' \dots out | \varphi^{\dagger}(x) | \vec{p}_2 \dots in \rangle (\overset{\leftarrow}{\Box} + m^2) f_{p_1}(x)$$

$$+ \ disconnected \ piece. \tag{10.38}$$

Hence for the connected part we have *effectively*

$$a_{in}^\dagger(\vec{p}) = iZ^{-\frac{1}{2}} \int d^4x \varphi^\dagger(x)(\overleftarrow{\Box} + m^2) f_p(x) . \qquad (10.39)$$

This completes the first step in the reduction process.

Second step in the LSZ-reduction process

We now treat in the same way the particle with momentum p_1' in the *out* state. Using this time

$$a_{out}(\vec{p}_1') = Z_\varphi^{-\frac{1}{2}} \int_{t'=+\infty} d^3x' f_{\vec{p}_1'}^*(x') i \overleftrightarrow{\partial_0'} \varphi(x') , \qquad (10.40)$$

we have in the integrand of (10.38),

$$\langle \vec{p}_1'...out|\varphi^\dagger(x)|\vec{p}_2 \cdots in \rangle = \langle \vec{p}_2' \cdots out|a_{out}(\vec{p}_1')\varphi^\dagger(x)|\vec{p}_2 \cdots in \rangle$$

$$= iZ_\varphi^{-\frac{1}{2}} \int_{t'=\infty} d^3x' f_{\vec{p}_1'}^*(x') \overleftrightarrow{\partial_0'} \langle \vec{p}_2' \cdots out|\varphi(x')\varphi^\dagger(x)|\vec{p}_2...in \rangle . \qquad (10.41)$$

In order to separate the disconnected piece we must move $\varphi(x')$ to the right of $\varphi^\dagger(x)$. To this end we introduce the *time-ordering* operation T for the product of two (bosonic) operators at different times,

$$T\varphi(x)\varphi^\dagger(y) = \theta(x^0 - y^0)\varphi(x)\varphi^\dagger(y) + \theta(y^0 - x^0)\varphi^\dagger(y)\varphi(x) . \qquad (10.42)$$

Since the integral in (10.41) is to be evaluated at time $t' = \infty$, we can rewrite it as

$$\langle \vec{p}_2' \cdots out|a_{out}(\vec{p}_1')\varphi(x)|p_2...in \rangle = Z_\varphi^{-1/2} \int_{t'=\infty} d^3x' \qquad (10.43)$$

$$\times f_{\vec{p}_1'}^*(x') i\overleftrightarrow{\partial_0'} \langle \vec{p}_2'...out|T\varphi(x')\varphi^\dagger(x)|\vec{p}_2...in \rangle .$$

Now, from (10.31) and (10.34) we have for the integrand,

$$\int_{t'=\infty} d^3x' f_{\vec{p}'}^*(x') \overleftrightarrow{\partial_0'} \langle \quad \rangle = \int_{t'=-\infty} d^3x' f_{\vec{p}'}^*(x') \overleftrightarrow{\partial_0'} \langle \quad \rangle$$

$$+ \int d^4x' f_{\vec{p}'}^*(x')(\overrightarrow{\Box}' + m^2)\langle \quad \rangle .$$

From the first term on the r.h.s. of this equation we obtain

$$i \int_{t'=-\infty} d^3x' f_{\vec{p}_1'}^*(x') \overleftrightarrow{\partial_0'} \langle \vec{p}_2'...out|T\varphi(x')\varphi^\dagger(x)|\vec{p}_2...in \rangle$$

$$= i \int_{t'=-\infty} d^3x' f_{\vec{p}_1'}^*(x') \overleftrightarrow{\partial_0'} \langle \vec{p}_2' \cdots out|\varphi^\dagger(x)\varphi(x')|\vec{p}_2...in \rangle$$

$$= Z_\varphi^{1/2} \langle \vec{p}_2' \cdots out|\varphi^\dagger(x)a_{in}(\vec{p}_1')|\vec{p}_2...in \rangle$$

$$= Z_\varphi^{1/2} \sum_{k=2} \langle \vec{p}_2' \cdots out|\varphi(x_1)|\vec{p}_2...\hat{p}_k...in \rangle \delta^3(\vec{p}_k - \vec{p}_1')$$

so that (10.41) may also be written in the form

$$\langle \vec{p}_1'...out|\varphi^\dagger(x)|\vec{p}_2...in\rangle = iZ_\varphi^{-1/2}\int d^4x' f_{\vec{p}_1}^*(x')(\overrightarrow{\Box}' + m^2)\langle \vec{p}_2'...out|T\varphi(x')\varphi^\dagger(x)|\vec{p}_2...in\rangle$$

$$+ \; disconnected \; piece \tag{10.44}$$

where the last term corresponds to an interaction leaving always one of the particles *untouched*, which therefore does not participate in the scattering process.

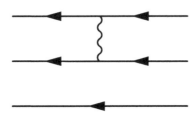

Fig. 10.1. 3-particle scattering with one disconnected piece.

Hence, substituting (10.44) into (10.39), we arrive at

$$\langle \vec{p}_1'\cdots out|\vec{p}_1...in\rangle = (iZ_\varphi^{-1/2})^2 \int d^4x_1 \int d^4x_1' f_{\vec{p}_1'}^*(x_1')(\overrightarrow{\Box}_1' + m^2)$$

$$\times \langle \vec{p}_2'...out|T\varphi(x_1')\varphi^\dagger(x_1)|p_2...in\rangle(\overleftarrow{\Box}_1 + m^2)f_{\vec{p}_1}(x_1)$$

$$+ disconnected \; piece \;,$$

or for the connected part we have *effectively*

$$a_{out}(\vec{p}') = Z^{-\frac{1}{2}}\int d^4x' f_{\vec{p}'}(x')(\overrightarrow{\Box}_1' + m^2)\varphi(x') \;. \tag{10.45}$$

This completes the second part of the reduction process.

     Dropping always the contribution from the disconnected terms, we obtain, after repeating this reduction process a sufficient number of times,

$$\langle 0|a_{out}(\vec{p}_1')\cdots a_{out}(\vec{p}_m')a_{in}^\dagger(\vec{p}_n)\cdots a_{in}^\dagger(\vec{p}_1)|0\rangle =$$

$$+ (iZ_\varphi^{-1/2})^{n+m}\int \prod_k d^4x_k' \int \prod_\ell d^4x_l \prod_{k=1}^m f_{\vec{p}_k'}^*(x_k')(\overrightarrow{\Box}_1' + m^2)$$

$$\times \; \langle \Omega|T(\varphi(x_1')...\varphi(x_m')\varphi^\dagger(x_n)...\varphi^\dagger(x_1))|\Omega\rangle \prod_{l=1}^n (\overleftarrow{\Box}_l + m^2)f_{\vec{p}_l}(x_l)$$

$$+ \; disconnected \; piece \;, \tag{10.46}$$

where the time-ordering operation for the product of n fields is defined in analogy to (10.42) by

$$T(\varphi(z_1)...\varphi(z_n)) = \theta(z_1^0 - z_2^0)\theta(z_2^0 - z_3^0)\cdots\theta(z_{n-1}^0 - z_n^0)\varphi(z_1)\cdots\varphi(z_n)$$

$$+ \; permutations \;.$$

Expression (10.46) is the desired LSZ reduction formula for the S-matrix elements.

The LSZ reduction formula for the Dirac and Maxwell fields

The result (10.46) for the S-matrix elements describing the scattering of particles is readily generalized to include the scattering of antiparticles as well. It may be written in the form

$$< out|\vec{p}_1' \cdots \vec{q}_m'|\vec{p}_1 \cdots \vec{q}_n|in >=< 0|a_{out}(\vec{p}_1') \cdots b_{out}(\vec{q}_{m'}')|b_{in}^\dagger(\vec{q}_m)...a_{in}^\dagger(p_1)|0 > \tag{10.47}$$

where we have the following identifications for scalar particles:

for an incoming particle:

$$a_{in}^\dagger(\vec{p}) \to iZ_\varphi^{-1/2} \int d^4x \varphi^\dagger(x)(\overleftarrow{\Box} + m^2)f_{\vec{p}}(x)$$

for an incoming antiparticle:

$$b_{in}^\dagger(\vec{q}) \to iZ_\varphi^{-1/2} \int d^4y f_{\vec{q}}^*(y)(\overrightarrow{\Box} + m^2)\varphi(y) \tag{10.48}$$

for an outgoing particle:

$$a_{out}(\vec{p}\,') \to iZ_\varphi^{-1/2} \int d^4x' f_{\vec{p}'}^*(x')(\overrightarrow{\Box}' + m^2)\varphi(x')$$

for an outgoing antiparticle:

$$b_{out}(\vec{q}\,') \to iZ_\varphi^{-1/2} \int d^4y' \varphi^\dagger(y')(\overleftarrow{\Box}' + m^2)f_{\vec{q}'}(y').$$

Note the equality between incoming (outgoing) particles and outgoing (incoming) antiparticles, a property leading to the so-called "crossing symmetry" of Feynman amplitudes. The differential operators appearing in these substitution rules appear originally outside the time-ordering operation. It turns out, however, that the *contact terms* arising from the commutation of these differential operators past the time-ordering operator do not contribute to the (on-shell) S-matrix elements, so that the result is in fact also correct in the form (10.48).

The reduction for a general process in QED is obtained by proceeding as in the scalar case. We recall that we have the following Fourier decompositions for the Dirac fields, and the Maxwell fields in the Coulomb gauge, which replace (10.28):

$$\psi_{\substack{in\\out}}(x)_\alpha = \int d^3p \sum_\sigma \left[u_\alpha(x;\vec{p},\sigma)a_{\substack{in\\out}}(\vec{p},\sigma) + v_\alpha(x;\vec{p},\sigma)b_{\substack{in\\out}}^\dagger(\vec{p},\sigma)\right]$$

$$\vec{A}_{\substack{in\\out}}(x) = \int d^3k \sum_{\lambda=\pm1} \left[\vec{\epsilon}(x;\vec{k},\lambda)c_{\substack{in\\out}}(\vec{k},\lambda) + \vec{\epsilon}^*(x;k,\lambda)c_{\substack{in\\out}}^\dagger(\vec{k},\lambda)\right],$$

where $u(x; \vec{p}, \sigma)$, $v(y; \vec{p}, \sigma)$ and $\epsilon^\mu(z; \vec{k}, \lambda)$ denote the free-particle plane waves

$$u_\alpha(x; \vec{p}, \sigma) = \frac{1}{(2\pi)^{3/2}} \sqrt{\frac{m}{\omega_p}} U(\vec{p}, \sigma) e^{-ip\cdot x} \tag{10.49}$$

$$v_\alpha(x; \vec{p}, \sigma) = \frac{1}{(2\pi)^{3/2}} \sqrt{\frac{m}{\omega_p}} V(\vec{p}, \sigma) e^{ip\cdot x} \tag{10.50}$$

$$\bar{\epsilon}(x; \vec{k}, \lambda) = \frac{1}{(2\pi)^{3/2}} \frac{1}{\sqrt{2|\vec{k}|}} \bar{\epsilon}(\vec{k}, \lambda) e^{-ik\cdot x} \tag{10.51}$$

with $\bar{\epsilon}(\vec{k}, \lambda)$ the polarization tensor (5.12). Making use of the orthogonality properties (6.16) and (5.13) we have (remember that $\epsilon^0(\vec{k}, \lambda) = 0$)

$$\int d^3x \, u^\dagger(x; \vec{p}', \sigma') u(x; \vec{p}, \sigma) = \delta^3(\vec{p}' - \vec{p}) \delta_{\sigma'\sigma}$$

$$\int d^3x \, v^\dagger(x; \vec{p}', \sigma') v(x; \vec{p}, \sigma) = \delta^3(\vec{p}' - \vec{p}) \delta_{\sigma'\sigma}$$

$$\int d^3x \, u^\dagger(x; \vec{p}', \sigma') v(x; \vec{p}, \sigma) = 0$$

$$\int d^3x \, \epsilon^{\mu *}(x; \vec{k}', \lambda')(-i\partial_0) \epsilon_\mu(x; \vec{k}, \lambda) = \delta^3(\vec{k}' - \vec{k}) \delta_{\lambda'\lambda} \ ,$$

which now replace the orthogonality relations (10.29) of the scalar case. Using these orthogonality relations, we have

$$a_{in}^\dagger(\vec{p}, \sigma) = Z_\psi^{-1/2} \int_{t=-\infty} d^3x \, \bar{\psi}(x) \gamma^0 u(x; \vec{p}, \sigma)$$

$$b_{in}^\dagger(\vec{p}, \sigma) = Z_\psi^{-1/2} \int_{t=-\infty} d^3x \, \bar{v}(x; \vec{p}, \sigma) \gamma^0 \psi(x)$$

$$c_{in}^\dagger(\vec{k}, \lambda) = Z_A^{-1/2} \int_{t=-\infty} d^3x \, \epsilon^\mu(x; \vec{k}, \lambda)(-i \overset{\leftrightarrow}{\partial_0}) A_\mu(x)$$

which now replace (10.30), and

$$a_{out}(\vec{p}, \sigma) = Z_\psi^{-1/2} \int_{t=+\infty} d^3x \, \bar{u}(x; \vec{p}, \sigma) \gamma^0 \psi(x)$$

$$b_{out}(\vec{p}, \sigma) = Z_\psi^{-1/2} \int_{t=+\infty} d^3x \, \bar{\psi}(x) \gamma^0 v(x; \vec{p}, \sigma)$$

$$c_{out}(\vec{k}, \lambda) = Z_A^{-1/2} \int_{t=+\infty} d^3x \, \epsilon^{\mu *}(x; \vec{k}, \lambda)(-i \overset{\leftrightarrow}{\partial_0}) A_\mu(x)$$

which now replace (10.40).

Following the above steps for the reduction process, we make again use of (10.31) as well as the equations of motion (in correspondence to (10.34))

$$(i\gamma \cdot \partial - m) u(x; \vec{p}, \sigma) = 0$$
$$(i\gamma \cdot \partial + m) v(x; \vec{p}, \sigma) = 0$$
$$\partial_\mu \epsilon^\mu(x; \vec{k}, \lambda) = 0 \ .$$

Thus, for example, using (10.31) we have

$$a_{in}^\dagger(\vec{p},\sigma) - a_{out}^\dagger(\vec{p},\sigma) = Z^{-\frac{1}{2}} \int d^4x \partial_0(\bar\psi(x)\gamma^0 u(x;\vec{p},\sigma)$$

$$= Z^{-\frac{1}{2}} \int d^4x \partial_0 \bar\psi(x)\gamma^0 u(x;\vec{p},\sigma) + \int d^4x \bar\psi(x)(-im + \vec\gamma \cdot \vec\partial)u(x;\vec{p},\sigma)$$

$$= iZ^{-\frac{1}{2}} \int d^4x \bar\psi(x)(-i \overleftarrow{\partial\!\!\!/} - m)u(x;\vec{p},\sigma)$$

where a partial integration in the spatial coordinates has been performed.

Omitting the disconnected part arising from $a_{out}(\vec{p},\sigma)$, and proceeding in a similar way for the gauge field, we arrive at a reduction formula of the form of (10.48) with the following identifications:

for an incoming electron:

$$a_{in}^\dagger(\vec{p},\sigma) \to iZ_\psi^{-1/2} \int d^4x\ \bar\psi(x)(i\overleftarrow{\partial\!\!\!/} + m)u(x;\vec{p},\sigma) \tag{10.52}$$

for an incoming positron:

$$b_{in}^\dagger(\vec{q},\sigma) \to iZ_\psi^{-1/2} \int d^4y\ \bar{v}(y;\vec{q},\sigma)(i\overrightarrow{\partial\!\!\!/} - m)\psi(y) \tag{10.53}$$

for an incoming photon:

$$c_{in}^\dagger(\vec{k},\lambda) \to iZ_A^{-1/2} \int dz^4 A^\mu(z) \overleftarrow{\Box}\ \epsilon_\mu(z;\vec{k},\lambda) \tag{10.54}$$

for an outgoing electron:

$$a_{out}(\vec{p'},\sigma') \to -iZ_\psi^{-1/2} \int d^4x\ \bar{u}(x;\vec{p'},\sigma')(i\overrightarrow{\partial\!\!\!/} - m)\psi(x) \tag{10.55}$$

for an outgoing positron:

$$b_{out}(\vec{q'},\sigma') \to -iZ_\psi^{-1/2} \int d^4y\ \bar\psi(y)(i\overleftarrow{\partial\!\!\!/} + m)v(y;\vec{p'},\sigma') \tag{10.56}$$

for an outgoing photon:

$$c_{out}(\vec{k'},\lambda') \to iZ_A^{-1/2} \int d^4z\ \epsilon_\mu^*(z,\vec{k'};\lambda')\overrightarrow{\Box} A_\mu(z). \tag{10.57}$$

Comparison of the substitution law for $a_{in}^\dagger(\vec{p},\sigma)$ and $b_{out}(\vec{p},\sigma)$ shows that they are mapped into each other by the reversal of the 4-momentum. Thus oriented external lines labeled by 4-momentum p^μ can be viewed as describing either incoming particles with momentum p^μ, or as outgoing antiparticles with momentum $-p^\mu$. Of course due attention has to be paid to the respective (on-shell) wave functions.

Chapter 11

Perturbation Theory

The exact computation of vacuum expectation values of field operators is only possible for some particular model quantum field theories such as the *Thirring model* as well as QED and *chiral gauge theories in 1+1 dimensions*.[1] In all these examples the (1+1)-dimensionality of space-time plays a crucial role. The S-matrix, on the other hand, can be computed exactly for a much larger class of *two*-dimensional quantum field theories, including the *Gross–Neveu, non-linear sigma* and CP^n models. The reason is that in two space-time dimensions the allowed scattering processes are strongly restricted by the existence of an infinite number of conservation laws, which make the theory *integrable*, as far as *on-shell* processes are concerned. In 3+1 dimensions we can compute the Green functions and S-matrix only perturbatively as some power series expansion in the coupling constant or as a "loop" expansion in powers of Planck's constant $\hbar$.

In the following we show how to arrive at an expansion of the Green functions in powers of the interaction strength. We again do this for the scalar field, for the sake of simplicity. Each term in such an expansion may then be computed in terms of a set of (Feynman) rules from the corresponding (Feynman) diagrams.

11.1 Interaction picture and U-matrix

The interacting fields satisfy the Hamilton equations

$$\partial_0 \varphi(x) = i[H, \varphi(x)]$$
$$\partial_0 \pi(x) = i[H, \pi(x)] \ ,$$

where H is a functional of $\varphi(x)$ and the momentum conjugate to $\varphi(x) : H = H[\varphi, \pi]$. We shall write this Hamiltonian as the sum of a "free" part and an interacting part:

$$H[\varphi, \pi] = H_0[\varphi, \pi] + H_{int}[\varphi, \pi] \ .$$

---

[1]E. Abdalla, M.C.B. Abdalla and K.D. Rothe, *Non-perturbative methods in 2-dimensional Quantum Field Theory* (World Scientific, first edition 1991 and second edition 2001).

Note that although H itself is time-independent, H_0 and H_{int} separately depend on the time labelling the fields. The interacting fields carry the time dependence associated with H:

$$\varphi(\vec{x},t) = e^{iHt}\varphi(\vec{x},0)e^{-iHt}, \quad \pi(\vec{x},t) = e^{iHt}\pi(\vec{x},0)e^{-iHt} \ .$$

We now suppose there exists an operator $V(t)$ with the property, that

$$\varphi_I(\vec{x},t) = V(t)\varphi(\vec{x},t)V^{-1}(t) \ , \tag{11.1}$$

satisfies the free field equation

$$\partial_0\varphi_I(x) = i\left[H_0[\varphi_I, \pi_I], \varphi_I(x)\right] \ . \tag{11.2}$$

The fields $\varphi_I(x)$ are called *interaction picture* fields.[2] The following operator first transforms the Heisenberg field at time t to time $t = 0$, and then lets it evolve in time according to the free field Hamiltonian H_0:

$$V(t) = e^{iH_0t}e^{-iHt} \ , \tag{11.3}$$

where

$$H = H[\varphi, \pi], \quad H_0 = H_0[\varphi_I, \pi_I] \ .$$

Note that both, H and H_0 are now time independent.

We wish to represent the operator (11.3) entirely in terms of the free fields of the interaction picture. To this end it is convenient to derive a differential equation for $V(t)$ in terms of the free fields.

Returning to the defining Eq. (11.1) and differentiating this equation with respect to time, one has ($x^0 = t$)

$$\begin{aligned}
\partial_0\varphi_I(\vec{x},t) &= \dot{V}(t)\varphi(\vec{x},t)V^{-1}(t) - V(t)\varphi(\vec{x},t)V^{-1}(t)\dot{V}(t)V^{-1}(t) \\
&\quad + V(t)\partial_0\varphi(\vec{x},t)V^{-1}(t) \\
&= [\dot{V}(t)V^{-1}(t), \ \varphi_I(x)] + iV(t)[H[\varphi,\pi], \varphi(x)]V^{-1}(t) \\
&= i\left[H_0[\varphi_I(t), \pi_I(t)], \varphi_I(x)\right] + \left[\dot{V}(t)V^{-1}(t) + iH_{int}[\varphi_I(t), \pi_I(t)], \varphi_I(x)\right] \ .
\end{aligned} \tag{11.4}$$

Correspondingly we also have

$$\partial_0\pi_I(\vec{x},t) = i\left[H_0[\varphi_I(t), \pi_I(t)], \pi_I(x)\right] + \left[\dot{V}(t)V^{-1}(t) + iH_{int}[\varphi_I(t), \pi_I(t)], \pi_I(x)\right] \ . \tag{11.5}$$

Since the fields in the interaction picture should satisfy the equations of motion (11.2), we must require

$$\dot{V}(t)V^{-1}(t) = -iH_{int}[\varphi_I(t), \ \pi_I(t)] \equiv -iH_I(t) \ , \tag{11.6}$$

implying the differential equation

$$i\partial_i V(t) = H_I(t)V(t) \ . \tag{11.7}$$

This equation has a non-trivial solution, thus demonstrating the existence of an operator with the property (11.1).

[2]F.J. Dyson, *Phys. Rev.* **75**, 486 (1949) 1736; J.D. Bjorken and S.D. Drell, *Relativistic Quantum Fields* (McGraw-Hill, 1965).

11.2 Interaction picture representation of Green functions

Equation (11.7) can be solved in terms of the interaction picture fields, once the initial condition is specified. It is however a different quantity, satisfying the same differential equation, which shall be of actual interest to us. Indeed, substituting the inverse of the isomorphism (11.1) for the fields in a Green function we obtain

$$< \Omega | T\varphi(x_1)...\varphi(x_n) | \Omega > = \qquad\qquad (11.8)$$
$$< \Omega | TV^{-1}(t_1)\varphi_I(x_1)V(t_1)V^{-1}(t_2)\varphi_I(x_2)V(t_2) \cdots V^{-1}(t_n)\varphi_I(x_n)V(t_n) | \Omega > ,$$

where $|\Omega >$ is the physical ground state. It is convenient to define the new operator

$$U(t,t') = V(t)V^{-1}(t') , \qquad\qquad (11.9)$$

with the obvious property

$$U(t,t'')U(t'',t') = U(t,t') .$$

From (11.9) and (11.3) we have

$$U(t,t') = e^{iH_0 t}e^{-iH(t-t'))}e^{-iH_0 t'} . \qquad\qquad (11.10)$$

In terms of this operator, Eq. (11.8) now takes the form

$$< \Omega | \mathbf{T}\varphi(x_1)...\varphi(x_n) | \Omega > = \qquad\qquad (11.11)$$
$$< \Omega | V^{-1}(T)\mathbf{T}\left[U(T,t_1)\varphi_I(x_1)U(t_1,t_2)...U(t_{n-1},t_n)\varphi_I(x_n)U(t_n,-T) \right] V(-T) | \Omega >,$$

where we take T to be large enough, so that $x_i^0 \, \epsilon(-T,T)$. The U-operator evidently satisfies the same differential equation with respect to t as $V(t)$.

$$i\partial_t U(t,t') = H_I(t)U(t,t') , \qquad\qquad (11.12)$$

with the boundary condition

$$U(t,t) = 1 .$$

Because of the time-ordering operation, (11.11) can be compactly written as

$$< \Omega | \mathbf{T}\varphi(x_1)...\varphi(x_n) | \Omega > = < \Omega | V^{-1}(T)\mathbf{T}\left[\varphi_I(x_1) \cdots \varphi_I(x_n)U(T,-T) \right] V(-T) | \Omega > .$$
$$(11.13)$$

Some formal considerations

The above formula involves the ground state $|\Omega >$ of the full Hamiltonian, whose energy we denote by E_0:

$$H[\varphi, \pi] |\Omega > = E_0 |\Omega > .$$

On the other hand, the operators $\varphi_I(x)$ are defined with respect to the vacuum of the interaction picture which we gauge by requiring

$$H_0[\varphi_I, \pi_I] |0 > = 0 .$$

Recall that both $H[\varphi, \pi]$ and $H_0[\varphi_I, \pi_I]$ are time independent. In the following we denote them by H and H_0 respectively.

We want to express (11.13) in terms of the "bare" vacuum $|0>$. To this end we note that completeness of the states allows us to write (we assume an energy gap)

$$V^{-1}(-T)|0> = e^{-iHT}e^{iH_0T}|0> = e^{-iE_0T}|\Omega><\Omega|0>$$
$$+ \sum_{E_n>E_0} e^{-iE_nT}|E_n><E_n|0> .$$

We are interested in the limit where $T \to \infty$. In this limit the oscillations associated with the exponentials increase with increasing energy. Making use of the Riemann–Lebesgue Lemma, we conclude that the sum over intermediate states is dominated in this limit by the first term involving the ground state:

$$V^{-1}(-T)|0> \to |\Omega> C(-T) , \qquad (11.14)$$

where

$$C(-T) =<\Omega|0> e^{-iE_0T} .$$

Inverting this relation we have

$$V(-T)|\Omega> \to |0> \frac{1}{C(-T)} .$$

Notice that the left-hand side of this equation is precisely one of the factors appearing on the right in (11.13). In a completely analogous way one finds

$$<\Omega|V^{-1}(T) \to \frac{1}{C^*(T)} <0| , \qquad (11.15)$$

which is the other factor on the left in (11.13). Hence we have

$$<\Omega|T\varphi(x_1)...\varphi(x_n)|\Omega> = \frac{<0|\mathbf{T}\{\varphi_I(x_1)\cdots\varphi_I(x_n)U(T,-T)\}|0>}{C^*(T)C(-T)} .$$

Now, from (11.14) and (11.15) we have

$$C^*(T)C(-T) =<0|V(T)V^{-1}(-T)|0> =<0|U(T,-T)|0>$$

since the ground state $|0>$ is assumed to be normalized to one. We thus finally have

$$<\Omega|T\varphi(x_1)...\varphi(x_n)|\Omega> = \frac{<0|\mathbf{T}\{\varphi_I(x_1)\cdots\varphi_I(x_n)U(T,-T)\}|0>}{<0|U(T,-T)|0>} . \qquad (11.16)$$

It remains to calculate $U(T,-T)$.

Calculation of $U(T, -T)$

In order to solve Eq. (11.12) we make use of the following.

Proposition:

The solution of equation

$$\frac{d}{dt}F(t) = G(t)F(t), \quad F(t_0) = 1 \tag{11.17}$$

with F and G non-commuting operators is given by

$$F(t) = T \exp \int_{t_0}^{t} dt' G(t') \, ,$$

where T denotes *time ordering* of the exponential as defined by the series

$$T \, e^{\int_{t_0}^{t} dt' G(t')} = 1 + \sum_{n=1}^{\infty} \int_{t_0}^{t} dt_1 \int_{t_0}^{t_1} dt_2 ... \int_{t_0}^{t_{n-1}} dt_n G(t_1)G(t_2)...G(t_n) \, . \tag{11.18}$$

Proof:

The proof is trivial. Differentiating the series (11.18) with respect to t we evidently obtain

$$\frac{d}{dt} T \, e^{\int_{t_0}^{t} dt' G(t')} = G(t) \left[1 + \sum_{n=1}^{\infty} \int_{t_0}^{t} dt_1 \int_{t_0}^{t_1} dt_2 ... \int_{t_0}^{t_{n-1}} dt_n G(t_1)...G(t_n) \right] \, .$$

Our proposition demonstrates the following important result: The exponential solution $F(t) = exp \int_{t_0}^{t} dt' G(t')$ of the linear differential Eq. (11.17) in the case where G is a *c-number*, is replaced by the corresponding "time-ordered exponential" in the case where $G(t)$ is an operator! This is an important and general result.
    Comparing (11.12) with (11.17), we conclude that

$$U(t, t') = T \, e^{-i \int_{t'}^{t} dt'' H_I(t'')} \, . \tag{11.19}$$

Let us finally observe that we can rewrite the time-ordered exponential (11.18) in a more convenient form. To this end we observe that we can rewrite for instance the second-order term in (11.18) as

$$\int_{t_0}^{t} dt_1 \int_{t_0}^{t_1} dt_2 G(t_1)G(t_2) = \frac{1}{2!} \int_{t_0}^{t} dt_1 \int_{t_0}^{t} dt_2$$
$$(\theta(t_1 - t_2)G(t_1)G(t_2) + \theta(t_2 - t_1)G(t_2)G(t_1))$$
$$= \frac{1}{2!} \int_{t_0}^{t} dt_1 \int_{t_0}^{t} dt_2 \, T(G(t_1)G(t_2)) \, ,$$

where T denotes the time-ordered product already introduced earlier,

$$T(G(t_1)G(t_2)) = \theta(t_1 - t_2)G(t_1)G(t_2) + \theta(t_2 - t_1)G(t_2)G(t_1) \, .$$

It is obvious how this result generalizes to arbitrary order, and we have

$$T\, e^{\int_{t_0}^t dt'\, G(t')} = 1 + \sum_{n=1}^{\infty} \frac{1}{n!} \int_{t_0}^t dt_1 ... \int_{t_0}^t dt_n\ T(G(t_1)G(t_2)...G(t_n))\ .$$

In this way the solution of (11.17) with $G(t)$ operator-valued resembles the solution in the case where $G(t)$ is a c-number function.

Using (11.19), and taking the limit $T \to \infty$, we finally obtain from (11.16) the desired result:

$$< \Omega|T\varphi(x_1)...\varphi(x_n)|\Omega > = \frac{< 0|T\varphi_I(x_1)...\varphi_I(x_n)e^{-i\int_{-\infty}^{\infty} dt H_I(t)}|0 >}{< 0|e^{-i\int_{-\infty}^{\infty} dt H_I(t)}|0 >}\ . \tag{11.20}$$

By expanding the exponential in powers of $H_I(t)$ we obtain a perturbative representation of the Green functions:

$$< \Omega|T\varphi(x_1)...\varphi(x_n)|\Omega > = \frac{< 0|T\varphi_I(x_1)...\varphi_I(x_n)\sum_{n=0} \frac{(-i)^n}{n!}\left(\int_{-\infty}^{\infty} d^4y H_I(y)\right)^n|0 >}{< 0|U(\infty, -\infty)|0 >} \tag{11.21}$$

where the normalization factor has the corresponding perturbative expansion

$$< 0|U(\infty, -\infty)|0 > = < 0|\left\{1 + \sum_{n=1}^{\infty} \frac{(-i)^n}{n!}T\left(\int_{-\infty}^{\infty} d^4y H_I(y)\right)^n\right\}|0 >\ . \tag{11.22}$$

Here $\mathcal{H}_I(y)$ is the interaction Hamiltonian density as defined by

$$\mathcal{H}_I(t) = \int d^3y \mathcal{H}_{int}(\varphi_I(\vec{y}, t), \pi_I(\vec{y}, t))\ . \tag{11.23}$$

Hence each term in the expansion (11.21) is explicitly computable, since it involves only the (free) interaction picture fields. The interaction Hamiltonian density $\mathcal{H}_I(y)$ is understood to be *normal ordered* with respect to these fields in order to allow for a lower bound of the ground-state energy (for an appropriate sign of the coupling constant). The computation of each term in the expansion (11.21) is then based on three theorems due to *Wick*, which we discuss next.

11.3   Wick theorems

For a formulation of the *Wick theorems* we need to introduce the notion of a *simple contraction* as well as *time-ordered contraction* of two free operators.

Simple contraction

Consider the product $A(x)B(y)$ of two *free* field operators. We write this product as the sum of a "normal-ordered" product and a "simple contraction", which we denote by a lower square bracket:

$$A(x)B(y) =: A(x)B(y): +\underline{A(x)B(y)}. \tag{11.24}$$

The first term stands here for the "normal ordered product"

$$: A(x)B(y) := A^{(+)}(x)B^{(+)}(y) + A^{(-)}(x)B^{(-)}(y)$$
$$+ A^{(-)}(x)B^{(+)}(y) + (-1)^{\epsilon(A)\epsilon(B)}B^{(-)}(y)A^{(+)}(x) \, ,$$

where $A^{(+)}(x)$ and $A^{(-)}(x)$ (etc.) denote the *positive* and *negative* frequency parts of $A(x)$ in the decomposition $A(x) = A^{(+)}(x) + A^{(-)}(x)$ in terms of destruction and creation operators, and the positive frequency parts are arranged to stand to the right of the negative energy parts, taking account of statistics. The second term is a c-number. We have

$$: A(x)B(y) := (-1)^{\epsilon(A)\epsilon(B)} : B(y)A(x) : \, , \tag{11.25}$$

where $\epsilon(Q)$ denotes the Grassman signature of the operator Q. Since

$$< 0| : A(x)B(y) : |0 >= 0$$

it follows that

$$< 0|\underbrace{A(x)B(y)}|0 >=< 0|A(x)B(y)|o > \, ,$$

and since $\underbrace{A(x)B(y)}$ is a c-number, we further have,

$$\underbrace{A(x)B(y)} =< 0|A(x)B(y)|0 > \, .$$

In particular, since $A^{(-)}(x)B^{(+)}(x)$ is already in normal-ordered form, we have

$$\underbrace{A^{(-)}(x)B^{(+)}(y)} = 0.$$

On the other hand,

$$A^{(+)}(x)B^{(-)}(y) = (-1)^{\epsilon(A)\epsilon(B)}B^{(-)}(y)A^{(+)}(x) + \underbrace{A^{(+)}(x)B^{(-)}(y)},$$

or

$$\underbrace{A^{(+)}(x)B^{(-)}(y)} = [A^{(+)}(x), B^{(-)}(y)]$$

where the bracket stands for the generalized commutator including statistics,

$$[A(x), B(y)] = A(x)B(y) - (-1)^{\epsilon(A)\epsilon(B)}B(y)A(x) \, . \tag{11.26}$$

Time-ordered contraction:

We repeat the above discussion for the case of the time-ordered product of two field operators, defined in general by

$$TA(x)B(y) = \theta(x^0 - y^0)A(x)B(y) + (-1)^{\epsilon(A)\epsilon(B)}\theta(y^0 - x^0)B(y)A(x) \, , \tag{11.27}$$

where $\epsilon(A)$ and $\epsilon(B)$ denote again the *Grassman signature* of the operators A and B. From (11.27), (11.25) and $\theta(x) + \theta(-x) = 1$, we have

$$T(A(x)B(y)) =: A(x)B(y) : +\overline{A(x)B(y)} \, ,$$

i.e.

$$A(x)B(y) =< 0|TA(x)B(y)|0 > . \tag{11.28}$$

It is thus convenient to define a new contraction, the time-ordered contraction of two operators, represented by an upper bracket:

$$\overline{A(x)B(y)} = \theta(x^0 - y^0)\underline{A(x)B(y)} + (-1)^{\epsilon(A)\epsilon(B)}\theta(y^0 - x^0)\underline{B(y)A(x)} .$$

Time-ordered contractions thus behave with respect to the exchange of the two operators like T-products.

We are now ready to state Wick's theorems:[3]

Wick theorem No. 1

The product of (free) operators is equal to the corresponding normal-ordered product plus the normal products including all possible contractions.

$$A_1...A_n =: A_1...A_n : + \sum_{i,j} : A_1...\underline{A_i...A_j}...A_n : A_1...A_i...A_j...A_n :$$

$$+ \sum_{i,j,k,l} : A_1...\underline{A_i...A_j}...\underline{A_k...A_l}...A_n :$$

Here the terms involving contractions are understood to be computed by first bringing the respective pairs next to each other by taking due account of the signature of the fields. This applies as well to the following theorems.

Corollary to Wick theorem No. 1

If an operator product is itself a normal-ordered product of operators, then all contractions of the operators *within* this normal product are to be omitted.

Wick theorem No. 2

The T-product of (free) operators is equal to the sum of the corresponding normal-ordered product and the normal ordered products including all possible time-ordered contractions:

$$T(A_1...A_n) =: A_1...A_n : + \sum_{i,j} : A_1...\overline{A_i...A_j}...A_n : A_1...A_i...A_j...A_n :$$

$$+ \sum_{i,j,k,l} : A_1...\overline{A_i...A_j...A_k...A_l}...A_n : \tag{11.29}$$

Corollary to Wick's theorem No. 2

If one of the operators in the T-product of (free) operators is itself a normal product of operators, then all time-ordered contractions of the operators *within* this normal ordered product are to be omitted.

---

[3]G.C. Wick, *Phys. Rev.* **80** (1950) 208.

Wick theorem No. 3

The vacuum expectation value of the time-ordered product of $n + 1$ free operators $A, B_1, B_2, ...B_n$ is equal to the sum of the n possible vacuum expectation values of the time-ordered products which one obtains by contracting the operator A with each of the n operators $B_i, i = 1, ...n$:

$$< 0|TAB_1...B_n|0 >= \sum_{i=1}^{n} < 0|T\overbrace{AB_1...B_i}...B_n|0 > .$$

The proof of this theorem is based on the fact that in a vacuum expectation value of such an operator product only those terms survive, in which *all* of the operators, including A, have been contracted. Theorem 3 implies that in the computation of the individual terms in the expansion (11.21) of a general Green function, all of the fields will have to be contracted in order to give a non-vanishing contribution. Hence each term in the expansion (11.21) will in general involve a multiple integral over a product of pairwise time-ordered contractions represented by the vacuum expectation value of two free time-ordered fields according to (11.28). The following section is thus devoted to a discussion of the various 2-point functions associated with the bosonic, fermionic, and radiation fields.

11.4   2-point functions

As we have seen, the simple and time-ordered Wick contractions are c-number functions, generally referred to as *2-point* functions. They represent the basic building blocks of any perturbative calculation, and hence will be discussed in the following sequel. We do this in detail for the case of the real scalar free field, limiting ourselves to a statement of the results for the case of fermions and the radiation field.

Basic definitions

In the case of a *real* scalar field $\phi(x)$, the following 2-point functions play a role in perturbative QFT:[4]

$$< 0|[\phi(x), \phi(0)]|0 >, = i\Delta(x)$$
$$< 0|\phi^{(+)}(x)\phi^{(-)}(0)|0 >, = i\Delta^{(+)}(x)$$
$$-\theta(x^0) < 0|[\phi(x), \phi(0)]|0 > = i\Delta_{ret}(x) = -i\theta(x^0)\Delta(x) \qquad (11.30)$$
$$\theta(-x^0) < 0|[\phi(x), \phi(0)]|0 > = i\Delta_{adv}(x) = i\theta(-x^0)\Delta(x)$$
$$< 0|T\phi(x)\phi(0)|0 > = i\Delta_F(x) .$$

Of these the first two 2-point functions are evidently solutions of the (homogeneous) Klein–Gordon equations:

$$(\Box + m^2)\Delta(x) = 0, \quad (\Box + m^2)\Delta^{(+)}(x) = 0 . \qquad (11.31)$$

---

[4] We have used translational invariance.

The remaining three are the Green functions corresponding to different boundary conditions:

$$(\Box + m^2)\Delta_{ret/adv}(x) = \delta^4(x)$$
$$(\Box + m^2)\Delta_F(x) = -\delta^4(x). \tag{11.32}$$

Equations (11.32) follow from the commutation relations

$$[\phi(x), \phi(0)]_{ET} = 0 , \quad [\phi(x), \partial_0\phi(y)]_{ET} = i\delta^3(x - y)$$

and the property

$$\partial_0\theta(x^0) = \delta(x^0)$$

of the Heaviside step function. For $\Delta_F(x)$,

$$\partial_0 < 0|T\phi(x)\phi(0)|0 > = < 0|T\partial_0\phi(x)\phi(0)|0 >$$
$$\partial_0^2 < 0|T\phi(x)\phi(0)|0 > = < 0|[\partial_0\phi(x)\phi(0)]|0 > \delta(x^0) + < 0|T\partial_0^2\phi(x)\phi(0)|0 >$$
$$= -i\delta^4(x) + < 0|T(\vec{\nabla}^2 - m^2)\phi(x)\phi(0)|0 > ,$$

where we have used $(\partial_0^2 - \vec{\nabla}^2 + m^2)\phi(x) = 0$.

Fourier representations

We now seek the Fourier representations of the various 2-point functions. For the function $\Delta(x)$ we had already found (see Eq. (7.34))

$$\Delta(x) = -\frac{i}{(2\pi)^3} \int \frac{d^3q}{2\omega_q} \left[e^{-i\omega_q x^0} - e^{i\omega_q x^0} \right] e^{i\vec{q}\cdot\vec{x}}$$
$$= -\frac{i}{(2\pi)^3} \int d^4q\epsilon(q^0)\delta(q^2 - m^2)e^{-iq\cdot x} .$$

Recalling the definition of the *epsilon* function

$$\epsilon(q^0) = \theta(q^0) - \theta(-q^0) ,$$

we may separate $\Delta(x)$ into its positive and negative frequency parts

$$\Delta(x) = \Delta^{(+)}(x) + \Delta^{(-)}(x) ,$$

where

$$\Delta^{(+)}(x) = -\frac{i}{(2\pi)^3} \int d^4q\theta(q^0)\delta(q^2 - m^2)e^{-iq\cdot x}$$
$$\Delta^{(-)}(x) = \frac{i}{(2\pi)^3} \int d^4q\theta(-q^0)\delta(q^2 - m^2)e^{-iq\cdot x} . \tag{11.33}$$

Note that

$$\Delta^{(-)}(x) = -\Delta^{(+)}(-x).$$

We further have from (11.33)

$$\Delta^{(+)}(x) = -\frac{i}{(2\pi)^3} \int \frac{d^3q}{2\omega_q} e^{-i\omega_q x^0 + i\vec{q}\cdot\vec{x}} \ .$$

Recalling further the following property of the step function $\theta(x) + \theta(-x) = 1$ we also have the following decomposition into an *advanced* and *retarded* part:

$$\Delta(x) = \Delta_{adv}(x) - \Delta_{ret}(x) \ .$$

Making use of the Fourier representation

$$\theta(x^0) = -\frac{1}{2\pi i} \int_{-\infty}^{\infty} dq^0 \frac{e^{-iq^0 x^0}}{q^0 + i\epsilon} \ , \quad \theta(-x^0) = \frac{1}{2\pi i} \int_{-\infty}^{\infty} dq^0 \frac{e^{-iq^0 x^0}}{q^0 - i\epsilon} \quad (11.34)$$

of the θ-function, where ϵ is an infinitesimal parameter, we have,

$$\Delta_{adv}(x) = \theta(-x^0)\Delta(x)$$
$$= \int \frac{dq^0}{2\pi i} \frac{e^{-iq^0 x^0}}{q^0 - i\epsilon} \times \frac{-i}{(2\pi)^3} \int \frac{d^3q}{2\omega_q} \left(e^{-i\omega_q x^0} - e^{i\omega_q x^0} \right) e^{i\vec{q}\cdot\vec{x}}$$
$$= -\frac{1}{(2\pi)^4} \int \frac{d^4q}{2\omega_q} \left(\frac{1}{q^0 - \omega_q - i\epsilon} - \frac{1}{q^0 + \omega_q - i\epsilon} \right) e^{-iq\cdot x} \ ,$$

where a shift in the q^0-integration has been made.

Proceeding in the same way for Δ_{ret}, and combining denominators we may summarize the results in the compact form

$$\Delta_{adv}(x) = -\int \frac{d^4q}{(2\pi)^4} \frac{e^{-iq\cdot x}}{q^2 - m^2 - iq^0\epsilon}$$
$$\Delta_{ret}(x) = -\int \frac{d^4q}{(2\pi)^4} \frac{e^{-iq\cdot x}}{q^2 - m^2 + iq^0\epsilon} \ .$$

Note that the *epsilon* prescriptions correspond to the integration contours shown in Fig. 11.1.

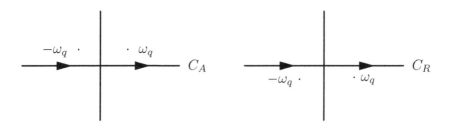

Fig. 11.1. Integration contour and pole location for $\Delta_{adv}(x)$ and $\Delta_{ret}(x)$.

In order to get the Fourier representation of the time-ordered 2-point function $\Delta_F(x)$, we write it as

$$\Delta_F(x) = \theta(x^0)\Delta^{(+)}(x) + \theta(-x^0)\Delta^{(+)}(-x) \, ,$$

and proceed as before. Making use again of the Fourier representation (11.34) of the step function, we have

$$\Delta_F(x) = \frac{i}{2\pi i} \int_{-\infty}^{\infty} dq^0 \int \frac{d^3q}{(2\pi)^3} \frac{1}{2\omega_q} \left[\frac{e^{-i(q^0+\omega_q)x^0+i\vec{q}\cdot\vec{x}}}{q^0 + i\epsilon} - \frac{e^{-i(q^0-\omega_q)x^0+i\vec{q}\cdot\vec{x}}}{q^0 - i\epsilon} \right] \, ,$$

or making a shift in the q^0-integration variable

$$\Delta_F(x) = \int \frac{d^4q}{(2\pi)^4} \frac{1}{2\omega_q} \left(\frac{1}{q^0 - \omega_q + i\epsilon} - \frac{1}{q^0 + \omega_q - i\epsilon} \right) e^{-iq^0x^0+i\vec{q}\cdot\vec{x}} \, .$$

An algebraic combination of the two terms results in

$$\Delta_F(x) = \int \frac{d^4q}{(2\pi)^4} \frac{e^{-iqx}}{q^2 - m^2 + i\epsilon} \, .$$

We check the result by observing that indeed $(\Box + m^2)\Delta_F(x) = -\delta^4(x)$. The *epsilon* in the denominator again tells us that the integration in the q^0-plane is to be performed along the contour C_F with pole locations shown in the figure below.

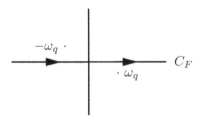

Fig. 11.2. Integration contour and pole locations for $\Delta_F(x)$.

It is evident from the Fourier representations above that the 2-point functions $\Delta_{ret}(x)$, $\Delta_{adv}(x)$ and $\Delta_F(x)$ satisfy the equations for a Green function of the Klein–Gordon operator. The function $\Delta_F(x)$ is called the *Feynman propagator* for a scalar field and will play a central role in the perturbative expansion of a general n-point "Green function", as is evident from the Wick theorems 2 and 3 of the previous section.

The above functions are singular on the light cone and exhibit logarithmic cuts in the variable x^2. The respective Fourier integrals may be computed to give[5] the following explicit representations of the 2-point functions $\Delta(x)$, $\Delta^{(\pm)}(x)$ and $\Delta_F(x)$

---

[5]N.N. Bogoljubov and O.V. Sirkov, *Quantenfelder* (Physik Verlag, 1984), p. 281.

$$\Delta(x) = -\frac{1}{2\pi}\epsilon(x^0)\delta(x^2) + \frac{m}{4\pi\sqrt{x^2}}\epsilon(x^0)\theta(x^2)J_1(m\sqrt{x^2})$$

$$\Delta^{(\pm)}(x) = \frac{1}{8\pi}\epsilon(x^0)\delta(x^2)$$

$$+ \frac{m\theta(x^2)}{8\pi\sqrt{x^2}}\left[\epsilon(x^0)J_1(m\sqrt{x^2}) \pm iN_1(m\sqrt{x^2}) \pm \frac{im\theta(-x^2)}{4\pi^2\sqrt{-x^2}}K_1(m\sqrt{-x^2})\right]$$

$$\Delta_F(x) = \frac{im}{4\pi^2}\frac{K_1(m\sqrt{-x^2 + i\epsilon})}{\sqrt{-x^2 + i\epsilon}}. \tag{11.35}$$

The corresponding expressions for the retarded and advanced Green function then follow from their definition (11.30). The functions $J_1(x), N_1(x)$ and $K_1(x)$ are the *Bessel, Neumann* and *Bessel* functions of the *second kind* (Hankel function for imaginary argument), respectively.

The explicit expressions (11.35) show that the 2-point functions are singular on the light-cone. Indeed, we have for small argument[6]

$$J_1(z) = \frac{z}{2} + O(z^3)$$

$$N_1(z) = -\frac{2}{\pi}\frac{1}{z} + \frac{z}{\pi}\left(\ln\frac{z}{2} + 1\right) + O(z^3)$$

$$K_1(z) = \frac{1}{z} + \frac{z}{2}\left(\ln\frac{z}{2} + 1\right) + O(z^3).$$

Introducing these expansions into (11.35) we have in particular for $\Delta(x)$,

$$\Delta(x) = -\frac{1}{2\pi}\epsilon(x^0)\delta(x^2) + \frac{m^2}{8\pi}\epsilon(x^0)\theta(x^2) + O(x^3).$$

The functions (11.35) belong to the class of *generalized functions* or *distributions* and are strictly defined only under an integral when acting on sufficiently smooth *test functions* $\zeta(x)$

$$f[\zeta] = \int dx f(x)\zeta(x).$$

Quite generally the behaviour of Green functions on the light cone determines the behaviour of scattering processes at high energies. The singular behaviour for $x^2 \to 0$ also lies at the root of the need for *renormalization* of quantum field theories, as we shall see in Chapter 16.

Causality

We have seen that the 2-point function $\Delta(x)$ is causal in the sense that it vanishes for space-like distances, which cannot be connected by a light signal:

$$\Delta(x) = 0, \quad x^2 < 0.$$

---

[6]See Magnus, Oberhettinger and Soni, *"Formulas and Theorems for the Special Functions of Mathematical Physics"*.

This property we had already established on the basis of Lorentz invariance and equal-time commutation relations. It no longer holds for the 2-point functions $\Delta^{(\pm)}(x)$ and $\Delta_F(x)$. We nevertheless have that both of these functions are exponentially damped for large space-like distances if the boson is massive:

$$\Delta_F(x) = \frac{im}{4\pi^2}\frac{K_1(m|x|)}{|x|}, \quad x^2 < 0 .$$

Pauli has given an interpretation for this acausal behaviour: Suppose we measure the position of an electron at the point x at time t with a shutter. At the *same time* another observer at another position y makes a measurement to see whether he finds the electron there. According to the acausal behaviour of $\Delta_F(x-y)$ he will find an electron with a certain probability. This, however, reflects the fact that in the course of the measurement observer 2 creates an electron-positron pair from the vacuum, of which the positron annihilates the electron measured by observer 1.

Table of 2-point functions of free scalar, fermion and Maxwell fields

Proceeding as in Chapter 7 we further obtain the 2-point functions for the fermion field and the gauge fields of QED in the Coulomb gauge. All results are summarized in the following table.

(I) *Scalar field*

$$< 0|[\phi(x), \phi(0)]|0 >= i\Delta(x)$$
$$< 0|\phi(x)\phi(0)|0 >= \underline{\phi(x)\phi(0)} = i\Delta^{(+)}(x)$$

$$< 0|T\phi(x)\phi(0)|0 >= \overline{\phi(x)\phi(0)} == i\Delta_F(x)$$

with

$$\Delta_F(x) = \int \frac{d^4q}{(2\pi)^4}\frac{e^{-iqx}}{q^2 - m^2 + i\epsilon}, \quad (\Box + m^2)\Delta_F(x) = -\delta^4(x)$$

(II) *Fermion field*

$$< 0|[\psi_\alpha(x), \bar{\psi}_\beta(0)]|0 >= iS(x)_{\alpha\beta} = (i\slashed{\partial} + m)_{\alpha\beta}i\Delta(x) \quad (11.36)$$
$$< 0|\psi_\alpha(x)\bar{\psi}_\beta(0)|0 >= \underline{\psi_\alpha(x)\bar{\psi}_\beta(0)} = iS^{(+)}(x)_{\alpha\beta} = (i\slashed{\partial} + m)_{\alpha\beta}i\Delta^{(+)}(x)$$

$$< 0|T\psi_\alpha(x)\bar{\psi}_\beta(0)|0 >= \overline{\psi_\alpha(x)\bar{\psi}_\beta(0)} = iS_F(x)_{\alpha\beta} = (i\slashed{\partial} + m)_{\alpha\beta}i\Delta_F(x)$$

with

$$S_F(x) = \int \frac{d^4p}{(2\pi)^4}\frac{(\slashed{p} + m)}{p^2 - m^2 + i\epsilon}e^{-ipx}$$

$$(i\slashed{\partial} - m)S_F(x) = \delta^4(x) .$$

Here [,] denotes the generalized commutator (7.27).

(III) *Gauge-field in Coulomb gauge*

$$< 0|[A^i(x), A^j(0)]|0 >= \left(\delta^{ij} - \frac{\partial^i \partial^j}{\Delta}\right) iD(x)$$

$$< 0|TA^i(x)A^j(0)|0 >= iD_F^{ij}(x) = \left(\delta^{ij} - \frac{\partial^i \partial^j}{\Delta}\right) iD_F(x)$$

where

$$D_F^{ij}(x) = \int \frac{d^4k}{(2\pi)^4} \frac{\delta^{ij} - \frac{k^i k^j}{\vec{k}^2}}{k^2 + i\epsilon} e^{-ikx} \ , \quad \partial_i D_F^{ij}(x) = 0 \qquad (11.37)$$

$$\Box D_F^{ij}(x) = - \left(\delta^{ij} - \frac{\partial^i \partial^j}{\Delta}\right) \delta^4(x) \,.$$

Notice that in the Coulomb gauge $\partial_i A^i(x) = 0$, which is reflected by the equations above.

11.5   Feynman Diagrams for QED

According to formulae (10.25), the calculation of the differential cross-section for electron-electron (Moeller) scattering, electron-positron (Bhabha) scattering, and electron-photon (Compton) scattering, as well as generalizations thereof involving arbitrary numbers of electrons, positrons and photons in the final state, require the calculation of the transition amplitude from the LSZ reduction formula (10.48).

A general $2 \to 2$ scattering process is represented diagrammatically in the figure below.

Fig. 11.3. A general $2 \to 2$ scattering process.

The arrows indicate the flow of momentum. With every *ingoing* electron (*outgoing* positron) is associated a field $\bar{\psi}(x)$, whereas with every *outgoing* electron (*ingoing* positron) is associated a field $\psi(x)$. Note that there is the same number n of fields $\psi(x)$ and $\bar{\psi}(x)$, reflecting conservation of charge.

We see from (11.21) that a general n-point Green function G can be written as the ratio of two vacuum expectation values of time ordered products. For the sake of clarity, let us write symbolically

$$G = \frac{\hat{G}}{V} \ , \qquad (11.38)$$

where V stands for "vacuum". We have from (11.21) and (11.22),

$$\hat{G} =< 0|T \left\{\varphi_I(x_1)...\varphi_I(x_n)\left[1 + \sum_{n=1}^{\infty} \frac{(-i)^n}{n!} \left(\int_{-\infty}^{\infty} d^4z \mathcal{H}_I(z)\right)^n\right]\right\}|0 > \qquad (11.39)$$

and

$$V =< 0|T \left\{ 1 + \sum_{n=1}^{\infty} \frac{(-i)^n}{n!} \left(\int_{-\infty}^{\infty} d^4 z \mathcal{H}_I(z) \right)^n \right\} |0 > . \tag{11.40}$$

We now give a diagrammatic interpretation of V for the case of QED.

Diagrammatic representation of V

For Electrodynamics (note that $\mathcal{H}_I$ is normal ordered)

$$\mathcal{H}_I(x) = -\mathcal{L}_I(x) = e_0 : \bar{\psi}_I(x) \gamma_\mu \psi_I(x) A_I^\mu(x) : , \tag{11.41}$$

where e_0 is the (unrenormalized, negative) bare electron charge.

Consider V to lowest order $O(e_0^2)$:

$$V \simeq 1 + \frac{(-ie_0)^2}{2!} \int d^4 z_1 \int d^4 z_2$$
$$\langle 0|T : \bar{\psi}_I(z_1) \gamma_\mu \psi_I(z_1) A_I^\mu(z_1) :: \bar{\psi}_I(z_2) \gamma_\nu \psi_I(z_2) A_I^\nu(z_2) : |0 \rangle$$

or doing the contractions

$$V^{(2)} = 1 - \tfrac{1}{2}(-ie_0)^2 \int d^4 z_1 \int d^4 z_2 tr[\gamma^\mu i S_F(z_1 - z_2) \gamma^\nu i S_F(z_2 - z_1)] i D_F(z_1 - z_2)_{\mu\nu} .$$

Note that an extra minus sign emerged as a result of the reordering of the fermion fields when computing the contractions. Note further that the fermionic part could be written as a trace. Both are characteristic for every *fermion loop*. Diagrams which are not connected to any external line are called vacuum diagrams. Examples are shown in the figures below.

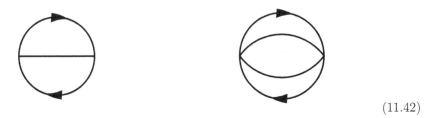

$$\tag{11.42}$$

Fig. 11.4. Second order vacuum graphs in ϕ^3 and ϕ^4 theory.

We now show that these vacuum diagrams cancel to all orders in the definition (11.38) of G.

Cancellation of vacuum diagrams

Consider as an example the 4-point function $< \Omega|T \psi(x_1) \psi(x_2) \bar{\psi}(y_1) \bar{\psi}(y_2)|\Omega > $ in QED. According to (11.39),

$$\hat{G}(x_1, x_2; y_1, y_2) = \langle 0|T \left\{ \psi_I(x_1) \psi_I(x_2) \bar{\psi}_I(y_1) \bar{\psi}_I(y_2) e^{-i \int d^4 z \mathcal{H}_I(z)} \right\} |0 \rangle$$

or, conveniently written using Wick's theorems,

$$\tilde{G}(x_1, x_2; y_1, y_2) = \sum_{p=0}^{\infty} \sum_{l=0}^{p} \frac{(-i)^p}{p!} \binom{p}{l} \tag{11.43}$$

$$\times \int d^4 z_1 ... d^4 z_l \langle 0|T\psi_I(x_1)\psi_I(x_2)\bar{\psi}_I(y_1)\bar{\psi}_I(y_2)\mathcal{H}_I(z_1)...\mathcal{H}_I(z_l)|0\rangle_{PC}$$

$$\times \int dz_{l+1} ... dz_p \langle 0|T\mathcal{H}_I(z_{l+1})...\mathcal{H}_I(z_p)|0\rangle \ ,$$

where the subscript PC stands for *"partially connected"*, i.e. diagrams involving no (disconnected) *vacuum* diagrams. The binomial coefficient

$$\binom{p}{l} = \frac{p!}{\ell!(p-\ell)!}$$

is the number of ways one can select l interaction Hamiltonians from a total of p.
We may rewrite the summation on the r.h.s. of (11.43) as follows:

$$\text{rhs} = \sum_{l=0}^{\infty} \frac{(-i)^l}{l!} \int d^4 z_1 ... d^4 z_l \langle 0|T\psi_I(x_1)\psi_I(x_2)\bar{\psi}_I(y_1)\bar{\psi}_I(y_2)\mathcal{H}_I(z_1)...\mathcal{H}_I(z_l)|0\rangle_{PC}$$

$$\times \sum_{p=l}^{\infty} \frac{(-i)^{p-l}}{(p-l)!} \int dz_{l+1} ... dz_p \langle 0|T\mathcal{H}_I(z_{l+1})...\mathcal{H}_I(z_p)|0\rangle \ .$$

The last factor can be rewritten as $(p = l + r)$

$$\sum_{r=0}^{\infty} \frac{(-i)^r}{r!} \int dz_1 ... dz_r \langle 0|T\mathcal{H}_I(z_1)...\mathcal{H}_I(z_r)|0\rangle = \langle 0|Te^{-i\int d^4 z \mathcal{H}_I(z)}|0\rangle = V \ .$$

Hence V factorizes out from $\hat{G}$, and G in (11.38) becomes

$$G(x_1, x_2; y_1, y_2) = \sum_{l} \frac{(-i)^l}{l!} \int d^4 z_1 ... \int d^4 z_l$$

$$\times \langle 0|\psi_I(x_1)\psi_I(x_2)\bar{\psi}_I(y_1)\bar{\psi}_I(y_2)\mathcal{H}_I(z_1)...\mathcal{H}_I(z_l)|0\rangle_{PC} \ .$$

This explicitly demonstrates the cancellation of vacuum graphs *to all orders*. We now take account of this from the outset in order to compute the fermion and photon propagator in second order of QED.

Second order Fermion propagator of QED

Making use of Wick's theorems we have to second order in the coupling constant for the fermion 2-point function

$$< \Omega|T\psi(x)_\alpha\bar{\psi}(y)_\beta|\Omega > \simeq < 0|T\psi_I(x)_\alpha\bar{\psi}_I(y)_\beta|0 > + \frac{(-ie_0)^2}{2!} < 0|T\psi_I(x)_\alpha\bar{\psi}_I(y)_\beta$$

$$\times \int d^4 z_1 : \bar{\psi}_I(z_1)\slashed{A}_I(z_1)\psi_I(z_1) : \int d^4 z_2 : \bar{\psi}_I(z_2)\slashed{A}_I(z_1)\psi_I(z_2) : |0 > \ . \tag{11.44}$$

In order to write down an explicit expression, we first reorder the fields by bringing contracted pairs adjacent to each other in standard form, keeping track of the number of permutations involved, and the corresponding Grassman signature. All fields must be contracted in order to give a non-vanishing result. Representing the time-ordered contractions by

$$\overbrace{\psi_\alpha^I(x)\bar{\psi}_\beta^I(y)} \;=\; \alpha, x \;\longleftarrow\; \beta, y \tag{11.45}$$

$$\overbrace{A_I^\mu(x)A_I^\nu(y)} \;=\; \mu, x \;\sim\!\sim\!\sim\; \nu, y \tag{11.46}$$

Fig. 11.5. Graphical representation of contractions.

we thus obtain

$$< \Omega|T\psi_\alpha(x)\bar{\psi}_\beta(y)|\Omega >\equiv iS_F'(x-y) = iS_F(x-y) + \tag{11.47}$$

$$(-ie_0)^2\int d^4z_1 d^4z_2 [iS_F(x-z_1)\gamma_\mu iS_F(z_1-z_2)\gamma_\nu iS_F(z_2-y)]_{\alpha\beta} iD_F^{\mu\nu}(z_1-z_2) + \cdots$$

as represented by the sum of diagrams

$$x,\alpha \;\longleftarrow\!\!\bigotimes\!\!\longleftarrow\; y,\beta \;\simeq\; \longleftarrow\!\!\!\longleftarrow \;+\; \longleftarrow\!\!\!\overset{\frown}{\longleftarrow}\!\!\!\longleftarrow \tag{11.48}$$

Fig. 11.6. Electron propagator to second order.

Notice that the combinatorial factor $\frac{1}{2!}$ in (11.21) has been omitted since there are 2! contractions leading to identical results.

Second order photon propagator of QED

In a similar way we have to second order for the 2-point function of the gauge field, after cancellation of the contribution of the vacuum graphs,

$$< \Omega|TA^\mu(x)A^\nu(y)|\Omega >=< 0|A_I^\mu(x)A_I^\nu(y)|0 > + \frac{(-ie_0)^2}{2!} < 0|TA_I^\mu(x)A_I^\nu(y)$$

$$\times \int d^4z_1 d^4z_2 : \bar{\psi}_I(z_1)\slashed{A}_I(z_1)\psi_I(z_1) :: \bar{\psi}_I(z_2)\slashed{A}_I(z_2)\psi_I(z_2) : |0 > + \cdots \tag{11.49}$$

or explicitly

$$< \Omega|TA^\mu(x)A^\nu(y)|\Omega > \equiv iD'_F(x-y) \simeq iD_F^{\mu\nu}(x-y) + \tag{11.50}$$

$$+ (-ie_0)^2 \int d^4z_1 \int d^4z_2 iD_F^{\mu\lambda}(x-z_1)tr\left[\gamma_\lambda iS_F(z_1-z_2)\gamma_\rho iS_F(z_2-z_1)\right] iD_F^{\rho\nu}(z_2-y).$$

The result has the diagrammatic representation

$$x,\mu \qquad y,\nu \simeq \qquad + \qquad \tag{11.51}$$

Fig. 11.6. Photon propagator to second order.

Topologically equivalent diagrams

The perturbative results obtained by performing Wick contractions keeping track of the Grassman signature of the fields, performing the space-time integrations dictated by the *LSZ* reduction formula, and attaching the respective momentum-space wave functions associated with the external states involved in the process can be resumed by a set of rules, called *Feynman rules*. They consist in first drawing all diagrams contributing to the perturbative expansion of the relevant Green function in *configuration space* up to the desired order in the coupling constant e_0, taking care of including only *topologically inequivalent* diagrams. In order to decide whether two diagrams are topologically equivalent or not, we must first label the fermion lines by arrows indicating the direction of flow of (*negative*) electron charge. Conservation of charge requires that the direction of flow of this charge is always the same as we follow through a fermionic "path". Since the gauge field is electrically neutral, there is no arrow to be associated with a photon line. We then label all external lines in configuration space. Thus, for example, the diagrams

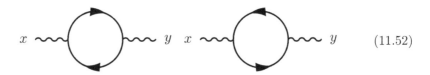

$$x \qquad y \quad x \qquad y \tag{11.52}$$

Fig. 11.7(a). Topologically equivalent diagrams.

are topologically *equivalent* since they are the same under reflection about the *xy*-axis. On the other hand, the diagrams

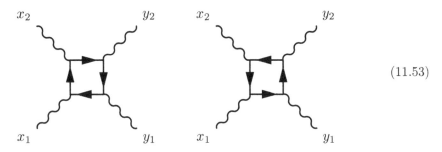

$$(11.53)$$

Fig. 11.7(b). Topologically inequivalent diagrams.

are topologically *inequivalent*. Note that in these diagrams we have not labeled the corresponding *vertices*. For such an n-point 1-loop diagram there exist $n!$ possible reorderings of these vertex labels for a diagram of order n. Since the corresponding coordinates are integrated over, every reordering will give the same result. Remembering that from the expansion of the exponential in (11.20) we also have to include a factor $1/n!$, we can ignore this factor if we do not count diagrams which only differ in their labeling of the vertices.

As one easily convinces oneself, every closed fermion loop involves an odd number of permutations of the fermion fields in the computation of the contractions so that a *minus sign* will have to be associated with every fermion loop. Gauge field loops do not occur because of the linear dependence on $A^\mu(x)$ of the interaction Hamiltonian.

11.6   Furry's theorem

The following theorem allows us to discard from the outset an in principle infinite set of Feynman diagrams.

Furry's theorem

All Feynman diagrams containing a fermion loop with an odd number of vertices can be omitted.

Proof:

Furry's theorem is a consequence of the charge conjugation invariance of QED. Under charge conjugation the gauge field transforms as follows:

$$U_c A^\mu(x) U_c^{-1} = -A^\mu(x) \ .$$

Making use of the charge conjugation invariance of the physical vacuum, we thus have

$$< \Omega | T A^{\mu_1}(x_1)...A^{\mu_n}(x_n) | \Omega > = (-1)^n < \Omega | T A^{\mu_1}(x_1)...A^{\mu_n}(x_n) | \Omega > \ ,$$

or

$$< \Omega | T A^{\mu_1}(x_1) \cdots A^{\mu_n}(x_n) | \Omega > = 0, \quad n \text{ odd} \ .$$

It will be useful to understand this result diagrammatically. Recall that we had introduced in (4.13) a matrix c with the property,

$$c\vec{\sigma}c^{-1} = -\vec{\sigma}^* .$$

Correspondingly we defined a matrix C in (4.17) with the property

$$C\gamma^{\mu}C^{-1} = \gamma^{\mu T} ,$$

where T denotes the transpose. With the definition $\mathbb{C} = \gamma^5 C$, it then follows that

$$\mathbb{C}\gamma^{\mu}\mathbb{C}^{-1} = -\gamma^{\mu T} , \tag{11.54}$$

implying for the fermion propagator

$$\mathbb{C}S_F(x)\mathbb{C}^{-1} = S_F^T(-x) . \tag{11.55}$$

Now, for every fermion loop contributing to a general diagram there will be a corresponding fermion loop with the fermion charge flowing in the opposite direction. Isolating the loop-part, we have for a loop with n vertices,

$$\text{loop1} = \int tr\left[\gamma^{\mu_n}S_F(z_n - z_{n-1})\gamma^{\mu_{n-1}}S_F(z_{n-1} - z_{n-2})\cdots\gamma^{\mu_1}S_F(z_1 - z_n)\right] ,$$

and for the charge flow in the opposite direction,

$$\text{loop2} = \int tr\left[S_F(z_n - z_1)\gamma^{\mu_1}\cdots S_F(z_{n-2} - z_{n-1})\gamma^{\mu_{n-1}}S_F(z_{n-1} - z_n)\gamma^{\mu_n}\right] ,$$

where the integrals stand for the integration over the space-time coordinates labelling the vertices. Rewriting the trace in the second integrand as $tr\,\mathbb{C}[\cdots]\mathbb{C}^{-1}$ will not change its value. Making use of (11.54) and (11.55) and adding the two loop contributions, we have

$$(1 + (-1)^n)tr[\gamma^{\mu_n}S_F(z_n - z_{n-1})\gamma^{\mu_{n-1}}S_F(z_{n-1} - z_{n-2})\cdots\gamma^{\mu_1}S_F(z_1 - z_n)],$$

showing that the two contributions add to *zero* for n odd.

11.7 Going over to momentum space

Having written down all topologically inequivalent diagrams with their respective labelling of electron charge flow, we go over to momentum space by replacing the space-time 2-point functions by their respective Fourier transforms. For each argument of the 2-point functions there will be a momentum in Fourier space associated with the particles participating in the interaction. Translational invariance will imply an *energy-momentum conservation* delta function at each vertex. We thus label the internal lines of the diagram by the momentum to be integrated over. There will survive an *overall energy-momentum conserving delta function* of the form

$$G(p_1 \cdots p_n) = (2\pi)^4 \delta^4(p_1 + p_2 + \ldots + p_n)G'(p_1 \cdots p_n) \tag{11.56}$$

where we have taken all external momenta to be incoming. We now exemplify the transition to momentum space for the case of the electron 2-point function.

Electron and photon 2-point function in momentum space

Consider the 2-point function (11.47). Taking the Fourier transform we have

$$\int d^4x \int d^4y \, e^{ip\cdot x} e^{iq\cdot y} S'_F(x-y) = (2\pi)^4 \delta^4(p+q) S'_F(p)$$

where

$$S'_F(p) = \int d^4z \, e^{ip\cdot z} S'_F(z) \; .$$

Hence $S'_F(p)$ corresponds in this case to G' in (11.56), where the overall delta-function has been factored out.

Going to momentum space we have from (11.47) for the fermion propagator in second order perturbation theory

$$\int \frac{d^4p}{(2\pi)^4} e^{-ip(x-y)} i S'_F(p) = \int \frac{d^4p}{(2\pi)^4} e^{-ip(x-y)} i S_F(p) \; +$$

$$+ \int d^4z_1 \int d^4z_2 \int \frac{d^4q_1}{(2\pi)^4} \frac{d^4q_2}{(2\pi)^4} \frac{d^4p}{(2\pi)^4} \int \frac{d^4k}{(2\pi)^4}$$

$$\times \, e^{-iq_1(x-z_1)} e^{-iq_2(z_1-z_2)} e^{-ip(z_2-y)} e^{-ik(z_1-z_2)}$$

$$\times \, [iS_F(q_1)(-ie_0\gamma_\mu) iS_F(q_2)(-ie_0\gamma_\nu) iS_F(p)] \, iD_F^{\mu\nu}(k) \; .$$

The z_1 and z_2 integrations implement 4-momentum conservation at each vertex:

$$\int d^4z_1 \int d^4z_2 \Rightarrow (2\pi)^4 \delta^4(q_1 - q_2 - k)(2\pi)^4 \delta^4(p - q_2 - k))$$

$$= (2\pi)^8 \delta^4(q_1 - p) \delta^4(p - q_2 - k) \; .$$

This leaves us with the exponential factor $e^{-iq_1 x} e^{ipy}$. Doing the integration over q_1 and q_2 we are left with

$$\int \frac{d^4p}{(2\pi)^4} e^{-ip(x-y)} i S'_F(p) = \int \frac{d^4p}{(2\pi)^4} e^{-ip(x-y)} \Big\{ iS_F(p) +$$

$$\int \frac{d^4k}{(2\pi)^4} [iS_F(p)(-ie_0\gamma_\mu) iS_F(p-k)(-ie_0\gamma_\nu) iS_F(p)] \, iD_F^{\mu\nu}(k) \Big\} + \cdots \; ,$$

or equivalently

$$iS'_F(p) = iS_F(p) + (-ie_0)^2 \int \frac{d^4k}{(2\pi)^4} [iS_F(p)(\gamma_\mu) iS_F(p-k)(\gamma_\nu) iS_F(p)] \, iD_F^{\mu\nu}(k) + \cdots$$

$$\tag{11.57}$$

For (11.50) we obtain correspondingly

$$iD'^{\mu\nu}_F(k) = iD^{\mu\nu}_F(k) + (-ie_0)^2 \int \frac{d^4p}{(2\pi)^4} iD^{\mu\lambda}(k) tr[\gamma_\lambda iS_F(k-p)\gamma_\rho iS_F(p)] iD^{\rho\nu}(k) + \cdots$$

$$\tag{11.58}$$

11.8 Momentum space Feynman rules for QED

There exist two ways of formulating the Feynman rules.

(A) Write down the Feynman diagram in question in momentum space, labelling all (oriented) lines by their momenta, taking account of energy-momentum conservation at each vertex. In the case of fermions, the direction of the flow of electron (positron) charge is taken to be along (opposite to) the direction of the momentum flow. An example is given by the self-energy diagrams

Fig. 11.8. Second and fourth order electron self-energy diagrams.

The Feynman rules for evaluating such a diagram are given by

1. With each oriented fermion propagator associate a factor

$$iS_F(p)_{\alpha\beta} = \left(\frac{i}{\not{p} - m + i\epsilon}\right)_{\alpha\beta}$$

2. With each photon propagator associate a factor (see (13.98)),

$$iD_F^{\mu\nu}(k) = -\frac{ig^{\mu\nu}}{k^2 + i\epsilon} + \text{gauge terms} \tag{11.59}$$

3. With each vertex associate a factor

$$-ie_0\gamma^\mu_{\alpha\beta}$$

4. For each loop momentum ℓ perform an integration with the measure

$$\int \frac{d^4l}{(2\pi)^4}$$

5. Introduce a minus sign for every independent fermion loop.

The results are the momentum-space Green functions G' in (11.56), with G stripped off the factor expressing overall energy-momentum conservation.

(B) A second possibility is to proceed as follows:

1. Label all internal and external lines by a momentum p_i, without taking account of energy-momentum conservation.

2. With each oriented fermion propagator associate a factor

$$iS_F(p)_{\alpha\beta} = \left(\frac{i}{\not{p} - m + i\epsilon}\right)_{\alpha\beta}$$

3. With each photon propagator associate a factor

$$iD_F^{\mu\nu}(k) = -\frac{ig^{\mu\nu}}{k^2 + i\epsilon} + \text{gauge terms}$$

4. With each vertex associate a factor

$$-ie_0\gamma_{\alpha\beta}^\mu(2\pi)^4\delta(p' + k - p)$$

5. For each internal momentum p_i perform the integration with measure

$$\int \frac{d^4p_i}{(2\pi)^4}$$

6. Introduce a minus sign for every independent fermion loop.

The results are the momentum-space Green functions G in (11.56) *including* the factor expressing overall energy-momentum conservation.

11.9 Moeller scattering

We now show for the case of electron–electron elastic scattering, how to compute the corresponding cross-section using the LSZ formalism of Chapter 10.

Moeller scattering in space-time

We first return to a space-time description of the Green function for electron–electron elastic scattering, in order to exemplify how to take account of the Fermi statistics involved.

Let us compute explicitly the T-matrix for electron–electron scattering in second order of perturbation theory. This requires us to calculate $G^{(2)}(x_1', x_2'; x_1, x_2)_C$, where the subscript stands for "connected". We have (we omit the subscript "I" for "interaction picture")

$$G^{(2)}(x_1', x_2'; x_1, x_2)_{\beta_1\beta_2\alpha_1\alpha_2} = \int d^4z_1 \int d^4z_2 \, \langle 0|T\psi_{\beta_1}(x_1')\psi_{\beta_2}(x_2')\bar\psi_{\alpha_1}(x_1)\bar\psi_{\alpha_2}(x_2)$$

$$\times : \bar\psi(z_1)(-ie_0)\not{A}(z_1)\psi(z_1) :: \bar\psi(z_2)(-ie_0)\not{A}(z_2)\psi(z_2) : |0\rangle_C \, .$$

Taking account of the Fermi statistics and performing the time-ordered contractions we have two distinct contractions (see Fig. 11.9)

$$G^{(2)}(x_1', x_2'; x_1, x_2)_{\beta_1\beta_2\alpha_1\alpha_2} = \int d^4z_1 \int d^4z_2 iD_F(z_2 - z_1)_{\mu_1\mu_2}$$

$$\times [iS_F(x_2' - z_2)(-ie_0\gamma^{\mu_2})iS_F(z_2 - x_2)]_{\beta_2\alpha_2}$$

$$\times [iS_F(x_1' - z_1)(-ie_0\gamma^{\mu_1})iS_F(z_1 - x_1)]_{\beta_1\alpha_1}$$

$$- (x_1' \leftrightarrow x_2', \beta_1 \leftrightarrow \beta_2) \, , \tag{11.60}$$

where the minus sign in the "exchange" term arises as a result of the Fermi statistics.

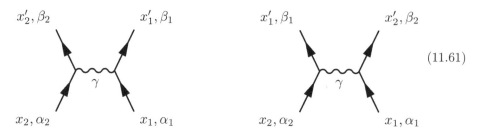

$$(11.61)$$

Fig. 11.9. Moeller scattering in space-time.

The T-matrix

Recalling that

$$S_F(x) = \int \frac{d^4q}{(2\pi)^4} S_F(q)e^{-iqx} \,, \quad D^{\mu\nu}(z) = \int \frac{d^4k}{(2\pi)^4} D_F^{\mu\nu}(k)e^{-ik\cdot z} \,,$$

we obtain for the connected Green function $G'^{(2)}$

$$
\begin{aligned}
G(x_1', x_2'; x_1, x_2)_{\beta_1\beta_2\alpha_1\alpha_2} = &\int d^4z_1 \int d^4z_2 \int \frac{d^4q_1'}{(2\pi)^4}\frac{d^4q_2'}{(2\pi)^4} \int \frac{d^4q_1}{(2\pi)^4}\frac{d^4q_2}{(2\pi)^4} \int \frac{d^4k}{(2\pi)^4} \\
&\times [iS_F(q_2')(-ie_0)\gamma^{\mu_2}iS_F(q_2)]_{\beta_2\alpha_2}e^{-iq_2'\cdot(x_2'-z_2)}e^{-iq_2\cdot(z_2-x_2)} \\
&\times [iS_F(q_1')(-ie_0)\gamma^{\mu_1}iS_F(q_1)]_{\beta_1\alpha_1}e^{-iq_1'\cdot(x_1'-z_1)}e^{-iq_1\cdot(z_1-x_1)} \\
&\times iD_F(k)_{\mu_1\mu_2}e^{-ik\cdot(z_2-z_1)} \\
&- (x_1' \leftrightarrow x_2', \beta_1 \leftrightarrow \beta_2) \,.
\end{aligned}
$$
$$(11.62)$$

According to the LSZ substitution rules (10.52) and (10.55) this Green function is to be multiplied from the right by

$$K_{\alpha_1\beta_1}^{(r)} = [iZ_\psi^{-\frac{1}{2}}(i\overleftarrow{\partial}_1 + m)u(x_1; \vec{p}_1, \sigma_1)]_{\alpha_1}[iZ_\psi^{-\frac{1}{2}}(i\overleftarrow{\partial}_2 + m)u(x_2; \vec{p}_2, \sigma_2)]_{\alpha_2}$$

and from the left by

$$K_{\beta_1\beta_2}^{(\ell)} = [-iZ_\psi^{\frac{1}{2}}\bar{u}(x_1'; \vec{p}_1', \sigma_1')(i\overrightarrow{\partial}_1' - m)]_{\beta_1}[-iZ_\psi^{-\frac{1}{2}}\bar{u}(x_2'; \vec{p}_2', \sigma_2')(i\overrightarrow{\partial}_2' - m)]_{\beta_2} \,.$$

Integration over z_1 and z_2 leads to the delta functions

$$\int d^4z_1 : \quad (2\pi)^4\delta^4(k + p_1' - p_1) \,, \qquad \int d^4z_2 : \quad (2\pi)^4\delta^4(k + p_2 - p_2') \,.$$

Subsequent integration over the space-time coordinates as implied by (10.52) and (10.55), will then produce four additional delta functions:

$$\int d^4x_1' : \quad (2\pi)^4\delta^4(q_1' - p_1') \,, \qquad \int d^4x_2' \quad (2\pi)^4\delta^4(q_2' - p_2')$$

$$\int d^4x_1 : \quad (2\pi)^4\delta^4(q_1 - p_1) \,, \qquad \int d^4x_2 \quad (2\pi)^4\delta^4(q_2 - p_2).$$

There remain five momentum integrations, leaving us with a δ function expressing overall energy-momentum conservation. The $\mathcal{T}$-matrix is defined in (10.12). Taking out the δ-function we obtain

$$
-i\mathcal{T}(\vec{p}_1', \vec{p}_2', \vec{p}_1, \vec{p}_2) = \left[\frac{-i}{(2\pi)^{\frac{3}{2}}}\right]^4 [\sqrt{Z_\psi}]^{-4} \sqrt{\frac{m}{\omega_{p_1}}} \sqrt{\frac{m}{\omega_{p_2}}} \sqrt{\frac{m}{\omega_{p_1'}}} \sqrt{\frac{m}{\omega_{p_2'}}} iD_F^{\mu_1\mu_2}(p_1' - p_1)
$$
$$
\times \bar{U}(\vec{p}_1', \sigma_1')(\not{p}_1' - m)\left[iS_F(p_1')(-ie_0\gamma_{\mu_1})iS_F(p_1)\right](\not{p}_1 - m)U(\vec{p}_1, \sigma_1)
$$
$$
\times \bar{U}(\vec{p}_2', \sigma_2')(\not{p}_2' - m)\left[iS_F(p_2')(-ie_0\gamma_{\mu_2})iS_F(p_2)\right](\not{p}_2 - m)U(\vec{p}_2, \sigma_2)
$$
$$
- (\vec{p}_1' \leftrightarrow \vec{p}_2', \sigma_1' \leftrightarrow \sigma_2'). \tag{11.63}
$$

We see that the LSZ rules correspond in the present case to the following substitution rules for the external legs of the Green function:

- for every *incoming* external fermion line:

$$
iS_F(p)_{\alpha\beta} \rightarrow i\frac{Z_\psi^{-\frac{1}{2}}}{(2\pi)^{\frac{3}{2}}}\sqrt{\frac{m}{\omega_p}}U_\beta(\vec{p}, \sigma) . \tag{11.64}
$$

- for every *outgoing* external fermion line:

$$
iS_F(p')_{\alpha\beta} \rightarrow i\frac{Z_\psi^{-\frac{1}{2}}}{(2\pi)^{\frac{3}{2}}}\sqrt{\frac{m}{\omega_{p'}}}\bar{U}_\alpha(\vec{p}', \sigma') , \tag{11.65}
$$

and completing these rules for the photon,

- for every *incoming* photon:

$$
iD_F^{\mu\nu}(k) \rightarrow i\frac{Z_A^{-\frac{1}{2}}}{(2\pi)^{\frac{3}{2}}}\frac{\epsilon^\nu}{2k_0^2} , \tag{11.66}
$$

- for every *outgoing* photon:

$$
iD_F^{\mu\nu}(k) \rightarrow i\frac{Z_A^{-\frac{1}{2}}}{(2\pi)^{\frac{3}{2}}}\frac{\epsilon^{*\mu}}{2k_0^2} . \tag{11.67}
$$

To order $O(e_0^2)$ we may set $Z_\psi = 1$ (see Chapter 16). The result (11.63) thus reads[7]

$$
\mathcal{T}_{fi} = (-ie_0)^2\sqrt{\frac{m}{(2\pi)^3\omega_1}\frac{m}{(2\pi)^3\omega_2}\frac{m}{(2\pi)^3\omega_1'}\frac{m}{(2\pi)^3\omega_2'}}
$$
$$
\times [\bar{U}(\vec{p}_1', \sigma_1')\gamma^\mu U(\vec{p}_1, \sigma_1)\left(\frac{g_{\mu\nu}}{(p_1' - p_1)^2} + \text{gauge}\right)\bar{U}(\vec{p}_2', \sigma_2')\gamma^\nu U(\vec{p}_2, \sigma_2)
$$
$$
- (\vec{p}_1' \leftrightarrow \vec{p}_2', \sigma_1' \leftrightarrow \sigma_2')] . \tag{11.68}
$$

---

[7]See (13.98) for the choice of photon propagator.

The gauge-dependent terms in the photon propagator are proportional to either k^μ or k^ν or $k^\mu k^\nu$, where $k = (p_1' - p_1) = -(p_2' - p_2)$ is the momentum associated with the photon line. Since we have the *on-shell* properties

$$\bar{U}(\vec{p}_1', \sigma_1)(\slashed{p}_1' - \slashed{p}_1)U(p_1, \sigma_1) = 0$$
$$\bar{U}(p_2', \sigma_2')(\slashed{p}_2' - \slashed{p}_1)U(p_1, \sigma_1) = 0 ,$$

it follows that the gauge-dependent terms in the photon propagator do not contribute, as promised. The result is represented in the figure below.

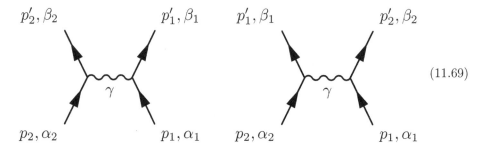

(11.69)

Fig. 11.10. Moeller scattering in momentum space.

11.10 The Moeller differential cross-section

To order $O(e_0^2)$ we can set the bare charge $e_0 = e$, $\alpha = e^2/4\pi = 1/137$ and $m_0 = m$, the physical mass.

For the corresponding differential cross-section we need to compute the absolute square of this amplitude. We shall do this for the case of *unpolarized* electrons, by averaging over the spins of the initial state, and summing over the spins of the final state. Recalling our definition of the Stapp $\mathcal{M}$-functions (10.22), we have

$$\overline{|T|^2} = \frac{1}{4} \sum_{\sigma_1, \sigma_2} \sum_{\sigma_1', \sigma_2'} | <\vec{p}_1'\sigma_1', \vec{p}_2'\sigma_2'|T|\vec{p}_1\sigma_1, \vec{p}_2\sigma_2 > |^2$$

$$= \frac{1}{(2\pi)^3 2\omega_1} \frac{1}{(2\pi)^3 2\omega_2} \frac{1}{(2\pi)^3 2\omega_1'} \frac{1}{(2\pi)^3 2\omega_2'} \overline{|\mathcal{M}|^2}$$

and recalling (4.43),

$$\sum_\sigma U_\alpha(\vec{p}, \sigma)\bar{U}_\beta(\vec{p}, \sigma) = \Lambda(p)_{\alpha\beta} = \left(\frac{\slashed{p} + m}{2m}\right)_{\alpha\beta} ,$$

one finds

$$\overline{|\mathcal{M}|^2} = \frac{1}{4}(2m)^4 e^4 \Big\{ tr(\gamma^\mu \Lambda(p_1)\gamma^\nu \Lambda(p_1')) tr(\gamma_\mu \Lambda(p_2)\gamma_\nu \Lambda(p_2')) \frac{1}{[(p_1' - p_1)^2]^2}$$

$$- tr(\gamma^\mu \Lambda(p_1)\gamma^\nu \Lambda(p_2')\gamma_\mu \Lambda(p_2)\gamma_\nu \Lambda(p_1')) \frac{1}{(p_1' - p_1)^2(p_2' - p_1)^2} + (p_1' \leftrightarrow p_2') \Big\} .$$

The traces are easily evaluated. We have[8]

$$tr(\gamma^\mu \Lambda(p_1)\gamma^\nu \Lambda(p_1')) = \frac{1}{m^2}(p_1^\mu p_1'^\nu - g^{\mu\nu} p_1 \cdot p_1' + p_1^\nu p_1'^\mu + m^2 g^{\mu\nu}) \,,$$

implying

$$tr(\gamma^\mu \Lambda(p_1)\gamma^\nu \Lambda(p_1'))tr(\gamma_\mu \Lambda(p_2)\gamma_\nu \Lambda(p_2'))$$
$$= \frac{2}{m^4}[(p_1 \cdot p_2)^2 + (p_1 \cdot p_2')^2 + 2m^2(p_1 \cdot p_2' - p_1 \cdot p_2)]$$

and

$$tr[\gamma^\mu \Lambda(p_1)\gamma^\nu \Lambda(p_2')\gamma_\mu \Lambda(p_2)\gamma_\nu \Lambda(p_1')] = -\frac{2}{m^4}[(p_1 \cdot p_2')^2 - 2m^2(p_1 \cdot p_2)] \,,$$

where we have used $p_1 + p_2 = p_1' + p_2'$, implying $p_1 \cdot p_2 = p_1' \cdot p_2'$, and $p_1 \cdot p_1' = p_2 \cdot p_2'$. Therefore

$$\overline{|\mathcal{M}|^2} = 8e^4\left\{\frac{(p_1 \cdot p_2)^2 + (p_1 \cdot p_2')^2 + 2m^2(p_1 \cdot p_2' - p_1 \cdot p_2)}{[(p_1' - p_1)^2]^2}\right. \tag{11.70}$$
$$\left. +\frac{(p_1 \cdot p_2)^2 + (p_1 \cdot p_1')^2 + 2m^2(p_1 \cdot p_1' - p_1 \cdot p_2)}{[(p_2' - p_1)^2]^2} + 2\frac{(p_1 \cdot p_2)^2 - 2m^2(p_1 \cdot p_2)}{(p_1' - p_1)^2(p_2' - p_1)^2}\right\} \,.$$

Differential cross-section in center of mass (cm)-system

The differential cross-section has the form given by (10.25). We shall compute it in the *cm* system of the initial particles as represented in the figure below.

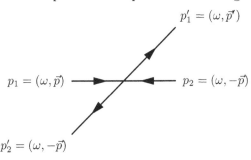

We may express the above invariant quantities in terms of the total *cm* energy and the *cm* scattering angle θ:

$$(p_1 + p_2)^2 \equiv s = E_{cm}^2 = 4\omega^2 \quad (\omega_1 = \omega_2 = \omega_1' = \omega_2' = \omega)$$
$$p_1 \cdot p_2 = \frac{s}{2} - m^2 = 2\omega^2 - m^2$$
$$p_1 \cdot p_1' = \omega^2 - \vec{p}^2 \cos\theta = \omega^2(1 - \cos\theta) + m^2 \cos\theta \tag{11.71}$$
$$p_1 \cdot p_2' = \omega^2(1 + \cos\theta) - m^2 \cos\theta \,.$$

8

$$tr(\gamma_\mu \gamma_\nu) = 4g_{\mu\nu} \rightarrow tr(\gamma_\mu \not{a}) = 4a_\mu$$
$$tr(\gamma_\mu \gamma_\nu \gamma_\lambda \gamma_\rho) = 4(g_{\mu\nu}g_{\lambda\rho} - g_{\mu\lambda}g_{\nu\rho} + g_{\mu\rho}g_{\nu\lambda})$$
$$tr(\gamma_\mu \not{a} \gamma_\nu \not{b}) = 4(a_\mu b_\nu - g_{\mu\nu}(a \cdot b) + b_\mu a_\nu)$$

Noting that

$$\sqrt{(s - 2m^2)^2 - 4m^2} = 2\sqrt{(p_1 \cdot p_2)^2 - m^4}$$

we have for the differential cross-section

$$d\sigma = \frac{(2\pi)^4 \delta(p_1' + p_2' - p_1 - p_2)}{\sqrt{4(p_1 \cdot p_2 - m^4)}} |\mathcal{M}|^2 \frac{d^3 p_1'}{(2\pi)^3 2\omega_1'} \frac{d^3 p_2'}{(2\pi)^3 2\omega_2'} . \qquad (11.72)$$

In the *cm* system,

$$\int \frac{d^3 p_2'}{(2\pi)^3 2\omega_2'} \delta^4(p_1' + p_2' - p_1 - p_2) \to \int \frac{d^3 p_2'}{(2\pi)^3 2\omega_2'} \delta^3(\vec{p}_1' + \vec{p}_2')\delta(2\omega_1' - 2\omega_1)$$

$$\to \frac{1}{(2\pi)^3 4\omega}\delta(\omega' - \omega) .$$

In order to do the $\vec{p}_1'$ integration, we note that it follows from $\omega^2 = \vec{p}\,^2 + m^2$ that

$$\omega d\omega = |\vec{p}| d|\vec{p}| .$$

Hence

$$|\vec{p}\,'|^2 d|\vec{p}\,'| = \sqrt{\omega'^2 - m^2}\omega' d\omega' ,$$

or

$$\frac{d^3 p_1'}{(2\pi)^3 2\omega_1'} = \frac{\sqrt{\omega'^2 - m^2}\omega' d\omega'}{(2\pi)^3 2\omega'} d\Omega' .$$

Furthermore

$$\sqrt{[(p_1 \cdot p_2)^2 - m^4]} = \sqrt{(2\omega^2 - m^2)^2 - m^4} = 2\omega\sqrt{\omega^2 - m^2} .$$

The remaining integration over $|\vec{p}_1|$ and ω' can now be trivially performed, and we obtain from (11.72)

$$d\sigma = \frac{1}{64} \frac{1}{(2\pi)^2 \omega^2} |\mathcal{M}|^2 d\Omega .$$

Finally, making use of (11.71) in (11.70) one obtains the *Moeller formula* ($\alpha = e^2/4\pi$)

$$\frac{d\sigma}{d\Omega} = \frac{\alpha^2 (2\omega^2 - m^2)^2}{4\omega^2 (\omega^2 - m^2)^2} \left[\frac{4}{\sin^4 \theta} - \frac{3}{\sin^2 \theta} + \frac{(\omega^2 - m^2)^2}{(2\omega^2 - m^2)^2} \left(1 + \frac{4}{\sin^2 \theta} \right) \right] .$$

In the *ultrarelativistic* limit, where $m/\omega \to 0$,

$$\frac{d\sigma}{d\Omega} \to \frac{\alpha^2}{\omega^2} \left(\frac{4}{\sin^4 \theta} - \frac{2}{\sin^2 \theta} + \frac{1}{4} \right) ,$$

whereas in the *nonrelativistic* limit we obtain[9] the *Mott* (1930) differential cross-

---

9

$$\sin \theta = 2\sin \frac{\theta}{2} \cos \frac{\theta}{2}$$

$$\frac{4}{\sin^4 \theta} - \frac{2}{\sin^2 \theta} = \frac{1}{4\sin^4 \frac{\theta}{2} \cos^4 z \frac{\theta}{2}} - \frac{1}{2\sin^2 \frac{\theta}{2} \cos^2 \frac{\theta}{2}}$$

$$= \frac{1 - 2\sin^2 \frac{\theta}{2} \cos^2 \frac{\theta}{2}}{4\sin^4 \frac{\theta}{2} \cos^4 \frac{\theta}{2}} = \frac{(\cos^2 \frac{\theta}{2} + \sin^2 \frac{\theta}{2}) - 2\sin^2 \frac{\theta}{2} \cos^2 \frac{\theta}{2}}{4\sin^4 \frac{\theta}{2} \cos^4 \frac{\theta}{2}}$$

section.[10]

$$\frac{d\sigma}{d\Omega} \to \left(\frac{\alpha}{m}\right)^2 \frac{1}{16v^4} \left(\frac{1}{\sin^4\frac{\theta}{2}} + \frac{1}{\cos^4\frac{\theta}{2}} - \frac{1}{\sin^2\frac{\theta}{2}\cos^2\frac{\theta}{2}}\right)$$

with $v = \frac{|\vec{p}|}{m}$. In the forward direction $\theta \to 0$ this expression reduces to the Rutherford cross-section,

$$\left(\frac{d\sigma}{d\Omega}\right)_{n.rel} \longrightarrow \left(\frac{\alpha}{m}\right)^2 \frac{1}{16v^2} \frac{1}{\sin^4\frac{\theta}{2}} \quad (Rutherford \ \ X\text{-}section) \, . \quad (11.73)$$

11.11 Compton scattering

As one further example, we consider the scattering of light (photon) off a free electron, as shown in Fig. 11.11. Topologically these diagrams are the same as in Fig. 11.10:[11]

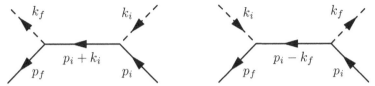

Fig. 11.11. Lowest order Compton scattering.

The corresponding differential cross-section now reads (see (10.25))

$$d\sigma_{fi} = \frac{(2\pi)^4\delta^4(p_f + k_f - p_i - k_i)|<\vec{p}_f,\sigma_f,\vec{k}_f|\mathcal{M}|\vec{p}_i,\sigma_i,\vec{k}_i>|^2}{2(s - m^2)} \frac{d^3k_f}{(2\pi)^32k_f^0} \frac{d^3p_f}{(2\pi)^32p_f^0}$$

with $\mathcal{M}$ the Stapp M-function. The transition amplitude is obtained from the LSZ formalism via the substitutions (11.64) to (11.67) in the corresponding *Green* function. For the Stapp-functions these substitutions take the simpler form:

- for every *incoming* external fermion propagator:

$$iS_F(p_i)_{\alpha\beta} \to iZ_\psi^{-\frac{1}{2}}U_\beta(\vec{p}_i,\sigma_i) \, .$$

- for every *outgoing* external fermion propagator:

$$iS_F(p_f)_{\alpha\beta} \to iZ_\psi^{-\frac{1}{2}}\bar{U}_\alpha(\vec{p}_f,\sigma_f) \, .$$

- for every *incoming* external photon propagator:

$$iD_F^{\mu\nu}(k_i) \to iZ_\psi^{-\frac{1}{2}}\epsilon^\nu(\vec{k}_i) \, .$$

[10]N.F. Mott and H.S.W. Messey, *Theory of Atomic Collisions* (Oxford, 1956).
[11]For technical reasons the photons are represented by dashed lines; the metafont program does allow for wiggly lines *with arrows* for aesthetical reasons.

- for every *outgoing* external photon propagator:

$$iD_F^{\mu\nu}(k_f) \rightarrow iZ_A^{-\frac{1}{2}}\epsilon^{*\mu}(k_f) \, .$$

Hence we have from Fig. 11.11, using the above rules,[12]

$$\mathcal{M} = e_0^2\bar{U}(p_f,\sigma_f)\left(\not{\epsilon}_f^*\frac{1}{\not{p}_i+\not{k}_i-m}\not{\epsilon}_i+\not{\epsilon}_i\frac{1}{\not{p}_i-\not{k}_f-m}\not{\epsilon}_f^*\right)U(p_i,\sigma_i).$$

Rationalizing the denominators; using $\not{a}\not{b} = 2b.a - \not{b}\not{a}$ and the Dirac equation of motion; going to the laboratory frame $\vec{p}_i = 0$ where $p_i.\epsilon_i = p_i.\epsilon_i^* = p_i.\epsilon_f = p_i.\epsilon_f^* = 0$; and, setting $k_i^2 = k_f^2$ $k_i \cdot \epsilon_i = k_f \cdot \epsilon_f = 0$ using (5.13), one obtains

$$\mathcal{M} = e_0^2\bar{U}(\vec{p}_f,\sigma_f)\left(\frac{\not{\epsilon}_f^*\not{k}_i\not{\epsilon}_i}{2p_i \cdot k_i} + \frac{\not{\epsilon}_i\not{k}_f\not{\epsilon}_f^*}{2p_i \cdot k_f}\right)U(\vec{p}_i,\sigma_i). \tag{11.74}$$

Note that unlike the amplitude (11.63) for Moeller scattering, this amplitude is symmetric under the exchange $(\not{\epsilon}_i, k_i) \leftrightarrow (\not{\epsilon}_f, k_f)$ corresponding to the Bose statistics of photons. Taking the absolute square of $\mathcal{M}$ and summing over the initial and final spins using (4.43) one obtains

$$\sum_{\sigma_f,\sigma_i}|\mathcal{M}|^2 = e_0^4 tr\left[\left(\frac{\not{\epsilon}_f^*\not{k}_i\not{\epsilon}_i}{2p_i \cdot k_i} + \frac{\not{\epsilon}_i\not{k}_f\not{\epsilon}_f^*}{2p_i \cdot k_f}\right)\frac{\not{p}_i+m}{2m}\left(\frac{\not{\epsilon}_i\not{k}_i\not{\epsilon}_f}{2p_i \cdot k_i} + \frac{\not{\epsilon}_f\not{k}_f\not{\epsilon}_i^*}{2p_i \cdot k_f}\right)\frac{\not{p}_i+m}{2m}\right]. \tag{11.75}$$

The trace can be computed using repeatedly the formula (4.45). The computation is very tedious and is just a matter of patience. We further simplify the calculation by taking the photons to be *linearly* polarized. The polarization tensors are then real, and the result of taking the trace is,[13]

$$\sum_{\sigma_f,\sigma_i}|\mathcal{M}|^2 = \frac{e_0^4}{2m^2}\left[\frac{\omega_f}{\omega_i} + \frac{\omega_i}{\omega_f} + 4(\epsilon_f \cdot \epsilon_i)^2 - 2\right]. \tag{11.76}$$

It remains to take care of the integration measure. We seek the differential cross-section for scattering a photon off an electron at rest into a solid angle $d\Omega$. Hence

$$d^3k_f = \omega_f^2 d\omega_f d\Omega_f$$

where $\omega_f = |\vec{k}_f|$. It remains to work out $\delta^4(p_f + k_f - p_i - k_i)\frac{d^3p_f}{(2\pi)^3 2\omega_f}$. Integration over $\vec{p}_f$ fixes the spatial momentum as a function of the incoming and outgoing photon-momentum:

$$\vec{p}_f = \vec{k}_i - \vec{k}_f$$

In the lab frame this leaves the remaining energy delta-function in the form

$$\delta(p_f^0 + \omega_f - m - \omega_i) = \delta(F(\omega_f; \omega_i, \cos\theta))$$

---

[12] A factor $-i$ has been taken out because of our definition of T.
[13] C. Itzikson and J.-B. Zuber, *Quantum Field Theory* (McGraw-Hill, 1980); S. Weinberg, *The Quantum Theory of Fields* (Cambridge University Press, 1996).

with

$$F(\omega_f; \omega_i, \cos\theta) = \left(\sqrt{\omega_i^2 - 2\omega_i\omega_f \cos\theta + \omega_f^2 + m^2} + \omega_f - m - \omega_i\right)$$

where θ is the angle between $\vec{k}_i$ and $\vec{k}_f$. Using the (familiar) formula,

$$\delta(F(\omega_f; \omega_i, \cos\theta)) = \frac{\delta(\omega_f - \omega_0)}{\left|\frac{d}{d\omega_f} F(\omega_f; \omega_i, \cos\theta)\right|_{\omega_f=\omega_0}} \tag{11.77}$$

with $\omega_f = \omega_0$ the (only) zero of $F(\omega_f; \omega_i, \cos\theta)$,

$$\omega_0 = \frac{\omega_i}{1 + \frac{\omega_i}{m}(1 - \cos\theta)},$$

we have

$$\delta(p_f^0 + \omega_f - m - \omega_i) = \frac{p_f^0 \omega_0}{m\omega_i}\delta(\omega_f - \omega_0(\theta)) .$$

From (11.77) we see that there is a shift in the frequency of the scattered photon, which carries the name of A.H. Compton, who discovered this shift in his X-ray studies in 1923.

Putting things together we have with (11.74),

$$d\sigma = \frac{(2\pi)^4|\mathcal{M}|^2}{2(s - m^2)}\left(\frac{\omega_f^2 d\omega_f d\Omega}{(2\pi)^3 2\omega_f}\right)\left(\frac{p_f^0 \omega_f}{m\omega_i}\delta(\omega_f - \omega_0)\right)\left(\frac{1}{(2\pi)^3 2p_f^0}\right) , \tag{11.78}$$

or integrating over ω_f,

$$\frac{d\sigma}{d\Omega} = \frac{\alpha_0^2}{2m^2}\left(\frac{\omega_f}{\omega_i}\right)^2\left[\frac{\omega_f}{\omega_i} + \frac{\omega_i}{\omega_f} + 4(\epsilon_f \cdot \epsilon_i)^2 - 2\right] \tag{11.79}$$

where we have made use of s defined in (10.24), which reduces in the laboratory frame to $s - m^2 = 2m\omega_i$. This is the *Klein–Nishina* scattering cross-section for scattering of photons off electrons at rest.

If the incident photons are unpolarized and the polarization of the final photons remains undetected, then we must average over the two helicities of the incoming and outgoing photons and divide by 2. We do the same for the electrons. Using (5.14) of Chapter 6 we have

$$\sum_{\lambda_f}\sum_{\lambda_i}(\vec{\epsilon}(k_f, \lambda_f) \cdot \vec{\epsilon}(k_i, \lambda_i))^2 = [(\hat{k}_f \cdot \hat{k}_i)^2 - \hat{k}_f^2 - \hat{k}_i^2 + 3] = (\cos^2\theta + 1) ,$$

or

$$\overline{\frac{d\sigma}{d\Omega}} = \frac{1}{4}\sum_{\lambda_f,\lambda_i}\sum_{\sigma_f,\sigma_i}\frac{d\sigma}{d\Omega} = \frac{\alpha^2}{2m^2}\left(\frac{\omega_f}{\omega_i}\right)^2\left[\frac{\omega_f}{\omega_i} + \frac{\omega_i}{\omega_f} - \sin^2\theta\right] . \tag{11.80}$$

Chapter 12

Parametric Representation of a General Diagram

In the previous chapter we have repeatedly used the representation of 1-loop diagrams in terms of integrals over the Feynman parameters ranging over the finite domain (0,1). The resulting integrals had the virtue of exhibiting explicitly the Lorentz-covariant structure with respect to the *external* momenta, the result having been of the form

$$I(p) = \int \prod_i d\alpha_i \; \delta \left(1 - \sum_i \alpha_i \right) \frac{N(\alpha; p)}{D(\alpha; p)} \; , \tag{12.1}$$

where $N(\alpha; p)$ was given by a (matrix valued) polynomial resulting from the spin of the electron. We now show how to obtain such a representation for a general Feynman diagram involving an arbitrary number of loops.[1]

12.1 Cutting rules for a general diagram

We shall avoid the complications introduced by the spin of a particle, by restricting ourselves to the case of spin-zero particles. Taking all external momenta to be incoming, a general Feynman diagram will be given in terms of the integral

$$I_G(p) = \int \prod_{i=1}^I \left[\frac{d^4 k_i}{(2\pi)^4} \left(\frac{i}{k_i^2 - m_i^2 + i\epsilon} \right) \right] \prod_{v=1}^V (2\pi)^4 \delta^4 \left(P_v - \sum_{l=1}^I \epsilon_{vi} k_i \right) \tag{12.2}$$

where our notation is as follows:

---

[1]We follow here C. Itzykson and J.-B. Zuber, *Quantum Field Theory* (McGraw-Hill Inc., 1980).

V = Number of vertices

I = Number of internal lines

L = Number of independent loops

P_v = External momentum *entering* a vertex v

k_l = Momentum associated with the lth *internal* line

$$\epsilon_{vi} \begin{cases} 1, & \text{if the momentum } k_i \text{ } \textit{leaves} \text{ the vertex } v \\ -1, & \text{if the momentum } k_i \text{ } \textit{enters} \text{ the vertex } v \\ 0, & \text{if the momentum } k_i \text{ neither leaves nor enters the vertex } v \end{cases}$$

The "incidence matrix" ϵ_{vi} evidently satisfies

$$\sum_{v=1}^{V} \epsilon_{vi} = 0 \ . \tag{12.3}$$

The delta-function in (12.2) implements energy-momentum conservation at each vertex. From (12.3) and the delta function in (12.2) follows

$$\sum_{v=1}^{V} P_v = 0 \ .$$

Since the delta functions at the vertices reduce the number of integrations by $V - 1$ (the last integration leaves an overall δ-function) we also have

$$L = I - (V - 1) \ . \tag{12.4}$$

In order to obtain the desired representation for (12.2) we begin by representing the individual Feynman propagators as

$$\frac{i}{k_i^2 - m_i^2 + i\epsilon} = \int_0^\infty d\beta_i e^{i\beta_i(k_i^2 - m_i^2 + i\epsilon)} \ . \tag{12.5}$$

For the energy-momentum conserving delta function we use the Fourier representation

$$(2\pi)^4 \delta^4 \left(P_v - \sum_{i=1}^{I} \epsilon_{vi} k_i \right) = \int d^4 y_v e^{-iy_v \cdot (P_v - \sum_i \epsilon_{vi} k_i)} \ . \tag{12.6}$$

Introduce (12.5) and (12.6) into (12.2) and perform the individual k_i-integrations by noting that

$$\int \frac{d^4 k_i}{(2\pi)^4} e^{i\beta_i \left(k_i^2 + \frac{1}{\beta_i} \sum_v y_v \epsilon_{vi} \cdot k_i \right)} = \left[\int \frac{d^4 k_i'}{(2\pi)^4} e^{i\beta_i k_i'^2} \right] e^{-i\beta_i \left(\frac{\sum_v y_v \epsilon_{vi}}{2\beta_i} \right)^2} \ .$$

With

$$\int_{-\infty}^\infty d^4 k e^{i\beta_i(k_0^2 - k_1^2 - k_2^2 - k_3^2)} = \left(\frac{\pi}{-i\beta_i} \right)^{1/2} \left(\frac{\pi}{i\beta_i} \right)^{3/2} = \frac{-i\pi^2}{\beta_i^2}$$

we have for (12.2),

$$I_G(p) = \int \prod_{v=1} d^4 y_v e^{-i \sum_v y_v \cdot P_v} \int_0^\infty \prod_{i=1}^I \left(d\beta_i \frac{e^{-i[\beta_i m_i^2 + (\sum_v y_v \epsilon_{vi})^2 / 4\beta_i]}}{i(4\pi\beta_i)^2} \right).$$

This integral is expected to contain an overall delta function expressing the conservation of the energy and momentum associated with the external momenta. In order to make this explicit, we perform the following shift in the integration variables $y_1...y_{v-1}$:

$$y_1 = z_1 + y$$
$$y_2 = z_2 + y$$

$$\cdot$$

$$\cdot$$

$$y_{v-1} = z_{v-1} + y$$
$$y_v = y.$$

The Jacobian of this transformation is *unity*. Making use of (12.3) we then have

$$I_G(p) = (2\pi)^4 \delta^4 \left(\sum_v P_v \right) \tilde{I}_G(p) , \tag{12.7}$$

where

$$\tilde{I}_G(p) = \int \prod_{v=1}^{V-1} d^4 z_v \int_0^\infty \prod_{i=1}^I \left[d\beta_i \frac{e^{-i\beta_i m_i^2}}{i(4\pi\beta_i)^2} \right] e^{-i \left(\sum_{i=1}^I \frac{\left(\sum_{v=1}^{V-1} z_v \epsilon_{vi}\right)^2}{4\beta_i} + \sum_{v=1}^{V-1} z_v \cdot P_v \right)}.$$

$$\tag{12.8}$$

The exponent in the integrand involves a quadratic form $q(z; \beta)$ in the integration variables z_v. Defining a $(V-1) \times (V-1)$ dimensional matrix $d_G(\beta)$ with matrix elements

$$[d_G(\beta)]_{v_1 v_2} = \sum_{i=1}^I \epsilon_{v_1 i} \frac{1}{\beta_i} \epsilon_{v_2 i} ,$$

we may write the expression in brackets in the exponential as

$$q(z, \beta) = \frac{1}{2} \sum_{v_1=1}^{V-1} \sum_{v_2=1}^{V-1} z_{v_1} \frac{d_G(\beta)}{2} z_{v_2} + \sum_{v=1}^{V-1} z_v \cdot P_v .$$

We now make use of the fact that the matrix $d_G(\beta)$ is *symmetric*, and hence may be diagonalized. Denoting the diagonal form of a symmetric matrix A_{ij} by a superscript D, we have quite generally $(x_i \to y_i, J_i \to \tilde{J}_i)$

$$\int \prod_{i=1}^n dx_i e^{-\frac{i}{2} \sum_{i,j} x_i A_{ij} x_j - i J_i x_i} = \int \prod_{i=1}^n dy_i e^{-\frac{i}{2} \sum_{i,j} y_i A_{ii}^D y_i - i \sum_i \tilde{J}_i y_i}$$

$$= \left(\prod_i \sqrt{\frac{2\pi}{iA_{ii}^D}} \right) e^{\frac{i}{2}\tilde{J}(A^D)^{-1}\tilde{J}} = (\sqrt{-i2\pi})^n \frac{e^{\frac{i}{2}JA^{-1}J}}{\sqrt{\det A}} \ .$$

Making use of this formula, and taking account of the fact that we are dealing with a four-dimensional integration,[2] we obtain from (12.8), after proper identifications,

$$\tilde{I}_G(p) = \int_0^\infty \left(\prod_{l=1}^I d\beta_i \right) \frac{e^{i\left(Q_G(p;\beta) - \sum_{i=1}^I \beta_i m_i^2 + i\epsilon \right)}}{\left[i(4\pi)^2 \right]^L \mathcal{P}_G(\beta)^2} \tag{12.9}$$

where

$$Q_G(p;\beta) = \sum_{v_1,v_2}^{V-1} d_G^{-1}(\beta)_{v_1 v_2} (P_{v_1} \cdot P_{v_2}) \tag{12.10}$$

and

$$\mathcal{P}_G(\beta) = \beta_1 \cdots \beta_I \det[d_G(\beta)] \ . \tag{12.11}$$

The computation of $\det d_G(\beta)$ requires the introduction of the notion of the *tree* graph corresponding to a given Feynman diagram.

Definition:

A tree graph corresponding to a general diagram G is a connected *subdiagram* (no vertex is isolated) of G which contains *all* vertices of G but *no* loops.

Example 1

With the box diagram

Fig. 12.1. Box diagram.

are associated four tree diagrams such as, for example,

Fig. 12.2. Tree diagram associated with box diagram.

Note that if one were to cut on further internal line, the tree diagram would separate into two disconnected pieces.

If we would reintroduce Planck's constant, the tree graphs would correspond to contributions of *zero* order in $\hbar$. Making use of (12.4) we see that G involves

---

[2] $n = V - 1$ and we have four copies $(y_1, \cdots y_4)$. Hence, with $A = d_G(\beta)$

$$\frac{(\sqrt{-i2\pi})^n}{\sqrt{\det A}} \to \left[(4\pi)^2 \right]^{V-1} \frac{1}{\left(\det d_G(\beta) \right)^2} \ .$$

Use further $L = I - (V - 1)$.

$I - L = V - 1$ internal lines, as one readily checks for the above diagrams. Since a tree graph has no loops, $L = I - (V - 1) = 0$, or the corresponding "incidence" matrix (ϵ_{vi}) for a tree diagram is thus a $V \times (V - 1)$ *matrix of rank* $(V - 1)$.

We are now in the position of stating[3] (without proof) the result for the determinant appearing in (12.11):

$$\det[d_G(\beta)] = \sum_{Tree} \left(\prod_{i \in Tree} \frac{1}{\beta_i} \right) . \tag{12.12}$$

Note that the result is given in terms of a sum over all tree diagrams associated with a given Feynman diagram, each term in the sum involving the product of Feynman parameters associated with the *internal* lines of the respective (connected) *tree* diagrams. It follows that

$$\mathcal{P}_G(\beta) := \beta_1 \cdots \beta_I \det d_G[\beta] = \sum_{Tree} \left(\prod_{i \notin Tree} \beta_i \right) \tag{12.13}$$

"$i \notin Tree$" refers to the internal lines of G to be cut in order to arrive at the tree diagram. Note that $\mathcal{P}_G(\beta)$ is a *homogeneous* polonomial of degree $I - (V - 1) = L$, with L the number of independent loops.

Next, one can show that (12.10) has the form

$$Q_G(p; \beta) = \frac{1}{\mathcal{P}_G(\beta)} \sum_C s_c \left(\prod_{i \in C} \beta_i \right) , \tag{12.14}$$

where the sum runs over all possible cuts C of $L + 1$ internal lines of the diagram separating this diagram into two *connected* tree diagrams $G_1(C)$ and $G_2(C)$, and where s_c denotes the *4-vector square* of the sum of *external momenta* entering $G_1(C)$(or, by energy-momentum conservation, equivalently $G_2(C)$):

$$s_c = \left(\sum_{v \in G_1(C)} P_v \right)^2 = \left(\sum_{v \in G_2(C)} P_v \right)^2 .$$

Finally, making use of the homogeneity properties of $Q_G(p, \beta)$ and $\mathcal{P}_G(\beta)$

$$Q_G(p, \lambda\beta) = \lambda Q_G(p, \beta) \tag{12.15}$$
$$\mathcal{P}_G(\lambda\beta) = \lambda^L \mathcal{P}_G(\beta) ,$$

we may reduce the integration in (12.9) to the range (0,1). To this end we introduce again the identity

$$1 = \int_0^\infty \frac{d\lambda}{\lambda} \delta \left(1 - \frac{1}{\lambda} \sum_{i=1}^I \beta_i \right)$$

---

[3] N. Nakanishi, *Graph Theory and Feynman Integrals* (Gordon and Breach, New York, 1970).

and rescale the integration variables as follows:

$$\beta_i = \lambda \alpha_i . \tag{12.16}$$

We then obtain from (12.9) and (12.15),

$$\tilde{I}_G(p) = \frac{1}{[i(4\pi)^2]^L} \int_0^1 \prod_{i=1}^{I} d\alpha_i \frac{\delta(1 - \sum_i \alpha_i)}{[\mathcal{P}_G(\alpha)]^2} \int_0^\infty \frac{d\lambda}{\lambda} \lambda^{\overbrace{I - 2L}^{2V-I-2}} e^{i\lambda[Q_G(p,\alpha) - \sum_i \alpha_i m_i^2 + i\epsilon]} . \tag{12.17}$$

Since $L = I - (V - 1) = I - L = V - 1$ it follows that[4]

$$\tilde{I}_G(p) = \frac{(i)^{3V-2I-3}}{[(4\pi)^2]^L} \Gamma(2V - I - 2) \int_0^1 \prod_{i=1}^{I} d\alpha_i \tag{12.18}$$

$$\times \frac{\delta(1 - \sum_1^I \alpha_i)}{[\mathcal{P}_G(\alpha)]^2 \left[Q_G(p,\alpha) - \sum_{i=1}^I \alpha_i m_i^2 + i\epsilon \right]^{2V-I-2}} .$$

The divergence of $\Gamma(2V - I - 2)$ for vanishing argument reflects the logarithmic divergence of the original integral (12.17) for $4L - 2I = 2(I - 2(V - 1))$. In case of such a divergence we must return to (12.17) and regularize the integral in λ by introducing a cutoff. Notice that now $\alpha \epsilon [0, 1]$ because of the δ-function.

Example 2

Consider the vertex diagram of ϕ^3 theory:

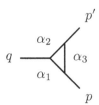

There are three tree graphs associated with it:

Fig. 12.3. Tree graphs of vertex diagram.

---

[4] We use the definition

$$\Gamma(n) = \int_0^\infty dx \, x^{n-1} e^{-x} .$$

Notice that $\Gamma[2V - I - 2] = \infty$ for $2V - I - 2 = -n (n = 0, 1, 2...)$. This reflects the divergence of the λ-integration at the lower endpoint, which in turn is a reflection of the UV divergence of the original integral!

We have correspondingly

$$\mathcal{P}(\alpha) = \alpha_1 + \alpha_2 + \alpha_3$$
$$Q(\alpha) = q^2 \alpha_1 \alpha_2 + m^2 \alpha_1 \alpha_3 + m^2 \alpha_2 \alpha_3 ,$$

so that we have the integral representation

$$\tilde{I}_{vertex} = \frac{1}{(4\pi)^2} \int_0^1 \prod_0 d\alpha_i \frac{\delta(\alpha_1 + \alpha_2 + \alpha_3 - 1)}{[q^2 \alpha_1 \alpha_2 + m^2 \alpha_1 \alpha_3 + m^2 \alpha_2 \alpha_3 - m^2 + i\epsilon]} . \qquad (12.19)$$

Example 3

Consider again the box-graph in Feynman parameter space (the labelling is chosen for later comparison)

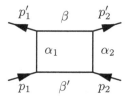

Fig. 12.4. Box-graph in Feynman-parameter space.

The corresponding integral is convergent, so that (12.18) applies. $\mathcal{P}_{box}(p)$ is computed from the sum of connected tree graphs of Example 1 to be

$$\mathcal{P}_{box}(\alpha, \beta) = \alpha_1 + \alpha_2 + \beta + \beta' \qquad (12.20)$$

and $Q_{box}(p)$ is computed from the (disconnected) tree graphs obtained by cutting one further *internal* line, to be

$$Q_{box}(p; \alpha, \beta, \beta') = \frac{\alpha_1 \alpha_2 s + \Delta(t; \alpha, \beta, \beta')}{\mathcal{P}_{box}(\alpha, \beta)} \qquad (12.21)$$

where

$$s = (p_1 + p_2)^2 , \quad t = (p_1' - p_1)^2$$

and

$$\Delta(t; \alpha, \beta, \beta') = \beta\beta' t + m_0^2 (\alpha_1 + \alpha_2)(\beta + \beta'). \qquad (12.22)$$

From (12.18) we thus have

$$\tilde{I}_{box} = \frac{i}{(4\pi)^2} \Gamma(2) \int_0^1 \prod_{i=1}^2 d\alpha_i (d\beta d\beta') \frac{\delta(\alpha_1 + \alpha_2 + \beta + \beta' - 1)}{[\alpha_1 \alpha_2 s + \Delta(t; \alpha, \beta, \beta') - m_0^2 + i\epsilon]} \qquad (12.23)$$

where we have set $\mathcal{P}_{box}(\alpha, \beta) = 1$ because of the delta function.

12.2   An alternative approach to cutting rules

The results (12.17) and (12.18) are well suited for getting a quick analytic expression for (12.1) when $N(\alpha; p) = 1$. The following approach also allows to include a polynomial $N(\alpha; p)$. The idea is based on the analogy to electrical circuits.[5]

---

[5] J.D. Bjorken and S.D. Drell, *Relativistic Quantum Fields* (McGraw-Hill Inc., 1965).

Consider the diagram

Fig. 12.5. Two current loops.

Let R be the number of independent loops (rungs) in a diagram. With the rth rung we associate a loop momentum l_r, which is eventually to be integrated over. Making a change of variable

$$k_i = K_i + \sum_{r=1}^{R} \eta_{ir} l_r \tag{12.24}$$

with K_i a function of the external variables p, p', and η_{ir} constants to be suitably chosen, we obtain for the integral (12.2),

$$I_G(p) \sim \int \frac{d^4 l_1 \cdots d^4 l_R}{(k_1^2 - m_1^2) \cdots (k_I^2 - m_I^2)}$$

$$= \int d^4 l_1 \cdots d^4 l_R \int_0^1 d\alpha_1 \cdots d\alpha_I \delta \left(1 - \sum_{j=1}^{I} \alpha_j \right)$$

$$\times \frac{(I-1)!}{[\sum_i (K_i^2 - m_i^2)\alpha_i + 2\sum_{j,r} K_j \alpha_j \eta_{jr} l_r + \sum_{jrr'} \alpha_j \eta_{jr} \eta_{jr'} l_r l_{r'}]^I}$$

where use has been made of the Feynman formula (15.4) (see Chapter 15 for a proof):

$$\prod_{j=1}^{I} \frac{1}{a_j + i\epsilon} = (I-1)! \int_0^1 d\alpha_1 ... d\alpha_I \frac{\delta(1 - \sum_j \alpha_j)}{(\sum_j \alpha_j a_j + i\epsilon)^I} . \tag{12.25}$$

Eliminating the mixed term by requiring

$$\sum_{j=1}^{I} K_j \alpha_j \eta_{jr} = 0, \quad r = 1 \cdots R \tag{12.26}$$

one finds, after integration over the loop momenta[6]

$$I_G(P) \sim \frac{\int_0^1 d\alpha_1 \cdots d\alpha_I \delta(1 - \sum_{j=1}^{I} \alpha_j)}{\Delta^2 [\sum_j (K_j^2 - m_j^2)\alpha_j]^{I-2R}} \tag{12.27}$$

where

$$\Delta = \det ||z||$$

with

$$z_{rr'} = \sum_{j=1}^{I} \eta_{jr} \eta_{jr'} \alpha_j .$$

---

[6] J.D. Bjorken and S.D. Drell, *Relativistic Quantum Fields* (McGraw-Hill, Inc.).

Since $R = L$ the result (12.27) agrees with (12.18).

In order to get a more detailed understanding of this result, let us return to our example of the triangle graph represented *on shell*, $p^2 = p'^2 = m^2$, by the integral

$$I(p) = \int \frac{d^4k}{(2\pi)^4} \frac{1}{(k^2 - 2p \cdot k)(k^2 - 2p' \cdot k)(k^2 - m^2) + i\epsilon}$$

$$= 2! \int \frac{d^4k}{(2\pi)^4} \frac{\int_0^1 d\alpha_1 \int_0^1 d\alpha_2 \int_0^1 d\alpha_3 \delta(1 - \alpha_1 - \alpha_2 - \alpha_3)}{[\alpha_1(k^2 - 2p \cdot k) + \alpha_2(k^2 - 2p' \cdot k) - \alpha_3(k^2 - m^2) + i\epsilon]^3}.$$

Our general result (12.27) evidently corresponds to a block-diagonalization of the denominator D, which, taking account of the delta function, we rewrite as

$$\begin{aligned}
D &= [k - (\alpha_1 p + \alpha_2 p')]^2 - (\alpha_1 p + \alpha_2 p')^2 - \alpha_3 m^2 \\
&= k'^2 - \alpha_1^2 m^2 - \alpha_2^2 m^2 - 2\alpha_1\alpha_2 p \cdot p' - \alpha_3 m^2 \\
&= k'^2 - \alpha_1^2 m^2 - \alpha_2^2 m^2 + \alpha_1\alpha_2(p - p')^2 - 2\alpha_1\alpha_2 m^2 - \alpha_3 m^2 \\
&= q^2\alpha_1\alpha_2 + \alpha_1\alpha_3 m^2 + \alpha_2\alpha_3 m^2 - m^2 + k'^2
\end{aligned}$$

where (k' corresponds to ℓ)

$$k' = k - (\alpha_1 p + \alpha_2 p') . \tag{12.28}$$

Integrating this result over k'^2 yields the result (12.19).

Let us take another look at this result from the point of view of (12.27).

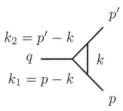

The above calculation suggests what change of variable (12.24) is to be performed:

$$\begin{aligned}
k_1 &\equiv p - k = K_1(q, p, p') - k' \\
k_2 &\equiv p' - k = K_2(q, p, p') - k' \\
k_3 &\equiv K_3(q, p, p') + k'
\end{aligned} \tag{12.29}$$

with

$$\begin{aligned}
K_1(q, p, p') &= p - \alpha_1 p - \alpha_2 p' \\
K_2(q, p, p') &= p' - \alpha_1 p - \alpha_2 p' \\
K_3(q, p, p') &= \alpha_1 p + \alpha_2 p'
\end{aligned} \tag{12.30}$$

and $\eta = 1$, $\ell = k'$. One checks that the mixed term vanishes, as required by (12.26):

$$\sum_i \alpha_j K_j(q, p, p') = 0.$$

Some further algebra shows that

$$\sum_{j=1}^{3}(K_j^2(q,p,p') - m^2)\alpha_i = \alpha_1 p^2 + \alpha_2 p'^2 - (\alpha_1 p + \alpha_2 p')^2 - m^2$$

$$= \alpha_1\alpha_2 q^2 + \alpha_1\alpha_3 m^2 + \alpha_2\alpha_3 m^2 - m^2 .$$

Putting things together and integrating over k (factor 2! cancels) we obtain (12.19).

The example demonstrates that, in general, the change of variable (12.24) one seeks may not be trivial. We shall make use of this change of variable when discussing the regularization and renormalization of the QED vertex function in Chapters 15 and 16, respectively.

12.3   4-point function in the ladder approximation

We now apply the results obtained in this chapter to the calculation of the asymptotic behaviour of the 4-point function of a ϕ^3-theory in the so-called ladder approximation, as given by the sum of diagrams

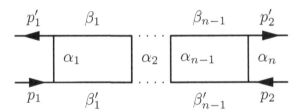

Fig. 12.6. Ladder diagram.

These diagrams are UV convergent, so that we can write for a general diagram the integral representation (12.18)

$$G_{lad}(p) = (-ig)^{2n} \tilde{I}_G^{(n)}(p) \tag{12.31}$$

with (recall definition (12.7))

$$\tilde{I}_{lad}^{(n)}(p) = \frac{i\Gamma(n)}{[(4\pi)^2]^{n-1}} \int \prod_{i}^{n} d\alpha_i \prod_{j-1}^{n-1} (d\beta_j d\beta_j') \tag{12.32}$$

$$\times \frac{\delta(\sum \alpha_i + \sum \beta_j + \sum \beta_j' - 1)}{[\mathcal{P}_{lad}(\alpha;\beta,\beta')]^2 [\mathcal{Q}_{lad}(p;\alpha;\beta,\beta') - m_0^2 + i\epsilon]^n} ,$$

where we have used $V = 2n, I = 3n - 2, L = n - 1$. Let us introduce again the two Lorentz invariants,

$$s = (p_1 + p_2)^2, \, t = (p_1' - p_1)^2 .$$

We shall be interested in studying the limit $s \to \infty$ of (12.32). For this, we shall need to exhibit only the s-dependence of the homogeneous polynomial (12.14); it is obtained by considering the cut

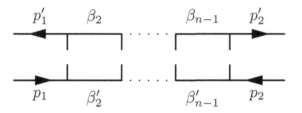

Fig. 12.7. Leading diagram for $s \to \infty$.

and has the form

$$Q_{lad}(p; \alpha, \beta, \beta') = \frac{\alpha_1 \alpha_2 \cdots \alpha_n s + \Delta(t; \alpha; \beta, \beta')}{\mathcal{P}_{lad}(\alpha; \beta, \beta')} .$$

The limit $s \to \infty$ is determined by the integration region around $\alpha_i = 0$. Hence, it is sufficient to consider the integral

$$\tilde{I}_{lad}^{(n)}(p) \to \frac{i}{\left[(4\pi)^2\right]^{n-1}} (n-1)! \int_0^1 \left(\prod_j d\beta_j d\beta_j' \right) \left(\mathcal{P}_{lad}(0; \beta, \beta') \right)^{n-2}$$

$$\times \delta \left(\sum_j \beta_j + \sum_j \beta_j' - 1 \right) \int_0^\epsilon \prod_i d\alpha_i \frac{1}{[\alpha_1 ... \alpha_n s + \bar{\Delta}(t; \beta, \beta')]^n} ,$$

with

$$\overline{\Delta}(t; \beta, \beta') = \Delta(t; 0; \beta, \beta') - m_0^2 \mathcal{P}_{lad}(0; \beta, \beta').$$

Note that the delta function no longer involves the α-*parameters*. One can do the α-integrations explicitly,[7] obtaining for (12.31) and $s \to \infty$, in the *leading log approximation*

$$G_{lad}^{(n)}(p) \to -ig^2 \frac{1}{s} \frac{(\ln s)^{n-1}}{(n-1)!} K_n(t) , \tag{12.33}$$

where

$$K_n(t) = \left(\frac{-g^2}{(4\pi)^2} \right)^{n-1} (n-2)! \int_0^1 \prod_{j=1}^{n-1} d\beta_j d\beta_j' \frac{\delta(\sum_j \beta_j + \sum \beta_j' - 1) \left[\mathcal{P}_{lad}(0; \beta, \beta') \right]^{n-2}}{\left[\overline{\Delta}(t; \beta, \beta') \right]^{n-1}} . \tag{12.34}$$

---

[7] See R.J. Eden, P. Landshoff, D.I. Olive and J.C. Polkinghorne, *The Analytic S-Matrix*.

For the computation of $\mathcal{P}_G(0, \beta, \beta')$ we only need the connected tree diagrams obtained from Fig. 12.6 by cutting $n-1$ internal β lines: with a cut of $n-1$ internal β-lines:

$$\mathcal{P}_{lad}(0, \beta, \beta') = \prod_{j=1}^{n-1}(\beta_j + \beta'_j) = \prod_{j=1}^{n-1} \mathcal{P}_j(0, \beta, \beta') \ .$$

The computation of $\Delta(t; 0, \beta, \beta')$ entering Q_G involves the sum of diagrams resulting from one further cut, which necessarily implies the presence of simultaneous cuts of a β_i and β'_i line. It is not hard to see that $\Delta(t, 0; \beta, \beta')$ has the form

$$\Delta(t, 0; \beta, \beta') = \sum_{j=1}^{n-1} \Delta_j(t; 0, \beta, \beta') \prod_{k \neq j} \mathcal{P}_k(0, \beta, \beta') \ , \tag{12.35}$$

where

$$\Delta_j(t; 0, \beta, \beta') = t\beta_j \beta'_j \ . \tag{12.36}$$

Regge behaviour

We now show that $K_n(t)$ factorizes. Make the change of variable (scale transformation)

$$\beta_j = \rho_j \bar{\beta}_j, \ \beta'_j = \rho_j(1 - \bar{\beta}_j) =: \rho_j \bar{\beta}'_j.$$

The Jacobian of the transformation is $J_j = \rho_j$. Hence

$$d\beta_j d\beta'_j = \rho_j d\rho_j d\bar{\beta}_j d\bar{\beta}'_j \delta(\bar{\beta}_j + \bar{\beta}'_j - 1)$$

$$\Delta_j(t; 0, \beta_j, \beta'_j) = \rho_j^2 \Delta_j(t, 0; \bar{\beta}_j, \bar{\beta}'_j) = t\rho_j^2 \bar{\beta}_j \bar{\beta}'_j$$

$$\mathcal{P}_{lad}(0, \beta, \beta') = \prod_{k=1}^{n-1} \rho_k \ ,$$

$$\frac{\Delta(t, 0; \beta, \beta')}{\mathcal{P}_{lad}(0, \beta, \beta')} = \sum_j \rho_j \Delta_j(t, 0; \bar{\beta}_j, \bar{\beta}'_j) \ .$$

We thus obtain for the integral in (12.34)

$$\mathcal{I} = \int_0^1 \prod_{j=1}^{n-1}(d\beta_j d\beta'_j) \frac{\delta(\sum \beta_j + \sum \beta'_j - 1)}{\mathcal{P}_{lad}(0; \beta, \beta') \left[\frac{\Delta(t;0;\beta,\beta')}{\mathcal{P}_{lad}(0;\beta,\beta')} - m_0^2 \right]^{n-1}}$$

$$= \int_0^1 \left[\prod_{k=1}^{n-1} d\bar{\beta}_k d\bar{\beta}'_k d\rho_k \right] \delta\left(\sum \rho_j - 1 \right) \frac{\prod_{j=1}^{n-1} \delta(\bar{\beta}_j + \bar{\beta}'_j - 1)}{\left[\sum_j^{n-1} \rho_j \Delta_j(t; 0; \bar{\beta}_j, \bar{\beta}'_j) - m_0^2 \right]^{n-1}} \ .$$

Making use of the inverse of the Feynman formula (12.25) in Chapter 15, we have

$$\int_0^1 \prod_{k=1}^{n-1} d\rho_k \frac{\delta(\sum_j \rho_j - 1)}{\left[\sum_j^{n-1} \rho_j \Delta_j(t; 0; \bar{\beta}_j, \bar{\beta}'_j) - m_0^2 \right]^{n-1}}$$

$$= \frac{1}{(n-2)!} \prod_{j=1}^{n-1} \left(\frac{i}{\Delta_j(t;0;\bar{\beta}_j,\bar{\beta}'_j) - m_0^2} \right) ,$$

and we finally arrive at

$$K_n(t) = \left[\frac{-g^2}{(4\pi)^2} \right]^{n-1} \prod_{j=1}^{n-1} \left\{ \int_0^1 d\bar{\beta}_j d\bar{\beta}'_j \frac{\delta(\bar{\beta}_j + \bar{\beta}'_j - 1)}{\Delta_j(t;0;\bar{\beta}_j,\bar{\beta}'_j) - m^2} \right\} .$$

Hence

$$K_n(t) = K^{n-1}(t)$$

with

$$K(t) = \left[\frac{-g^2}{(4\pi)^2} \right] \left\{ \int_0^1 d\bar{\beta}_j d\bar{\beta}'_j \frac{\delta(\bar{\beta}_j + \bar{\beta}'_j - 1)}{t\bar{\beta}_j\bar{\beta}'_j - m_0^2} \right\} . \tag{12.37}$$

Thus we have from (12.33) in the limit $s \to \infty$,

$$\tilde{I}_{lad}^{(n)}(s,t) \to -g^2 s^{-1} \frac{\left(K(t)\ln s \right)^{n-1}}{(n-1)!} .$$

In the leading log-approximation, the sum of the ladder diagrams is therefore given by the exponential series

$$G_{lad}^{(n)}(s,t) \sim -g^2 s^{-1} \sum_{n=1}^{\infty} \frac{1}{(n-1)!} \left(K(t)\ln s \right)^{n-1} ,$$

which can be summed to give

$$G_{lad}^{(n)}(s,t) \sim -g^2 s^{\alpha(t)} , \tag{12.38}$$

where

$$\alpha(t) = -1 + K(t) .$$

The function $\alpha(t)$ is called *Regge trajectory*. This sample calculation thus provides an explicit example for what has been termed *Regge behaviour* of scattering amplitudes.[8] Such Regge behaviour, and the associated *Regge poles*, have been the subject of intensive studies in the sixties, and developed into a field of research of its own, referred to as S-matrix theory.

---

[8]T. Regge, *Nuovo Cimento* **14**, (1959) 951; P.D.B. Collins, *An Introduction to Regge Theory and High Energy Physics* (Cambridge University Press, Cambridge, 1977).

Chapter 13

Functional Methods

The derivation of the Feynman rules in Chapter 11 has been based on operator methods in the interaction picture. The rules allow for the perturbative computation of the Green functions. The corresponding perturbative series is known to be at best an *asymptotic* series in the coupling constant with zero radius of convergence around the origin of the complex coupling-constant plane. It thus reflects the true content of the theory only as long as it provides a good approximation in the sense of an asymptotic series for sufficiently small coupling constant. It furthermore presupposes the existence of a *unique* vacuum given by the *Fock vacuum*. This appears to be the case for *Quantum Electrodynamics (QED)*, where perturbation theory provides an excellent description of the experimental results. As we know however, this reflects only partially, in general, a much more complex situation. Thus it is known that the *QCD* vacuum is highly non-trivial, and is in fact a condensate of quark–antiquark pairs, very much reminiscent of the superconducting ground state in superconductors known to be populated by spin-zero electron–electron (Cooper) pairs. Many of the modern developments in Quantum Field Theory relating to such non-perturbative phenomena would have been missed had one adhered to perturbation theory in the formulation of Chapter 11. In fact, for a very small class of models in 1+1 dimensions (Thirring model, QED_2) exact solutions have been constructed using operator methods; they exhibit non-trivial properties lying outside the scope of perturbation theory.[1] From here one learned that Quantum Field Theories in 3+1 dimensions can be expected to exhibit highly non-trivial non-perturbative features. It was only the *functional* approach to QFT which opened the possibility for studying such non-perturbative phenomena in "realistic" theories. This approach, dispensing of the notion of an underlying "Fock space", will be the subject of this chapter. We shall show how one recovers the perturbative series embodied by Eq. (11.21), and prepare the ground for new approximation schemes paying tribute to "non-perturbative" phenomena, as exemplified in Chapter 20.

---

[1] E. Abdalla, M. Christina B. Abdalla and K.D. Rothe, *Non-Perturbative Methods in 2 Dimensional Quantum Field Theory* (World Scientific, 1991, and extended version 2001).

13.1   The Generating Functional

For the sake of simplicity, we shall restrict our discussion in this section to the case of a self-interacting real scalar field $\phi(x)$. We define[2]

$$\mathcal{Z}[j] = <\Omega|T\; e^{i\int d^4x\phi(x)j(x)}|\Omega> \;, \tag{13.1}$$

where $j(x)$ is an external c-*number* source, and $\mathcal{Z}[0] = 1$. We define the functional derivative $\delta/\delta j(x)$ as an operation commuting with the integration $\int d^4x$ and having the property

$$\left[\frac{\delta}{\delta j(x)}, \; j(y)\right] = \delta^4(x-y) \;. \tag{13.2}$$

Thus

$$\frac{\delta}{\delta j(x)}\int d^4y j(y)\phi(y) = \int d^4y \left[\frac{\delta}{\delta j(x)}, j(y)\right]\phi(y) = \phi(x) \;.$$

From (13.1) we thus have,

$$\frac{\delta}{i\delta j(x)}\mathcal{Z}[j]|_{j=0} = <\Omega|\phi(x)|\Omega> \;,$$

or more generally

$$\left(\frac{\delta}{i\delta j(x_n)}\cdots\frac{\delta}{i\delta j(x_1)}\mathcal{Z}[j]\right)_{j=0} = <\Omega|T\phi(x_n)...\phi(x_1)|\Omega> \;. \tag{13.3}$$

Hence $\mathcal{Z}[j]$ plays the role of a generating functional for the n-point Green functions of QFT. Our next task thus consists in obtaining an explicit representation for this generating functional. We shall discuss two constructions, due to J. Schwinger and R. Feynman, respectively. Of these, the second one will provide the basis for the chapters to follow.

13.2   Schwinger's Construction of $\mathcal{Z}[j]$

Schwinger's construction of the generating functional is based on the (Euler–Lagrange) equations satisfied by the fields in question. We exemplify the construction for the case of a real scalar field satisfying the Euler–Lagrange equations of motion

$$(\Box + m^2)\phi(x) = J(\phi(x)) \tag{13.4}$$

associated with a lagrangian density

$$\mathcal{L} = \frac{1}{2}\partial_\mu\phi\partial^\mu\phi - \frac{1}{2}m^2\phi^2 + \mathcal{L}_I(\phi) \;, \tag{13.5}$$

with

$$J(\phi(x)) = \frac{\partial\mathcal{L}_I}{\partial\phi(x)} \;, \tag{13.6}$$

---

[2] J. Schwinger, *Quantum Electrodynamics* (Dover Press, New York, 1958).

some polynomials of the fields acting as the source of the interaction. From (13.1) we have (note that this time we have not set $j = 0$)

$$\frac{1}{i}\frac{\delta}{\delta j(x)}\mathcal{Z}[j] = <\Omega|T_{x^0}^\infty \phi(x)T_{-\infty}^{x^0}|\Omega>,$$

where we have introduced

$$T_a^b[j] = T e^{i\int_a^b dt L_j(t)},$$

with

$$L_j(t) = \int d^3x \mathcal{L}_j(x), \quad \mathcal{L}_j(x) = \phi(x)j(x)$$

and the time-ordering operation defined by

$$T_a^b[j] = \sum_{n=0}^{\infty}\frac{i^n}{n!}\int_{t_a}^{t_b} dt_1...dt_n T(L_j(t_1)...L_j(t_n))$$

$$= 1 + \sum_{n=1}^{\infty}i^n\int_{t_a}^{t_b} dt_1\int_{t_a}^{t_1} dt_2...\int_{t_a}^{t_{n-1}} dt_n L_j(t_1)L_j(t_2)...L_j(t_n).$$

It follows from here that

$$\frac{\partial}{\partial x^0}T_{x^0}^\infty = T_{x^0}^\infty(-iL_j(x^0))$$

and

$$\frac{\partial}{\partial x^0}T_{-\infty}^{x^0} = iL_j(x^0)T_{-\infty}^{x^0}.$$

We thus obtain

$$\frac{\partial}{\partial x^0} <\Omega|T_{x^0}^\infty \phi(x)T_{-\infty}^{x^0}|\Omega> = <\Omega|T_{x^0}^\infty \partial_0\phi(x)T_{-\infty}^{x^0}|\Omega>$$

$$+ i <\Omega|T_{x^0}^\infty[\phi(\vec{x},x^0),L_j(x^0)]T_{-\infty}^{x^0}|\Omega>.$$

Now,

$$[\phi(x),L_j(x^0)] = \int d^3y[\phi(\vec{x},x^0),\phi(x^0,\vec{y})]j(\vec{y},x^0) = 0,$$

so that

$$\frac{\partial}{\partial x^0} <\Omega|T_{x^0}^\infty \phi(x)T_{-\infty}^{x^0}|\Omega> = <\Omega|T_{x^0}^\infty \partial_0\phi(x)T_{-\infty}^{x^0}|\Omega>.$$

Differentiating once again and making use of the ETC

$$[L_j(x^0),\partial_0\phi(\vec{x},x^0)] = ij(x)$$

we similarly obtain

$$\frac{\partial^2}{\partial x_0^2} <\Omega|T_{x^0}^\infty \phi(x)T_{-\infty}^{x^0}|\Omega> = <\Omega|T_{x^0}^\infty \partial_0^2\phi(x)T_{-\infty}^{x^0}|\Omega> + j(x) <\Omega|T_{-\infty}^\infty|\Omega>,$$

or

$$(\Box_x + m^2)\frac{\delta}{i\delta j(x)}\mathcal{Z}[j] = <\Omega|T_{x^0}^{\infty}[(\Box_x + m^2)\phi(x)]T_{-\infty}^{x^0}|\Omega> +j(x)\mathcal{Z}[j] \ . \qquad (13.7)$$

We thus finally have with (13.4),

$$K_x \frac{\delta}{i\delta j(x)}\mathcal{Z}[j] = J\left(\frac{\delta}{i\delta j(x)}\right)\mathcal{Z}[j] + j(x)\mathcal{Z}[j] \qquad (13.8)$$

where

$$K_x = \Box_x + m^2 \ .$$

The non-interacting case

In the absence of an interaction, Eq. (13.8) reads

$$K_x \frac{\delta}{i\delta j(x)}\mathcal{Z}^{(0)}[j] = j(x)\mathcal{Z}^{(0)}[j] \ . \qquad (13.9)$$

The form of this equation suggests making the exponential Ansatz

$$\mathcal{Z}^{(0)}[j] = e^{iW^{(0)}[j]}$$

with the "initial condition"
$$W^{(0)}[0] = 0 \ .$$

This leads us to solve the equation

$$K_x \frac{\delta}{\delta j(x)}W^{(0)}[j] = j(x) \ , \qquad (13.10)$$

having the solution (see (11.32))

$$\frac{\delta}{\delta j(x)}W^{(0)}[j] = -\int d^4 y \Delta_F(x - y; m^2)j(y) \ .$$

Note that we have chosen for the Green function the Feynman prescription. The functional differential Eq. (13.10) has the solution

$$W^{(0)}[j] = -\frac{1}{2}\int d^4 x d^4 y \, j(x)\Delta_F(x - y; m^2)j(y) \ .$$

We thus conclude that for the non-interacting case

$$\mathcal{Z}^{(0)}[j] = e^{\frac{i}{2}\int d^4 x \int d^4 y \, j(x)K^{-1}(x-y)j(y)} \ ,$$

where $K^{-1}(z)$ is the inverse of the Klein–Gordon operator. Note that $\mathcal{Z}^{(0)}[0] = 1$ in accordance with $<\Omega|\Omega>= 1$.

We now turn to the case where an interaction is present.

The interacting case

For the case where $\mathcal{L}_I(\phi)$ does not vanish, we make the Ansatz

$$\mathcal{Z}[j] = N^{-1}e^{F[\frac{\delta}{i\delta j}]}\mathcal{Z}^{(0)}[j] ,$$

where N is a normalization constant to be determined below. Using (13.9) we obtain

$$K_x \frac{\delta}{i\delta j(x)}\mathcal{Z}[j] = N^{-1}e^{F[\frac{\delta}{i\delta j}]}K_x\frac{\delta}{i\delta j(x)}\mathcal{Z}^{(0)}[j]$$

$$= N^{-1}\left[e^{F[\frac{\delta}{i\delta j}]}, j(x)\right]\mathcal{Z}^{(0)}[j] + j(x)\mathcal{Z}[j] .$$

Comparing the right-hand side of this expression with the right-hand side of Eq. (13.8), we conclude that we must require

$$\left[e^{F[\frac{\delta}{i\delta j}]}, j(x)\right]\mathcal{Z}^{(0)}[j] = J\left(\frac{\delta}{i\delta j(x)}\right)e^{F[\frac{\delta}{i\delta j}]}\mathcal{Z}^{(0)}[j]. \tag{13.11}$$

Setting

$$\frac{\delta}{i\delta j(x)} = \pi_j(x),$$

we may rewrite (13.11) as

$$\left[e^{F[\pi_j]}, j(x)\right]\mathcal{Z}^{(0)}[j] = J(\pi_j)e^{F[\pi_j]}\mathcal{Z}^{(0)}[j].$$

Noting with (13.2) that

$$[j(x), \pi_j(y)] = i\delta^4(x - y)$$

we require

$$[e^{F[\pi_j]}, j(x)] = J(\pi_j)e^{F[\pi_j]} .$$

Expanding the exponential and $F[\pi_j]$ into a Taylor series, we see that this means,

$$-i\frac{\delta F[\pi_j]}{\delta\pi_j(x)} = J(\pi_j) ,$$

or recalling (13.6) we conclude,

$$F\left[\frac{\delta}{i\delta j}\right] = i\int d^4z\mathcal{L}_I\left(\frac{1}{i}\frac{\delta}{\delta j(z)}\right) \equiv iS_I\left[\frac{\delta}{i\delta j}\right] .$$

Hence

$$\mathcal{Z}[j] = N^{-1}e^{iS_I[\frac{\delta}{i\delta j}]}\mathcal{Z}^{(0)}[j]. \tag{13.12}$$

The so-far arbitrary normalization constant N is fixed by requiring $\mathcal{Z}[0] = 1$:

$$N = \left(e^{iS_I[\frac{\delta}{i\delta j}]}\mathcal{Z}^{(0)}[j]\right)_{j=0} . \tag{13.13}$$

Note that for an interaction lagrangian not involving any derivatives in the fields, we have $\mathcal{L}_I = -\mathcal{H}_I$. By expanding the exponential in (13.12) in powers of j and performing the differentiations with respect to the external source as in (13.3), one verifies that one generates the perturbative series (11.21) with N given by the sum over vacuum graphs.

13.3   Feynman Path-Integral

Schwinger's construction of the generating functional provides a closed formula from which one recovers the perturbative series representing a general n-point Green function by differentiating the generating functional n times with respect to the external source and then setting the source equal to zero, according to Eq. (13.3). We now derive an explicit representation of this generating functional in terms of an infinite dimensional (functional) integral. We shall do this first for the case of Quantum Mechanics of one degree of freedom and then extend the result to the case of an infinite number of degrees of freedom.[3]

Path-integral representation of free propagation kernel in QM

Consider the Schrödinger wave function in the coordinate representation, $\psi(q,t) = < q|\Psi(t) >$, where $|q>$ is an eigenstate of the position operator Q: $Q|q>= q|q>$. For simplicity we shall consider here the space to be one-dimensional. Using completeness of the states $|q>$,

$$\int dq|q><q| = 1 ,$$

and the fact that ($t' \geq t$; We introduce here $\hbar$.)

$$|\Psi(t')>= e^{-\frac{i}{\hbar}(t'-t)H_0}|\Psi(t)> ,$$

we have the identity

$$\psi(q',t') = \int dq K_0(q',t';q,t)\psi(q,t) ,$$

where $K_0(q',t';q,t)$ stands for

$$K_0(q',t';q,t) =< q',t'|q,t >=< q'|e^{-\frac{i}{\hbar}(t'-t)H_0}|q > \qquad (13.14)$$

with H_0 the Hamiltonian of a free particle

$$H_0 = \frac{p^2}{2m} .$$

The computation of the integral kernel proceeds by introducing a complete set of eigenstates of the momentum operators, and gives (see Chapter 1, Eq. (1.11))

$$K_0(q',t';q,t) = \left(\frac{m}{2\pi\hbar i(t'-t)} \right)^{1/2} e^{\frac{i}{\hbar}\frac{m}{2}\frac{(q'-q)^2}{t'-t}} . \qquad (13.15)$$

The kernel (13.14) has the interesting property of satisfying the following composition law:

$$K_0(q',t';q,t) = \int dq'' K_0(q',t';q'',t'')K_0(q'',t'';q,t) . \qquad (13.16)$$

---

[3]R.P. Feynman and A.R. Hibbs, *Quantum Mechanics and Path Integrals* (McGraw-Hill, 1965).

We make use of this composition law in order to obtain a *functional representation* of this kernel. To this end we subdivide the time interval $\Delta t = t' - t$ into N equal time intervals

$$\epsilon = (t' - t)/N \; ,$$

where we shall eventually let N tend to infinity. Making repeated use of the completeness of the states $|q >$, we have

$$< q'|e^{-\frac{i}{\hbar} H_0(t'-t)}|q > = \int \prod_{\ell=1}^{N-1} dq^{(\ell)} < q'|e^{-\frac{i}{\hbar} H_0 \epsilon}|q^{(N-1)} > \qquad (13.17)$$

$$\times < q^{(N-1)}|e^{-\frac{i}{\hbar} H_0 \epsilon}|q^{(N-2)} > \cdots < q^{(1)}|e^{-\frac{i}{\hbar} H_0 \epsilon}|q > \; ,$$

where, according to (13.15), we have, exactly,

$$< q^{(\ell+1)}|e^{-\frac{i}{\hbar} H_0 \epsilon}|q^{(\ell)} > = \left(\frac{m}{2\pi\hbar i\epsilon}\right)^{1/2} e^{\frac{i}{\hbar}\frac{m}{2}\frac{(q^{(\ell+1)}-q^{(\ell)})^2}{\epsilon}} \qquad (13.18)$$

for the individual factors. Using $H_0 = -L_0$, the following expression is short hand for the multiple integral (13.17) in the limit where $\epsilon \to 0$ with $N\epsilon = t' - t$ (it is now convenient to do some relabelling):

$$K_0(q_b, t_b; q_a, t_a) = \int_{q(t_a)=q_a}^{q(t_b)=q_b} Dq \, e^{\frac{i}{\hbar}\int_{t_a}^{t_b} dt \, L_0(q(t),\dot{q}(t))} \; , \qquad (13.19)$$

where we have made the identifications

$$\frac{1}{2}m\frac{(q^{(\ell+1)}-q^{(\ell)})^2}{\epsilon} \to \epsilon L_0(q^{(\ell)}, \dot{q}^{(\ell)}),$$

$$\sum_{\ell=0}^{N-1} \epsilon L_0(q^{(\ell)}, \dot{q}^{(\ell)}) \to \int_{t_a}^{t_b} dt L_0(q(t), \dot{q}(t)) \; ,$$

and where Dq is shorthand for the integration measure

$$Dq_i := \left(\frac{m}{2\pi i\hbar\epsilon}\right)^{\frac{N}{2}} \prod_{\ell=1}^{N-1} dq^{(\ell)} \qquad (13.20)$$

in the limit

$$N \to \infty, \; \epsilon \to 0, \quad N\epsilon = t_a - t_b = fixed \, . \qquad (13.21)$$

Checking the use of the completeness relation

In order to create some confidence in our procedure, let us verify that we recover from (13.17) the formerly derived result for the kernel (13.15). We have

$$K(b; a) = \lim_{\epsilon \to 0} \left(\frac{m}{2\pi i\hbar\epsilon}\right)^{1/2} \int [Ndx_{N-1}]... \int [Ndx_1] e^{\frac{i}{\hbar}\frac{m}{2}\sum_{i=1}^{N}\frac{(x_i-x_{i-1})^2}{\epsilon}} \qquad (13.22)$$

with

$$x_N = x_b, \quad x_0 = x_a$$

and

$$\mathcal{N} = \left(\frac{m}{2\pi i \hbar \epsilon}\right)^{\frac{1}{2}} .$$

Consider the exponential factors involving x_1. We have ($y = x_1 - x_0$),

$$\int [\mathcal{N} dx_1] e^{\frac{im}{2\hbar\epsilon}[(x_2-x_1)^2+(x_1-x_0)^2]} = \mathcal{N} \int dy e^{\frac{im}{2\hbar\epsilon}\{[(x_2-x_0)-y]^2+y^2\}}$$

$$= \mathcal{N} \int dy e^{\frac{im}{2\hbar\epsilon}\{2y^2-2y(x_2-x_0)+(x_2-x_0)^2\}}$$

$$= \mathcal{N} \int dy e^{\frac{im}{2\hbar\epsilon}\{2[y-\frac{1}{2}(x_2-x_0)]^2+\frac{1}{2}(x_2-x_0)^2\}}$$

$$= \mathcal{N} \int dz e^{\frac{im}{2\hbar\epsilon}[2z^2]} e^{\frac{im}{2\hbar 2\epsilon}(x_2-x_0)^2} .$$

Now

$$\int dz e^{ibz^2} = \left(\frac{i\pi}{b}\right)^{\frac{1}{2}} . \tag{13.23}$$

Hence

$$\int [\mathcal{N} dx_1] e^{\frac{im}{2\hbar\epsilon}[(x_2-x_1)^2+(x_1-x_0)^2]} = \left(\frac{1}{2}\right)^{\frac{1}{2}} e^{\frac{im}{2\hbar\cdot 2\epsilon}(x_2-x_0)^2} .$$

Repeating this process with the next adjacent integration variable x_2, we have

$$\int [\mathcal{N} dx_2] e^{\frac{im}{2\hbar\cdot\epsilon}(x_3-x_2)^2} e^{\frac{im}{2\hbar\cdot 2\epsilon}(x_2-x_0)^2} = \int [\mathcal{N} dy] e^{\frac{im}{2\hbar\cdot 2\epsilon}[3y^2-4y(x_3-x_0)+2(x_3-x_0)^2]}$$

$$= \int [\mathcal{N} dy] e^{\frac{i3m}{2\hbar\cdot 2\epsilon}[(y-\frac{2}{3}(x_3-x_0))^2+\frac{2}{9}(x_3-x_0)^2]}$$

$$= \left[\left(\frac{2i\pi\hbar 2\epsilon}{3m}\right)^{\frac{1}{2}} \mathcal{N}\right] e^{\frac{im}{2\hbar\cdot 3\epsilon}(x_3-x_0)^2]} ,$$

or finally

$$\left(\frac{1}{2}\right)^{\frac{1}{2}} \int [\mathcal{N} dx_2] e^{\frac{im}{2\hbar\cdot\epsilon}(x_3-x_2)^2} e^{\frac{im}{2\hbar\cdot 2\epsilon}(x_2-x_0)^2} = \left(\frac{1}{3}\right)^{\frac{1}{2}} e^{\frac{im}{2\hbar\cdot 3\epsilon}(x_3-x_0)^2]} ,$$

where we have again used (13.23).

Repeating this procedure $N - 1$ times, we finally obtain for the r.h.s. of (13.22),

$$\left(\frac{1}{\mathcal{N}}\right)^{1/2} \exp\left[\frac{im}{2\hbar\cdot N\epsilon}(x_b - x_a)^2\right] .$$

Hence, multiplying this result with the remaining factor $\mathcal{N}$, we finally obtain, after identifying $N\epsilon$ with $t_b - t_a$,

$$K(b; a) = \left(\frac{m}{2\pi i \hbar (t_b - t_a)}\right)^{1/2} e^{\frac{im}{2\hbar}\frac{(x_b-x_a)^2}{t_b-t_a}} , \tag{13.24}$$

where we have set $N\epsilon = t_b - t_a$. We have thus recovered the propagation kernel for the free particle motion. Notice that the singular ϵ dependence in (13.22) has been completely absorbed by the integration.

Semi-classical derivation for the free-particle kernel

In the case where an interaction is present, the above explicit calculation can in general no longer be performed. Approximate methods reminiscent of the WKB method in Quantum Mechanics may be of interest. Let us illustrate this for the case of the free particle with the lagrangian

$$L_0 = \frac{1}{2}m\dot{x}^2 \ .$$

Let x_{cl} be the solution of the associated Lagrange equation $\ddot{x}_{cl} = 0$. We make the change of variable

$$x = x_{cl} + y$$

subject to the boundary conditions

$$x_{cl}(t_a) = x_a \ , \quad x_{cl}(t_b) = x_b \ .$$

Then

$$y(t_a) = y(t_b) = 0 \ .$$

Now

$$x_{cl}(t) = x_a + \frac{x_b - x_a}{t_b - t_a}(t - t_a) \ .$$

We further have

$$S[x_{cl} + y] = \int_{t_a}^{t_b} dt L(\dot{x}_{cl} + \dot{y})$$

$$= \int_{t_a}^{t_b} dt \left[L(\dot{x}_{cl}) + \dot{y}\left(\frac{\partial L}{\partial \dot{x}}\right)_{x_{cl}} + \frac{1}{2}\dot{y}^2\left(\frac{\partial^2 L}{\partial \dot{x}^2}\right)_{x_{cl}} \right]$$

$$= S_{cl} + \frac{m}{2}\int_{t_a}^{t_b} dt \, \dot{y}^2 \tag{13.25}$$

since $\left(\frac{\partial L}{\partial \dot{x}}\right)_{cl} = const$ and $y(t_a) = y(t_b) = 0$. Hence we have from (13.19),

$$K(b; a) = F(t_b, t_a)e^{\frac{i}{\hbar}S_{cl}}$$

with

$$F(t_b, t_a) = e^{\frac{i}{\hbar}S_{cl}}\int_{y(t_a)=0}^{y(t_b)=0} \mathcal{D}y \, e^{\frac{i}{\hbar}\int_{t_a}^{t_b} dt \, \frac{m}{2}\dot{y}^2} \ .$$

Now

$$L_{cl} = \frac{m}{2}\dot{x}_{cl}^2 = \frac{m}{2}\frac{(x_b - x_a)^2}{(t_b - t_a)^2} \ ,$$

or

$$S_{cl} = \frac{m}{2} \frac{(x_b - x_a)^2}{t_b - t_a} \ .$$

Hence

$$K(b, a) = F(t_b, t_a) e^{\frac{im}{2\hbar} \frac{(x_b - x_a)^2}{t_b - t_a}}$$

to be compared with (13.24). To compute $F(t_b, t_a)$, we nevertheless need to do an explicit calculation. In many cases, one is however only interested in the limit $t_a \to -\infty$, $t_b \to \infty$. In that case, $F(t_b, t_a)$ just tends to an irrelevant number, which divides out after proper normalization of the ground state.

Path integral representation of kernel with interaction

Although in the interacting case we can no longer obtain a closed expression for the propagation kernel, we can still obtain for it a functional representation of the form of Eq. (13.19), with the free *action* replaced by the action of the fully interacting theory:

$$< q_b|e^{-\frac{i}{\hbar}H(t_b - t_a)}|q_a > = \int_{q_a}^{q_b} \mathcal{D}q e^{\frac{i}{\hbar} \int_{t_a}^{t_b} dt L(q(t), \dot{q}(t))} \ . \qquad (13.26)$$

From the mathematical standpoint, one considers the kernel (13.26) for time analytically continued to imaginary time, $t = -i\tau$:

$$< q_b|e^{-\frac{1}{\hbar}H(\tau_b - \tau_a)}|q_a > = \int_{q_a}^{q_b} \mathcal{D}q \, e^{-\frac{1}{\hbar} \int_{\tau_a}^{\tau_b} d\tau L_E(q(\tau), \dot{q}(\tau))} \ , \qquad (13.27)$$

with

$$L_E(q, \dot{q}) = \frac{1}{2} m \left(\frac{dq}{d\tau}\right)^2 + V(q) \ .$$

This leaves in (13.27) an exponentially damped exponential. We now provide a systematic derivation of this result. Corresponding to the composition property (13.16) one splits the (euclidean) time interval into N segments. Using the completeness of the eigenstate $|q >$ of the position operator, one then has analogous to (13.17),

$$< q'|e^{-\frac{1}{\hbar}H(\tau' - \tau)}|q > = \int \prod_{\ell=1}^{N-1} dq^{(\ell)} < q'|e^{-\frac{1}{\hbar}H\epsilon}|q^{(N-1)} > \times \qquad (13.28)$$

$$< q^{(N-1)}|e^{-\frac{1}{\hbar}H\epsilon}|q^{(N-2)} > \cdots < q^{(1)}|e^{-\frac{1}{\hbar}H\epsilon}|q > \ .$$

We take the Hamiltonian operator H to have the form

$$H = \frac{\hat{P}2}{2m} + V(\hat{Q}) \ ,$$

with the interaction part only depending on the position operator $\hat{Q}$. Making use of the *Baker–Campbell–Hausdorff* formula

$$e^A e^B = e^{A+B+\frac{1}{2}[A,B]+\cdots} \ ,$$

we may factorize $\exp(-\frac{1}{\hbar}H\epsilon)$ for ϵ small as follows

$$e^{-\frac{1}{\hbar}H\epsilon} \approx e^{-\frac{\epsilon}{\hbar}\frac{\hat{p}^2}{2m}}e^{-\frac{\epsilon}{\hbar}V(\hat{Q})} \,,$$

where we have neglected terms of order $O(\epsilon^2)$ in the exponent. We thus have

$$< q^{(\ell+1)}|e^{-\frac{\epsilon}{\hbar}H}|q^{(\ell)}>\approx< q^{\ell+1}|e^{-\frac{\epsilon}{\hbar}\frac{\hat{p}^2}{2m}}|q^{(\ell)}>e^{-\frac{\epsilon}{\hbar}V(q^{(\ell)})} \,.$$

The matrix element for real time has already been given in (13.18). We thus find (make substitution $i\epsilon \to \epsilon$)

$$< q^{(\ell+1)}|e^{-\frac{\epsilon}{\hbar}H}|q^{(\ell)}>\approx \left(\frac{m}{2\pi\hbar\epsilon}\right)^{1/2} e^{-\frac{m}{2\hbar}\frac{(q^{(\ell)}-q^{(\ell-1)})^2}{\epsilon}} e^{-\frac{1}{\hbar}\epsilon V(q^{(\ell)})} \,.$$

Substituting this expression into (13.28), we thus obtain

$$< q'|e^{-\frac{1}{\hbar}H(\tau'-\tau)}|q >\approx \left(\frac{m}{2\pi\hbar\epsilon}\right)^{N/2} \int \prod_{\ell=1}^{N-1} dq^{(\ell)} e^{-\frac{1}{\hbar}\sum_{\ell=0}^{N-1} \epsilon L_E(q^{(\ell)},\dot{q}^{(\ell)})} \,,$$

where

$$L_E(q^{(\ell)},\dot{q}^{(\ell)}) = \frac{m}{2}\left(\frac{q^{(\ell+1)}-q^{(\ell)}}{\epsilon}\right)^2 + V(q^{(\ell)}) \,.$$

Set

$$q^{(\ell)} \to q(t_\ell) \,.$$

Taking the limit (13.21), we obtain with the measure (13.20) $(i\epsilon \to \epsilon)$ after suitable relabelling, the announced result (13.27). Continuing back to real time we end up with (13.26).[4]

13.4 Path-integral representation of correlators in QM

In quantum field theory, the vacuum expectation value of time-ordered products of field operators (Green functions)

$$G(x_1, x_2..x_n) :=< \Omega|T\phi(x_1)\phi(x_2)...\phi(x_n)|\Omega > \qquad (13.29)$$

are of interest. These are known to be analytic functions in the complex t-plane, allowing for a continuation $t \to -i\tau$ in the time variables. The operators $\phi(x_i)$ appearing in (13.29) are related to the corresponding operators $\phi(\vec{x}, 0)$ in the Schrödinger picture by

$$\phi(\vec{x}, t) = e^{iHt}\phi(\vec{x}, 0)e^{-iHt} \,.$$

---

[4]For a non-trivial application to the (i) unharmonic oscillator, see I. Bender, D. Gromes and K.D. Rothe, *Nucl. Phys.* **B136** (1978) 259. (ii) Coulomb potential, see I.H. Duru and H. Kleinert, *Phys. Lett.* **84B** (1979) 185.

Continuing these operators formally to imaginary time, we define the "euclidean" operators

$$\hat{\phi}(\vec{x}, \tau) = e^{H\tau} \phi(\vec{x}, 0) e^{-H\tau} \ .$$

The QM analogon of the field $\phi(\vec{x}, t)$ is the infinite set of coordinates $Q_{\vec{x}}(t)$ labelled $\vec{x}$. In the following, we shall restrict ourselves to just one degree of freedom $Q(t)$. The generalization to an infinite number is then done in the following section. In the euclidean formulation, we then have

$$\hat{Q}(\tau) = e^{H\tau} \hat{Q}(0) e^{-H\tau} \ .$$

Actually these operators, taken by themselves, are ill defined if the spectrum of H is unbounded above (we shall always assume that the spectrum is bounded from below). However, for an operator product ordered according to *descending* times from left to right, the (euclidean) *correlator*

$$< q', \tau' | \hat{Q}(\tau_1) \hat{Q}(\tau_2) ... \hat{Q}(\tau_n) | q, \tau >, \quad \tau' > \tau_1 > \tau_2 > ... \tau_n > \tau$$

is well defined. Inserting a complete set of energy eigenstates to the left and right of the operators, we obtain from here

$$< q', \tau' | \hat{Q}(\tau_1) ... \hat{Q}(\tau_n) | q, \tau > = \sum_{\ell, \ell'} e^{-E_{\ell'} \tau'} e^{E_{\ell} \tau} \psi_{\ell'}(q') \psi_{\ell}^*(q) < E_{\ell'} | \hat{Q}(\tau_1) ... \hat{Q}(\tau_n) | E_{\ell} >,$$

where we have set

$$< q, \tau | E_\ell > = < q | e^{-H\tau} | E_\ell > =: \psi_\ell(q) e^{-E_\ell \tau}$$

with $\psi_\ell(q)$ eigenfunctions of the Hamiltonian corresponding to the eigenvalues E_ℓ. Assuming that there exists an energy gap between the ground state and the first excited state, we conclude that in the limit $\tau' \to \infty$ and $\tau \to -\infty$

$$< q', \tau' | \hat{Q}(\tau_1) ... \hat{Q}(\tau_n) | q, \tau > \longrightarrow e^{-E_0(\tau'-\tau)} \psi_0(q') \psi_0^*(q) < E_0 | \hat{Q}(\tau_1) ... \hat{Q}(\tau_n) | E_0 > \ . \tag{13.30}$$

In particular, replacing the operators $\hat{Q}(\tau)$ by unit operators, we have

$$< q', \tau' | q, \tau > \longrightarrow e^{-E_0(\tau'-\tau)} \psi_0(q') \psi_0^*(q) \ . \tag{13.31}$$

We thus conclude that $(\tau' > \tau_1 > \tau_2 > ... \tau_n > \tau)$

$$\frac{< q', \tau' | \hat{Q}(\tau_1) ... \hat{Q}(\tau_n) | q, \tau >}{< q', \tau' | q, \tau >} \longrightarrow < E_0 | \hat{Q}(\tau_1) ... \hat{Q}(\tau_n) | E_0 > \ . \tag{13.32}$$

Notice that for the formula (13.32) to be valid we only require the existence of an energy gap, and a non-vanishing projection of the states $| q >$ and $| q' >$ on the ground state $| E_0 >$.

In order to appreciate the importance of our *euclidean* discussion it is instructive to interrupt for a moment our discussion to consider as example the Minkowski functional integral for the quantum mechanical harmonic oscillator.

Digression: Harmonic oscillator

We wish to calculate $< x_b, T | x_a, -T >$ for the case of the harmonic oscillator described by the lagrangian

$$L = \frac{1}{2} m (\dot{x}^2 - \omega^2 x^2) .$$

Since this is the interacting case, we shall treat it using the semiclassical method described previously. In order to implement the boundary conditions it is again convenient to make the change of variable

$$x = x_{cl} + y ,$$

where x_{cl} is a solution of the classical equation of motion

$$\ddot{x}_{cl} + \omega^2 x_{cl} = 0 . \tag{13.33}$$

The most general solution of (13.33) is given by

$$x_{cl} = A \sin \omega t + B \cos \omega t .$$

Imposing the boundary conditions

$$x(-T) = x_a, \quad x(T) = x_b ,$$

this fixes A and B to be

$$A = \frac{x_b - x_a}{2 \sin \omega T} , \quad B = \frac{x_b + x_a}{2 \cos \omega T} .$$

Making use of this, we obtain for L_{cl},

$$L_{cl} = \frac{1}{2} m \omega^2 \left[\left(\frac{(x_a^2 + x_b^2) \cos(2\omega T) - 2 x_a x_b}{\sin^2(2\omega T)} \right) \cos(2\omega t) - \frac{(x_b^2 - x_a^2)}{\sin(2\omega T)} \sin(2\omega t) \right] .$$

Integrating over time from $-T$ to $+T$, we obtain for the corresponding action

$$S_{cl} = \frac{m\omega}{2 \sin(2\omega T)} [(x_a^2 + x_b^2) \cos(2\omega T) - 2 x_a x_b] .$$

We thus conclude,

$$< x_b, T | x_a, -T >= F(T; \omega) e^{\frac{im\omega}{2\hbar \sin(2\omega T)} [(x_a^2 + x_b^2) \cos(2\omega T) - 2 x_a x_b]} , \tag{13.34}$$

where

$$F(T, \omega) = \int_{y(-T)=0}^{y(T)=0} \mathcal{D}y \, e^{\frac{i}{\hbar} \int_{-T}^{T} dt \frac{m}{2} (\dot{y}^2 - \omega^2 y^2)} = \int_{y(-T)=0}^{y(T)=0} \mathcal{D}y \, e^{\frac{i}{\hbar} \int_{-T}^{T} dt \frac{m}{2} y \mathcal{D} y} , \tag{13.35}$$

with D the differential operator

$$D = -\frac{d^2}{dt^2} - \omega^2 .$$

Note again that the linear term in y is absent since S_{cl} corresponds to an extremum of the action.

Expand y in a complete set of orthonormal eigenfunctions of the operator D,

$$Du_n = \lambda_n u_n ,$$

satisfying the boundary condition

$$u_n(-T) = u_n(T) = 0 .$$

We have for the normalized eigenfunctions and eigenvalues

$$u_n(t) = \frac{1}{\sqrt{T}} \sin \frac{n\pi t}{T} \qquad \lambda_n = \left(\frac{n\pi}{T}\right)^2 - \omega^2 .$$

Make the expansion

$$y(t) = \sum_{n=0}^{\infty} a_n u_n(t) .$$

With the change of variable

$$\mathcal{D}y \rightarrow J \prod_n da_n ,$$

where J is the Jacobian of the transformation, we have for (13.35)

$$\int_{y(-T)=0}^{y(T)=0} \mathcal{D}y \, e^{i\frac{m}{2\hbar} \int_{-T}^{T} dt \, y D y} \rightarrow J \int \prod_{n=1}^{N} da_n e^{\frac{im}{2\hbar} \sum \lambda_n a_n^2}$$

$$= J \frac{1}{\left(\prod_{n=1}^{N} \frac{m}{2\hbar\lambda_n}\right)^{1/2}} = J' \prod_{n=1}^{N} \left(1 - \frac{\omega^2 T^2}{n^2 \pi^2}\right)^{-\frac{1}{2}}$$

where J' is independent of ω. Noting that

$$\lim_{N\to\infty} \prod_{n=1}^{N} \left(1 - \frac{\omega^2 T^2}{n^2 \pi^2}\right)^{-1/2} = \left[\frac{(\sin \omega T)}{\omega T}\right]^{-1/2}$$

we finally obtain

$$F(T,\omega) = J'(T) \left(\frac{\omega T}{\sin \omega T}\right)^{1/2} .$$

The function $J'(T)$ is determined by taking the limit $\omega \to 0$ and comparing with the free-particle case (1.11):

$$J'(T) = \left(\frac{m}{2\pi i\hbar T}\right)^{1/2} .$$

Hence

$$F(T,\omega) = \left(\frac{m\omega}{2\pi i\hbar \sin 2\omega T}\right)^{1/2} . \tag{13.36}$$

Summarizing we finally have for (13.34),

$$< x_b, T|x_a, -T >= \left(\frac{m\omega}{2\pi i\hbar \sin(2\omega T)}\right)^{1/2} e^{i\frac{m\omega}{\sin(2\omega T)}[(x_a^2+x_b^2)\cos(2\omega T)-2x_ax_b]} \,. \quad (13.37)$$

We now observe that this transition amplitude has no definite limit for $T \to \infty$. However, for imaginary time $t = -iT$, this result tends to

$$< x_b, -iT|x_a, iT > \to \left(\frac{m\omega}{2\pi i\hbar \sin 2\omega T}\right)^{1/2} e^{-\frac{m\omega}{2\hbar}(x_a^2+x_b^2)}$$
$$\to e^{-\omega T}\varphi_0(x_b)\varphi_0(x_a)$$

with

$$\varphi_0(x) = \left(\frac{m\omega}{\pi\hbar}\right)^{\frac{1}{4}} e^{-\frac{m\omega}{2\hbar}x^2}$$

the oscillator ground state wave function with ground state energy $\frac{1}{2}\omega$. One checks that $\int \phi_0^2 = 1$ corresponds to $< E_0|E_0 >= 1$. This result is thus of the form (13.30) with $\hat{Q} = 1$.

Back to (13.32)

We now return to our original discussion. In order to obtain a path-integral representation for the left-hand side of (13.32), we write

$$< q', \tau'|\hat{Q}(\tau_1) \cdots \hat{Q}(\tau_n)|q, \tau >=$$
$$< q'|e^{-H(\tau'-\tau_1)}\hat{Q}e^{-H(\tau_1-\tau_2)} \cdots e^{-H(\tau_{n-1}-\tau_n)}\hat{Q}e^{-H(\tau_n-\tau)}|q > ,$$

where $\hat{Q} = \hat{Q}(0)$. We proceed as before and repeatedly insert a complete set of eigenstates of the operator $\hat{Q}$, thus obtaining

$$< q', \tau'|\hat{Q}(\tau_1)...\hat{Q}(\tau_n)|q, \tau >= \int \prod_{\ell=1}^{n} dQ^{(\ell)}$$
$$\left[< q', \tau'|Q^{(1)}, \tau_1 > Q^{(1)} < Q^{(1)}, \tau_1|Q^{(2)}, \tau_2 > Q^{(2)}...Q^{(n)} < Q^{(n)}, \tau_n|q, \tau >\right] \,.$$

Making use of the path integral representation for the individual propagation kernels, and following the steps leading to (13.27), we may write the final result in the compact form

$$< q', \tau'|Q(\tau_1)...Q(\tau_n)|q, \tau >= \int_{Q(\tau)=q}^{Q(\tau')=q'} DQ\, Q(\tau_1)...Q(\tau_n)e^{-S_E[Q;\tau',\tau]} \,, \quad (13.38)$$

where $S_E[Q; \tau', \tau]$ stands for the *euclidean* action

$$S_E[Q; \tau', \tau] = \int_{\tau}^{\tau'} d\tau'' L_E(Q(\tau''), \dot{Q}(\tau'')) \,.$$

The path integral is calculated as implied by (13.38):

(i) Split the time interval $[\tau, \tau']$ into n sub-intervals $[\tau_{i+1}, \tau_i]$ corresponding to the n "insertions" of the Operators $\hat{Q}(\tau_i)$ at "times" $\tau_1 \cdots \tau_n$.

(ii) Divide each of the sub-intervals $[\tau_{i+1}, \tau_i]$ into n_i further sub-intervals.

(iii) Consider all paths $Q(\tau)$ starting at q at time τ and ending at q' at time τ'.

(iv) Weigh each path with the factor $Q(\tau_1)...Q(\tau_n) \exp(-S_E[Q; \tau', \tau])$. Sum the contributions over all paths by integrating over all possible values of the coordinates at the intermediate times, with the integration measure (13.20).

(v) Take the limit $\epsilon_i = (\tau_{i+1} - \tau_i)/n_i \to 0, n_i \to \infty$, keeping the product $n_i \epsilon_i = (\tau_{i+1} - \tau_i)$ fixed in each sub-sub-interval.

We may relax the condition $\tau_1 > \tau_2 > ... > \tau_n$. In that case the r.h.s. represents the matrix element of the time-ordered product of the n operators $\hat{Q}(\tau_i), i = 1, \cdots n$:

$$< q', \tau' | T\hat{Q}(\tau_1)...\hat{Q}(\tau_n) | q, \tau > = \int_{Q(\tau)=q}^{Q(\tau')=q'} \mathcal{D}Q \, Q(\tau_1)...Q(\tau_n) e^{-S_E[Q; \tau', \tau]} .$$

The right-hand side exhibits the symmetry with respect to the operators $Q(\tau_i)$ of the time-ordered product of the left-hand side. Taking now the limit $\tau \to -\infty, \tau' \to \infty$, we obtain (see (13.32))

$$< E_0 | T\hat{Q}(\tau_1)...\hat{Q}(\tau_n) | E_0 > = \frac{\int \mathcal{D}Q Q(\tau_1)...Q(\tau_n) e^{-S_E[Q]}}{\int \mathcal{D}Q \, e^{-S_E[Q]}} , \qquad (13.39)$$

where $S[Q]$ is now the euclidean action

$$S_E[Q] = \int_{-\infty}^{\infty} d\tau L_E(Q(\tau), \dot{Q}(\tau)) .$$

This is our final result. Note that the normalization factor of the measure in (13.20) drops out in the ratio (13.39).

13.5   Feynman path-integral representation in QFT

The extension of the above result to field theory is now obvious. To this end we first extend the result (13.39) to the case of a quantum-mechanical system of K degrees of freedom, with coordinates

$$Q \to \{Q_k\} = (Q_1....Q_K) .$$

We evidently have

$$< E_0 | T\hat{Q}_{k_1}(\tau_1)...\hat{Q}_{k_n}(\tau_n) | E_0 > = \frac{\int \mathcal{D}Q Q_{k_1}(\tau_1)...Q_{k_n}(\tau_n) e^{-S_E[Q]}}{\int \mathcal{D}Q e^{-S_E[Q]}} ,$$

where the integration measure now stands for

$$\mathcal{D}Q := \prod_{k=1}^{K} \prod_{l=1}^{N-1} dQ_k(\tau_l) .$$

We now consider the limit where this number K of degrees of freedom tends to infinity. Replacing the label k by a continuous label $\vec{x}$,

$$\hat{Q}_k(\tau) \to \hat{\phi}(\tau, \vec{x}) , \quad Q_k(\tau) \to \phi(\tau, \vec{x}) ,$$

we thus arrive at the following expression for the euclidean correlator in a scalar field theory:

$$< \Omega | T\hat{\phi}(x_1)...\hat{\phi}(x_n) | \Omega > = \frac{\int \mathcal{D}\phi \, \phi(x_1)...\phi(x_n) e^{-S_E[\phi]}}{\int \mathcal{D}\phi \, e^{-S_E[\phi]}} ,$$

where, in order to conform to QFT notation, we have set $|E_0 > = |\Omega >$.

The corresponding n-point Green function in Minkowski space is recovered by performing an analytic continuation in "time" to the real time axis ($\tau \to it$) with the (formal) result

$$< \Omega | T\hat{\phi}(x_1)...\hat{\phi}(x_n) | \Omega > = \frac{\int \mathcal{D}\phi \, \phi(x_1)...\phi(x_n) e^{iS[\phi]}}{\int \mathcal{D}\phi \, e^{iS[\phi]}} \qquad (13.40)$$

with

$$S[\phi] = \int d^4 y \mathcal{L}(y) ,$$

where

$$\mathcal{L}(y) \equiv \mathcal{L}(\phi(y), \partial_\mu \phi(y))$$

is now the Minkowski–Lagrange density defining the (scalar) field theory of interest. The right-hand side of (13.40) represents the "partially connected" n-point Green function.

Perturbation Theory recovered

We now illustrate how one recovers from (13.40) the perturbative series defined by (11.21). To this end we define the generating functional

$$Z[J] = \int \mathcal{D}\phi \, e^{i \int d^4 y (\mathcal{L}(y) + J(y)\phi(y))} . \qquad (13.41)$$

In particular,

$$Z[0] = \int \mathcal{D}\phi \, e^{i \int d^4 y \mathcal{L}(y)} .$$

The ratio $\frac{Z[J]}{Z[0]}$ is the path-integral version of Schwinger's generating functional $\mathcal{Z}[J]$, Eq. (13.1). The partially connected n-point Green functions of the theory are correspondingly obtained by functional differentiation of $\ln Z[J]$ with respect to the external source $J(x)$ and subsequently setting $J = 0$. Separating the action associated with the Lagrange density into its free and interacting part,

$$S[\phi] = S_0[\phi] + S_I[\phi] \qquad (13.42)$$

and in (13.41) making the expansion

$$e^{iS[\phi]} = e^{iS_0[\phi]} \sum_{n=0}^{\infty} \frac{i^n}{n!} \left(S_I[\phi] \right)^n \ ,$$

we may evidently write[5]

$$Z[J] = \sum_{n=0}^{\infty} \frac{i^n}{n!} \left(S_I \left[\frac{\delta}{i\delta J} \right] \right)^n Z^{(0)}[J] \ , \tag{13.43}$$

where

$$Z^{(0)}[J] = \int \mathcal{D}\phi \, e^{iS_0[\phi] + i \int d^4y \, J(y)\phi(y)} \ . \tag{13.44}$$

This is to be compared with the Schwinger result (13.12). For the case of a real scalar field, the <u>free</u> part of the action is purely quadratic in the fields having the form

$$S_0[\phi] = \int d^4z \frac{1}{2}(\partial_\mu \phi \partial^\mu \phi - m^2\phi^2) = -\int d^4x \frac{1}{2}\phi(\Box + m^2)\phi \tag{13.45}$$

$$= -\frac{1}{2}\int d^4x \int d^4y \phi(x) K(x,y)\phi(y) \ .$$

$$K(x,y) = (\Box_x + m^2)\delta^4(x-y) \ ,$$

$Z^{(0)}[J]$ is easily evaluated by rewriting the exponent in (13.44) as a quadratic form, leaving us with:

$$Z^{(0)}[J] = \int \mathcal{D}\phi \, e^{-i \int d^4z \int d^4z' [\frac{1}{2}(\phi - JK^{-1})(z)K(z,z')(\phi - K^{-1}J)(z') - \frac{1}{2}J(z)K^{-1}(z,z')J(z')]} \ ,$$

where

$$K^{-1}(x,y) = (\Box + m^2)^{-1}_{xy} = -\Delta_F(x-y) \ .$$

By making a shift, the ϕ integration will contribute a constant which drops out in the ratio $\mathcal{Z}[J] = \frac{Z[J]}{Z[0]}$. We thus have

$$\frac{Z[J]}{Z[0]} = \sum_{n=0}^{\infty} \frac{i^n}{n!} \left(S_I[\frac{\delta}{i\delta J}] \right)^n e^{-\frac{i}{2}\int d^4z \int d^4z' J(z)\Delta_F(z-z')J(z')}. \tag{13.46}$$

Performing the functional differentiations with respect to the external source and setting this source equal to zero at the end, reproduces the perturbative series of the scalar field theory in question.

---

[5] From classical mechanics we are familiar with the computation of $\delta S[q] \equiv S[q + \delta q] - S[q]$. Thus $\delta \int dt f(q(t)) \equiv \int dt [f(q(t) + \delta q(t)) - f(q(t))]$. With the "rule"

$$\left(\frac{\delta}{\delta J(x)} J(y) \right) = \delta^4(x-y)$$

we correspondingly have $\frac{\delta}{\delta J(x)} \int d^4y J(y)\phi(y) = \phi(x)$. This is a natural generalization of $\frac{\partial}{\partial x_i}x_j = \delta_{ij}$, i.e. $\frac{\partial}{\partial x_i}\sum_j x_j k_j = k_i$.

Example: ϕ^3 *theory*

In this case

$$S_I[\phi] = \frac{\lambda}{3!} \int d^4z \, \phi^3(z)$$

or

$$S_I \left[\frac{1}{i} \frac{\delta}{\delta J} \right] = \frac{\lambda}{3!} \int d^4z \left(\frac{1}{i} \frac{\delta}{\delta J(z)} \right)^3.$$

To second order

$$Z[J] = F^{(2)} \left[\frac{1}{i} \frac{\delta}{\delta J} \right] Z^{(0)}[J] + \ldots$$

where

$$F^{(2)} \left[\frac{1}{i} \frac{\delta}{\delta J} \right] = 1 + \frac{i}{1!} S_I \left[\frac{1}{i} \frac{\delta}{\delta J} \right] + \frac{i^2}{2!} \left(S_I \left[\frac{1}{i} \frac{\delta}{\delta J} \right] \right)^2.$$

Now

$$< \Omega|T\phi(x)\phi(y)|\Omega > = \frac{\left(\frac{\delta}{i\delta J(x)} \frac{\delta}{i\delta J(y)} Z[J] \right)_{J=0}}{Z[0]}$$

$$= \frac{1}{Z[0]} \left[F^{(2)} \left[\frac{\delta}{i\delta J} \right] \frac{\delta}{i\delta J(x)} \frac{\delta}{i\delta J(y)} Z^{(0)}[J] \right]_{J=0} + \cdots.$$

We have

$$\frac{\delta}{i\delta J(y)} Z^{(0)}[J] = -\frac{1}{2} \left(\int d^4\xi \, J(\xi)\Delta_F(\xi - y) + \int d^4\xi \, \Delta_F(y - \xi)J(\xi) \right) Z^{(0)}[J]$$

$$= - \left(\int d^4\xi \, J(\xi)\Delta_F(\xi - y) \right) Z^{(0)}[J]$$

and

$$\frac{\delta}{i\delta J(x)} \frac{\delta}{i\delta J(y)} Z^{(0)}[J] = i\Delta_F(x - y)Z^{(0)}[J]$$

$$+ \left[\int d^4\xi \, J(\xi)\Delta_F(\xi - y) \right] \left[\int d^4\xi' \, J(\xi')\Delta_F(\xi' - x) \right] Z^{(0)}[J].$$

Now apply $F^{(2)}$, divide by $Z[0]$ and set $J = 0$ at the end. The result is represented schematically by the following set of diagrams:

$$\left[F^{(2)} \left[\frac{\delta}{i\delta J} \right] \frac{\delta}{i\delta J(x)} \frac{\delta}{i\delta J(y)} Z^{(0)}[J] \right]_{J=0} =$$

$$\tag{13.47}$$

and

$$Z[0] = 1 + \bigominus \tag{13.48}$$

The last term in the sum (*tad-pole*) is eliminated by normal ordering. Regarding these diagrams as a power series in λ we see that the contribution of the vacuum diagrams (second term in the bracket) cancels in the ratio (13.46).

13.6   Path-Integral for Grassman-valued fields

So far we have considered quantum mechanical systems involving only bosonic degrees of freedom. But the fundamental matter fields in nature carry spin 1/2. In contrast to the bosonic case these fields anticommute in the limit $\hbar \to 0$, and hence become elements of a Grassmann algebra in this limit. We therefore expect that the path integral representation of Green functions built from fermion fields will involve the integration over anticommuting (Grassmann) variables.[6] We therefore begin with the introduction of some basic notions regarding the differentiation and integration of functions of Grassmann variables. The integration rules are then applied to calculate specific integrals, which will play an important role throughout this book. The results we obtain will give us a strong hint regarding the path integral representation of fermionic Green functions in theories of interest for elementary particle physics. We begin our discussion with some basic definitions.

Grassmann Algebra

The elements $\eta_1, ..., \eta_N$ are said to be the generators of a Grassmann algebra, if they anticommute among each other, i.e. if

$$\{\eta_i, \eta_j\} = \eta_i \eta_j + \eta_j \eta_i = 0, \quad i, j = 1, ..., N. \tag{13.49}$$

From here it follows that

$$\eta_i^2 = 0. \tag{13.50}$$

A general element of a Grassmann algebra is defined as a power series in the η_i's. Because of (13.50), however, this power series has only a finite number of terms:

$$f(\eta) = f_0 + \sum_i f_i \eta_i + \sum_{i \neq j} f_{ij} \eta_i \eta_j + ... + f_{12...N} \eta_1 \eta_2 ... \eta_N. \tag{13.51}$$

As an example consider the function

$$g(\eta) = e^{-\sum_{i,j=1}^{N} \eta_i A_{ij} \eta_j}.$$

---

[6]This section has been taken over from *Lattice Gauge Theories — An Introduction* by Heinz J. Rothe (World Scientific), with the author's permission. For a comprehensive discussion of the functional formalism for fermions the reader may consult the book by F. Berezin, *The Method of Second Quantization* (Academic Press, New York, London, 1966).

It is defined by the usual power series expansion of the exponential. Since the terms appearing in the sum — being quadratic in the Grassmann variables — commute among each other, we can also write $g(\eta)$ as the product

$$g(\eta) = \prod_{i,j} e^{-\eta_i A_{ij} \eta_j},$$

or, making use of (13.50),

$$g(\eta) = \prod_{\substack{i,j=1 \\ i \neq j}}^{N} (1 - \eta_i A_{ij} \eta_j).$$

Next we consider the following function of a set of $2N$-Grassmann variables which we denote by $\eta_1, .., \eta_N, \bar{\eta}_1, ..., \bar{\eta}_N$:

$$h(\eta, \bar{\eta}) = e^{-\sum_{ij} \bar{\eta}_i A_{ij} \eta_j}.$$

Proceeding as above, we now have that

$$h(\eta, \bar{\eta}) = \prod_{i,j=1}^{N} (1 - \bar{\eta}_i A_{ij} \eta_j).$$

Notice that in contrast to the previous case, this expression also involves diagonal elements of A_{ij}.

Integration over Grassmann variables

We now state the Grassmann rules for calculating integrals of the form

$$\int \prod_{i=1}^{N} d\eta_i f(\eta),$$

where $f(\eta)$ is a function whose general structure is given by (13.51). Since a given Grassmann variable can at most appear to the first power in $f(\eta)$, the following rules suffice to calculate an arbitrary integral:

$$\int d\eta_i = 0, \quad \int d\eta_i \eta_i = 1. \tag{13.52}$$

When computing multiple integrals one must further take into account that the integration measures $\{d\eta_i\}$ also anticommute among themselves, as well as with all η_j's

$$\{d\eta_i, d\eta_j\} = \{d\eta_i, \eta_j\} = 0, \quad \forall i, j.$$

These integration rules look indeed very strange. But, as we shall see soon, they are the appropriate ones to allow us to obtain a path-integral representation of fermionic Green functions. As an example, let us apply these rules to calculate the following integral:

$$I = \int \prod_{\ell=1}^{N} d\bar{\eta}_\ell d\eta_\ell e^{-\sum_{i,j=1}^{N} \bar{\eta}_i A_{ij} \eta_j}. \tag{13.53}$$

To evaluate (13.53), we first write the integrand in the form

$$e^{-\sum_{i,j} \bar{\eta}_i A_{ij} \eta_j} = \prod_{i=1}^{N} e^{-\bar{\eta}_i \sum_{j=1}^{N} A_{ij} \eta_j}.$$

Since $\bar{\eta}_i^2 = 0$, only the first two terms in the expansion of the exponential will contribute. Hence

$$e^{-\sum_{i,j} \bar{\eta}_i A_{ij} \eta_j} = \left(1 - \bar{\eta}_1 \sum_{i_1} A_{1 i_1} \eta_{i_1}\right)\left(1 - \bar{\eta}_2 \sum_{i_2} A_{2 i_2} \eta_{i_2}\right) \cdots \left(1 - \bar{\eta}_N \sum_{i_N} A_{N i_N} \eta_{i_N}\right).$$

$$\tag{13.54}$$

Now, because of the Grassmann integration rules (13.52), the integrand of (13.53) must involve the product of all the Grassmann variables. We therefore only need to consider in (13.54) the term

$$K(\eta, \bar{\eta}) \equiv \sum_{i_1,\dots,i_N} (\eta_{i_1} \bar{\eta}_1)(\eta_{i_2} \bar{\eta}_2) \dots (\eta_{i_N} \bar{\eta}_N) A_{1 i_1} A_{2 i_2} \dots A_{N i_N}, \tag{13.55}$$

where we have set $\bar{\eta}_k \eta_{i_k} = -\eta_{i_k} \bar{\eta}_k$ to eliminate the minus signs appearing in (13.54). The summation clearly includes only those terms for which all the indices $i_1, \dots, i_N$ are different. Now, the product of Grassmann variables in (13.55) is antisymmetric under the exchange of any pair of indices i_ℓ and $i_{\ell'}$. Hence we can write the expression (13.55) in the form

$$K(\eta, \bar{\eta}) = \eta_1 \bar{\eta}_1 \eta_2 \bar{\eta}_2 \dots \eta_N \bar{\eta}_N \sum_{i_1 \dots i_N} \epsilon_{i_1 i_2 \dots i_N} A_{1 i_1} A_{2 i_2} \dots A_{N i_N},$$

where $\epsilon_{i_1 i_2 \dots i_N}$ is the Levi–Civita-tensor in N-dimensions. Recalling the standard formula for the determinant of a matrix A, we therefore find that

$$K(\eta, \bar{\eta}) = (\det A)\eta_1 \bar{\eta}_1 \eta_2 \bar{\eta}_2 \dots \eta_N \bar{\eta}_N.$$

We now replace the exponential in (13.53) by this expression and obtain

$$I = \left[\prod_{i=1}^{N} \int d\bar{\eta}_i d\eta_i \eta_i \bar{\eta}_i\right] \det A = \det A.$$

Let us summarize our result for later convenience:

$$\int D(\bar{\eta}\eta) e^{-\sum_{i,j=1}^{N} \bar{\eta}_i A_{ij} \eta_j} = \det A, \quad D(\bar{\eta}\eta) = \prod_{\ell=1}^{N} d\bar{\eta}_\ell d\eta_\ell. \tag{13.56}$$

There is another important formula we shall need. It will allow us to calculate integrals of the type

$$I_{i_1 \cdots i_\ell i'_1 \cdots i'_\ell} = \int D(\bar{\eta}\eta)\eta_{i_1} \cdots \eta_{i_\ell}\bar{\eta}_{i'_1} \cdots \bar{\eta}_{i'_\ell} e^{-\sum_{i,j=1}^{N} \bar{\eta}_i A_{ij}\eta_j}. \tag{13.57}$$

Consider the following generating functional

$$Z[\rho, \bar{\rho}] = \int D(\bar{\eta}\eta)e^{-\sum_{i,j} \bar{\eta}_i A_{ij}\eta_j + \sum_i (\bar{\eta}_i\rho_i + \bar{\rho}_i\eta_i)}, \tag{13.58}$$

where all indices are understood to run from 1 to N, and where the "sources" $\{\rho_i\}$ and $\{\bar{\rho}_i\}$ are now also anticommuting elements of the Grassmann algebra generated by $\{\eta_i, \bar{\eta}_i, \rho_i, \bar{\rho}_i\}$. To evaluate (13.58) we first rewrite the integral as follows:

$$Z[\rho, \bar{\rho}] = \left[\int D(\bar{\eta}\eta)e^{-\sum_{i,j} \bar{\eta}'_i A_{ij}\eta'_j} \right] e^{\sum_{i,j} \bar{\rho}_i A_{ij}^{-1}\rho_i},$$

where

$$\eta'_i = \eta_i - \sum_k A_{ik}^{-1}\rho_k, \quad \bar{\eta}'_i = \bar{\eta}_i - \sum_k \bar{\rho}_k A_{ki}^{-1},$$

and A^{-1} is the inverse of the matrix A. Making use of the invariance of the integration measure under the above transformation[7] and of (13.56), we find that

$$Z[\rho, \bar{\rho}] = \det A\, e^{\sum_{i,j} \bar{\rho}_i A_{ij}^{-1}\rho_j}. \tag{13.59}$$

Notice that in contrast to the bosonic case, this generating functional is proportional to $\det A$.

Differentiation of Grassmann Variables

We now complete our discussion on Grassmann variables by introducing the concept of a partial derivative on the space of functions defined by (13.51). Suppose we want to differentiate $f(\eta)$ with respect to η_i. Then the rules are the following:

1. If $f(\eta)$ does not depend on η_i, then $\partial_{\eta_i} f(\eta) = 0$

2. If $f(\eta)$ depends on η_i, then the *left* derivative $\rightarrow \partial/\partial\eta_i$ is performed by first bringing the variable η_i (which never appears twice in a product!) all the way to the left, using the anticommutation relations (13.49), and then applying the rule

$$\frac{\vec{\partial}}{\partial\eta_i}\eta_i = 1.$$

Correspondingly, we obtain the *right* derivative $\overleftarrow{\partial}/\partial\eta_i$ by bringing the variable η_i all the way to the right and then applying the rule

$$\eta_i\frac{\overleftarrow{\partial}}{\partial\eta_i} = 1.$$

---

[7]This is ensured by the Grassmann integration rules.

Thus for example

$$\frac{\partial}{\partial \eta_i} \eta_j \eta_i = -\eta_j \quad (i \neq j),$$

or

$$\bar{\eta}_i \eta_j \frac{\overleftarrow{\partial}}{\partial \bar{\eta}_i} = -\eta_j.$$

Notice that, because of the peculiar definition of Grassmann integration, we have that

$$\int d\eta_i f(\eta) = \frac{\partial}{\partial \eta_i} f(\eta).$$

Hence integration over η_i is equivalent to partial differentiation with respect to this variable! Another property, which can be easily proved, is that

$$\left\{ \frac{\partial}{\partial \eta_i}, \frac{\partial}{\partial \eta_j} \right\} f(\eta) = 0.$$

Let us apply these rules to some cases of interest. Consider the function

$$E(\bar{\rho}) = e^{\sum_j \bar{\rho}_j \eta_j},$$

where $\{\eta_i, \bar{\rho}_i\}$ are the generators of a Grassmann algebra. If they were ordinary c-numbers then we would have that

$$\frac{\partial}{\partial \bar{\rho}_i} E(\bar{\rho}) = \eta_i E.(\bar{\rho}) . \tag{13.60}$$

This result is in fact correct. To see this, let us write $E(\bar{\rho})$ in the form

$$E(\bar{\rho}) = \prod_j (1 + \bar{\rho}_j \eta_j).$$

Applying the rules of Grassmann differentiation, we have that

$$\frac{\partial}{\partial \bar{\rho}_i} E(\bar{\rho}) = \eta_i \prod_{j \neq i} (1 + \bar{\rho}_j \eta_j).$$

But because of the appearance of the factor η_i we are now free to extend the product to include $i = j$. Hence we arrive at the naive result (13.60). It should, however, be noted, that the order of the Grassmann variables in $\sum_i \bar{\rho}_i \eta_i$ was important. By reversing this order we get a minus sign, and the rule is not the usual one! By a similar argument one finds that

$$e^{\sum_j \bar{\eta}_j \rho_j} \frac{\overleftarrow{\partial}}{\partial \rho_i} = \bar{\eta}_i e^{\sum_j \bar{\eta}_j \rho_j}.$$

Let us now return to the generating functional defined in (13.58). Proceeding as above one can easily show that

$$
I_{i_1...i_\ell;i'_1...i'_\ell} = \left[\frac{\partial}{\partial\bar{\rho}_{i_1}} \cdots \frac{\partial}{\partial\bar{\rho}_{i_\ell}} Z[\rho,\bar{\rho}] \frac{\overleftarrow{\partial}}{\partial\rho_{i'_1}} \cdots \frac{\overleftarrow{\partial}}{\partial\rho_{i'_\ell}} \right]_{\rho=\bar{\rho}=0}, \tag{13.61}
$$

where the left-hand side has been defined in (13.57). By making use of the explicit expression for Z given in (13.59), one can calculate the right-hand side of (13.61). Since we shall need this expression in later chapters, we will derive it here. To this effect we first rewrite (13.59) as follows

$$
Z[\rho,\bar{\rho}] = \det A \prod_i e^{\bar{\rho}_i \sum_j A_{ij}^{-1} \rho_j} = (\det A) \left(1 + \bar{\rho}_{i_1} \sum_{k_1} A_{i_1 k_1}^{-1} \rho_{k_1} \right) \tag{13.62}
$$

$$
\times \left(1 + \bar{\rho}_{i_2} \sum_{k_2} A_{i_2 k_2}^{-1} \rho_{k_2} \right) \cdots \left(1 + \bar{\rho}_{i_\ell} \sum_{k_\ell} A_{i_\ell k_\ell}^{-1} \rho_{k_\ell} \right) [\cdots],
$$

where the indices $k_1, ..., k_\ell$ are summed, and $[\cdots]$ stands for the remaining factors not involving the variables $\bar{\rho}_{i_1}, ..., \bar{\rho}_{i_\ell}$. The only terms which contribute to the left derivatives in (13.61) are those involving the product $\bar{\rho}_{i_1}...\bar{\rho}_{i_\ell}$. Furthermore since we will eventually set all "sources" ρ_i and $\bar{\rho}_i$ equal to zero, we can replace $[\cdots]$ by 1. The contribution in (13.62), which is relevant when computing (13.61), is therefore given by

$$
\tilde{Z}[\rho,\bar{\rho}] = \det A \sum_{\{k_i\}'} (\bar{\rho}_{i_1} A_{i_1 k_1}^{-1} \rho_{k_1})...(\bar{\rho}_{i_\ell} A_{i_\ell k_\ell}^{-1} \rho_{k_\ell}),
$$

where all k_i's are different, and the "prime" on $\{k_i\}'$ indicates that the k_i's take only values in the set $(i'_1, i'_2, ..., i'_\ell)$, labelling the right derivatives in (13.61). Thus we can write the above expression in the form

$$
\tilde{Z}[\rho,\bar{\rho}] = \det A \sum_P A_{i_1 i'_{P_1}}^{-1} A_{i_2 i'_{P_2}}^{-1} \cdots A_{i_\ell i'_{P_\ell}}^{-1} \bar{\rho}_{i_1} \rho_{i'_{P_1}} \bar{\rho}_{i_2} \rho_{i'_{P_2}} \cdots \bar{\rho}_{i_\ell} \rho_{i'_{P_\ell}}, \tag{13.63}
$$

where the sum extends over all permutations

$$
P : \begin{pmatrix} i'_1 & i'_2 & \cdots & i'_\ell \\ i'_{P_1} & i'_{P_2} & \cdots & i'_{P_\ell} \end{pmatrix}. \tag{13.64}
$$

Each of the products of Grassman variables appearing in the sum (13.63) can be put into the form

$$
F_{i_1 i_2...,i_\ell}^{i'_1 i'_2...i'_\ell} \equiv \bar{\rho}_{i_1} \rho_{i'_1} \bar{\rho}_{i_2} \rho_{i'_2} \cdots \bar{\rho}_{i_\ell} \rho_{i'_\ell}
$$

by using the anticommutation rules for Grassmann variables. It follows that

$$
\tilde{Z} = (\det A) \left[\sum_P (-1)^{\sigma_P} A_{i_1 i'_{P_1}}^{-1} \cdots A_{i_\ell i'_{P_\ell}}^{-1} \right] F_{i_1...i_\ell}^{i'_1...i'_\ell}(\rho,\bar{\rho}),
$$

where $(-1)^{\sigma_P}$ is the signum of the permutation (13.64). We now apply the left and right derivatives indicated in (13.61) to the above expression and obtain the following important result:

$$\int D(\bar{\eta}\eta)\eta_{i_1}...\eta_{i_\ell}\bar{\eta}_{i'_1}...\bar{\eta}_{i'_\ell}e^{-\sum_{i,j}\bar{\eta}_i A_{ij}\eta_j} = \xi_\ell(\det A)\sum_P(-1)^{\sigma_P}A^{-1}_{i_1 i'_{P_1}}..A^{-1}_{i_\ell i'_{P_\ell}},$$

$$(13.65)$$

where $\xi_\ell = (-1)^{\ell(\ell-1)/2}$. As a particular case of (13.65) we have that

$$\int D(\bar{\eta}\eta)\eta_i\bar{\eta}_j e^{-\sum_{i,j}\bar{\eta}_i A_{ij}\eta_j} = (\det A)A^{-1}_{ij}.$$

$$(13.66)$$

Let us define the 2-point correlation function

$$< \eta_i\bar{\eta}_j >= \frac{\int D(\bar{\eta}\eta)\eta_i\bar{\eta}_j e^{-\sum_{i,j}\bar{\eta}_i A_{ij}\eta_j}}{\int D(\bar{\eta}\eta)e^{-\sum_{i,j}\bar{\eta}_i A_{ij}\eta_j}}.$$

Then it follows from (13.56) and (13.66) that

$$< \eta_i\bar{\eta}_j >= A^{-1}_{ij}.$$

$$(13.67)$$

We shall refer to (13.67) as a *contraction*. The generalization of (13.67) to arbitrary "correlation" function is then given by

$$< \eta_{i_1}...\eta_{i_\ell}\bar{\eta}_{i'_1}...\bar{\eta}_{i'_\ell} >= \frac{\int D(\bar{\eta}\eta)\eta_{i_1}...\eta_{i_\ell}\bar{\eta}_{i'_1}\cdots\bar{\eta}_{i'_\ell}e^{-\sum_{i,j}\bar{\eta}_i A_{ij}\eta_j}}{\int D(\bar{\eta}\eta)e^{-\sum_{i,j}\bar{\eta}_i A_{ij}\eta_j}}.$$

$$(13.68)$$

Note that the euclidean correlation function contains no time-ordering operation.

13.7   Extension to Field Theory

From our discussion in Section 13.5, the formal extension of the generating functional (13.58) to Field Theory is obvious. We have for euclidean time τ,

$$Z[\rho,\bar{\rho}] = \int D(\bar{\eta}\eta)e^{-\int d^4z\int d^4z'\bar{\eta}(z)K(z,z')\eta(z')}e^{\int d^4z(\bar{\rho}(z)\eta(z)+\bar{\eta}(z)\rho(z))}.$$

One way of making the connection with a system with a discrete number of degrees of freedom is to suppose K to be an operator defined on a compact manifold $\mathcal{M}$, with a discrete eigenvalue spectrum and a complete set of eigenfunctions:

$$Ku_n = \lambda_n[K]u_n .$$

$$(13.69)$$

For the time being we shall assume that K has no zero eigenvalues.

Starting from

$$Z[0] = \int \mathcal{D}\bar{\eta}\mathcal{D}\eta\, e^{-\int \bar{\eta}K\eta}$$

$$(13.70)$$

we expand the fields $\eta(x)$, $\bar{\eta}(x)$ in terms of the above eigenfunctions:

$$\eta(x) = \sum_n a_n u_n(x),$$

$$\bar{\eta}(x) = \sum_n \bar{a}_n u_n^\dagger(x) ,$$

(13.71)

with a_n, $\bar{a}_n$, b_n, $\bar{b}_n$ Grassmann-valued c-number coefficients. Choosing u_n and v_m to be orthonormal with respect to a measure $d\mu(x)$ on $\mathcal{M}$,

$$\int d\mu(x) u_n^\dagger(x) u_{n'}(x) = \delta_{nn'}$$

(13.72)

we make the change of variable

$$\mathcal{D}\bar{\eta}\mathcal{D}\eta = J \prod_n d\bar{a}_n da_n ,$$

(13.73)

where J is the Jacobian of the transformation.

Introducing the expansions (13.71) in (13.70) and using (13.72), we obtain

$$Z[0] = J \int \prod_n d\bar{a}_n da_n e^{-\sum \lambda_n[K]\bar{a}_n a_n} .$$

Now,

$$e^{-\sum \lambda_n(K)\bar{a}_n a_n} = \prod_n e^{-\lambda_n[K]\bar{a}_n a_n} = \prod_n (1 - \lambda_n[K]\bar{a}_n a_n) .$$

The last equality only holds, because of the Grassmann properties $a_n^2 = \bar{a}_n^2 = 0$. Furthermore, according to the Berezin integration rules of the previous section,

$$\int \prod_n d\bar{a}_n da_n e^{-\sum \lambda_n[K]} = \prod_n \lambda_n[K] .$$

(13.74)

Hence we conclude,

$$Z[0] = J \prod_n \lambda_n[K] = J \det K.$$

(13.75)

The unknown constant J arising from the change of variable (13.73) will eventually drop out when calculating correlators such as (13.68).

Equation (13.75) corresponds to the identification made in (13.56). If there are zero modes present, then these should be omitted in (13.75), resulting in

$$Z[0] = J \prod_n{}' \lambda_n[K] = \det' K ,$$

(13.76)

where the "prime" indicates omission of zero modes. Hence $\det' K$ is just the subdeterminant corresponding to the non-vanishing eigenvalues of K. The existence of "zero-modes" is not seen in perturbation theory.[8]

The divergencies occurring in the loop-integrations have their reflection in the non-existence of the product (13.75) because of the (in general unbounded) growth of the eigenvalues with $n \to \infty$. Thus again, a regularization procedure has to be given for defining the determinant via (13.75).

13.8   Mathews–Salam representation of QED generating functional

We now make use of the results above in order to arrive at a representation of the generating functional of QED in terms of a path integral over gauge-field configurations *alone*. Returning to Minkowski space via the formal substitutions $\tau \to ix^0$, we have for the generating functional of the n-point Green-functions of fermions interacting with a gauge field A^μ:[9]

$$Z[\eta, \bar{\eta}; J] = \int [\mathcal{D}A] \int [\mathcal{D}\bar{\psi}][\mathcal{D}\psi]\, e^{i \int d^4z \mathcal{L}(z)} e^{i \int d^4z (\bar{\eta}\psi + \bar{\psi}\eta)(z)} e^{i \int d^4z J^\mu(z) A_\mu(z)} ,$$

$$(13.77)$$

where $\mathcal{L}(x)$ is the QED Lagrange density

$$\mathcal{L} = \mathcal{L}_G(A) + \bar{\psi}(i\not{D} - m)\psi, \quad \mathcal{L}_G(A) = -\frac{1}{4}F_{\mu\nu}F^{\mu\nu} , \qquad (13.78)$$

with D_μ the covariant derivative

$$iD_\mu = i\partial_\mu - e_0 A_\mu . \qquad (13.79)$$

Integrating over the fermions we have

$$Z[\eta, \bar{\eta}, J] = \int [\mathcal{D}A]\mathrm{Det}(i\not{D} - m)e^{i \int d^4z \mathcal{L}_G(A(z))} \qquad (13.80)$$

$$\times\, e^{i \int d^4z \int d^4z' \bar{\eta}(z)G(z,z';A)\eta(z')} e^{i \int d^4z J(z)_\mu A^\mu(z)} ,$$

where $G(x, y; A)$ is the inverse of the Dirac operator (fermionic Green function in an external field A^μ):

$$(i\not{D} - m)G(x, y; A) = \delta^{(4)}(x - y) . \qquad (13.81)$$

Let us consider the individual terms in (13.80). In order to simplify the presentation, we set $m = 0$. Then formally

$$i\, \not{\partial}G = 1 + e_0\, \not{A}G ,$$

---

[8]E. Abdalla, M. Christina B. Abdalla and K.D. Rothe, *Non-Perturbative Methods in 2 Dimensional Quantum Field Theory* (World Scientific, 1991, and extended version 2001).
[9]P.T. Mathews and A. Salam, *Phys. Rev.* **94** (1954) 185.

that is,

$$G = \frac{1}{i\,\partial\!\!\!/} + e_0 \frac{1}{i\,\partial\!\!\!/} A\!\!\!/ G \ .$$

Iterating this equation

$$G(z, z'; A) = \left[\frac{1}{i\,\partial\!\!\!/} + \frac{1}{i\,\partial\!\!\!/} e_0 A\!\!\!/ \frac{1}{i\,\partial\!\!\!/} + \frac{1}{i\,\partial\!\!\!/} e_0 A\!\!\!/ \frac{1}{i\,\partial\!\!\!/} e_0 A\!\!\!/ \frac{1}{i\,\partial\!\!\!/} + \cdots \right]_{z,z'}$$

where

$$\left(\frac{1}{i\,\partial\!\!\!/} \right)_{zz'} = S_F(z - z')$$

is the Feynman propagator satisfying

$$i\,\partial\!\!\!/ S_F(z) = \delta^4(z) \ .$$

In terms of Feynman diagrams

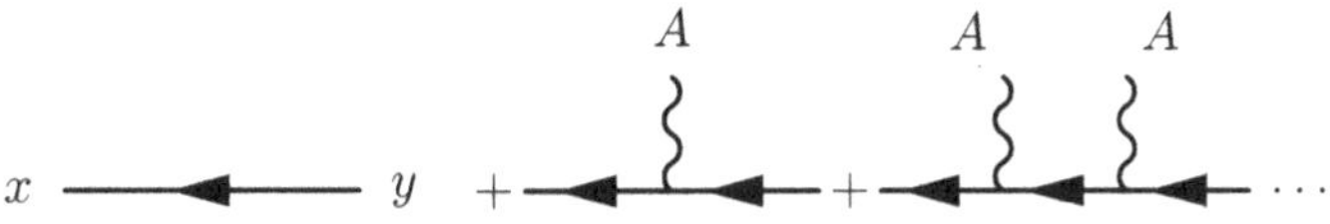

$$\tag{13.82}$$

Fig. 13.1. Green function in external gauge-field.

The diagrammatic sum represents the effect of repeated scatterings of virtual fermions in the external field A^μ, and cannot generally be performed in closed form (except in two space-time dimensions!). Explicitly it is represented by

$$iG(x, y; A)_{\alpha\beta} = iS_F(x - y)_{\alpha\beta} + \int d^4 z_1 [iS_F(x - z_1)(-ie_0\, A\!\!\!/(z_1))iS_F(z_1 - y)]_{\alpha\beta}$$

$$+ \int d^4 z_1 \int d^4 z_2 [iS_F(x - z_1)(-ie_0\, A\!\!\!/(z_1))iS_F(z_1 - z_2)(-ie_0\, A\!\!\!/(z_2))iS_F(z_2 - y)]_{\alpha\beta} + \cdots$$

In a similar spirit we now show that the logarithm of a functional determinant has a simple interpretation in terms of Feynman diagrams: it may be represented as an infinite sum of properly weighted one fermion-loop diagrams. This is easily seen on a formal level. Thus consider in general an hermitian operator D which we decompose into a "free" (D_0) and an "interacting" part (D_I) as follows:

$$D = D_0 + D_I \ .$$

Borrowing familiar formulas from matrix algebra, we write:

$$\omega_D \equiv \ln \det D = \operatorname{Tr} \ln(D_0 + D_I) \ ,$$

$$\omega_{\frac{D}{D_0}} = \ln \frac{\det D}{\det D_0} = \operatorname{Tr} \ln(1 + D_0^{-1} D_I) = \sum_{n=1}^{\infty} \frac{(-1)^{n+1}}{n} \operatorname{Tr}\left(\frac{1}{D_0} D_I \right)^n \tag{13.83}$$

where $D_0^{-1}D_I$ is to be regarded as a (non-local) operator, and where Tr denotes the trace with respect to "internal" indices as well as space-time. The functional ω_D possesses the obvious property, $\omega_{\frac{D}{D_0}} = \omega_D - \omega_{D_0}$.

The sum (13.83) has a simple graphical interpretation: it is given by the sum of (connected) 1-loop graphs as shown in Fig. 13.2, with the following "Feynman" rules:

(a) $-D_I$ for each vertex,

(b) D_0^{-1} for each propagator,

(c) (-1) overall factor for being a fermion loop,

(d) $\frac{(n-1)!}{n!} = \frac{1}{n}$ combinatorial factor for a nth order diagram.

As an example consider the euclidean Dirac operator of QED,

$$i\,\slashed{D} = (i\,\slashed{\partial} - e_0\,\slashed{A})$$

with $A^\mu(x)$ some arbitrarily chosen "external" gauge-field configuration. According to (13.83) we have

$$\ln\left[\frac{\det(i\,\slashed{\partial} - e_0\,\slashed{A})}{\det(i\,\slashed{\partial})}\right] = \sum_{n=1}^{\infty} \frac{(-1)^{n+1}}{n} Tr\left\{\frac{-e_0}{i\,\slashed{\partial}}\,\slashed{A}\right\}^n$$

$$= -\int d^4z_1\, tr\, iS_F(0)(-ie_0\,\slashed{A}(z_1)) \tag{13.84}$$

$$- \frac{1}{2}\int d^4z_1 \int d^4z_2\, tr\, [iS_F(z_1 - z_2)(-ie_0\,\slashed{A}(z_2))iS_F(z_2 - z_1)(-ie_0\,\slashed{A}(z_1))] + \cdots$$

This series has the graphical representation

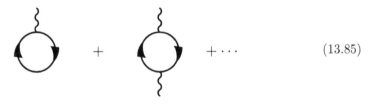

$$\tag{13.85}$$

Fig. 13.2. Functional determinant.

The functional

$$W[A] = -i\ln \det \slashed{D}$$

is called the 1-loop effective action. As we have learned, $W[A]$ is given by the sum of 1-loop *connected* diagrams shown in the figure above, whereas $e^{iW[A]}$ will also involve disconnected diagrams such as

$$(13.86)$$

Fig. 13.3. Disconnected 1-loop graphs in external field.

The effect of the A^μ integration in (13.80) is then to turn the external A^μ lines into photon propagators connecting these loops in all possible ways. This will lead to diagrams such as

Fig. 13.4. Second order vacuum graph.

The divergences occurring in the loop-integrations implicit in (13.84) have their reflection in the non-existence of the product (13.75) because of (in general, unbounded) growth of the eigenvalues with $n \to \infty$. Thus again, a regularization procedure has to be given for defining these integrals.

13.9 Faddeev–Popov quantization and α-gauges

We have seen in Sections 3 and 4 of Chapter 8 that gauge theories are first class systems requiring a choice of gauge for their quantization. The Faddeev–Popov method[10] provides a systematic procedure for implementing a particular gauge in a gauge theory. We illustrate it for the case of QED.

Starting point is the functional integral (we go to Minkowski space)

$$Z[0] = \int \mathcal{D}A \int \mathcal{D}\bar{\psi}\mathcal{D}\psi e^{iS[A,\psi,\bar{\psi}]}\,,\tag{13.87}$$

with

$$S[A,\psi,\bar{\psi}] = S_G[A] + S_F[A,\psi,\bar{\psi}]\,,$$

where S_F is the fermionic action, and S_G is the Maxwell action which, after a partial integration, can be written in the form

$$S_G[A] = \int d^4z A^\mu(z)(g_{\mu\nu}\Box - \partial_\mu\partial_\nu)A^\nu(z)\,.$$

[10]L.D. Faddeev and V.N. Popov, *Phys. Lett.* **B167** (1986) 225.

The differential operator appearing in $S_G[A]$ is not invertible, since it is transversal to the gradiant of any function, thus having an infinite set of zero-modes of the form $\partial^\mu \Lambda(x)$. In fact, the corresponding Lagrange density is invariant under the simultaneous $U(1)$ gauge transformations

$$A^\mu(x) \to A^\theta_\mu(x) = A_\mu(x) + \partial_\mu \theta(x)$$
$$\psi(x) \to \psi^\theta(x) = e^{-ie_0\theta(x)}\psi(x) .$$

(13.88)

All configurations $A^\theta_\mu, \psi^\theta$ and $\bar\psi^\theta$ are said to belong to the orbit of A^μ, ψ, ψ and are physically equivalent. The integration measure (Haar measure) $\mathcal{D}A$ is, however, gauge-invariant, and so is $\mathcal{D}\bar\psi \mathcal{D}\psi$. All configurations belonging to a given orbit thus contribute to the functional integral with the same weight, with the result that the A^μ-integral diverges.

To gain control of this infinity, we shall have to fix the gauge. The Faddeev–Popov method provides a systematic procedure for achieving this. To do this, we first have to choose a condition which selects just one configuration on each orbit. Familiar conditions are:

$$\vec\nabla \cdot \vec A = 0 \quad \text{(Coulomb gauge)}$$
$$\partial_\mu A^\mu = 0 \quad \text{(Lorentz gauge)} .$$

In general, we suppose such a condition to be local in x^μ:

$$f(A(x)) = 0.$$

(13.89)

Let $\Delta[A]$ be defined by

$$1 = \Delta[A] \int \mathcal{D}\theta \, \delta[f(A^\theta)] ,$$

(13.90)

where

$$\delta[f(A^\theta)] = \prod_x \delta\big(f(A^\theta(x))\big) .$$

It follows from the invariance of $\mathcal{D}\theta$ under the shift $\theta \to \theta + \theta'$, that

$$\Delta[A] = \Delta[A^\theta] ,$$

so that the identity (13.90) can also be written in the form

$$1 = \int \mathcal{D}\theta \, \Delta[A^\theta] \delta[f(A^\theta)] .$$

Introduce this identity into the functional integral and make use of the gauge invariance of the integration measure and of the action in the functional integral (13.87),

$$S[A, \psi, \bar\psi] = S[A^\theta, \psi^\theta, \bar\psi^\theta]$$

in order to write this functional integral in the form

$$Z[0] = \int \mathcal{D}\theta \int \mathcal{D}A^\theta \Delta[A^\theta] \delta[f(A^\theta)] \int \mathcal{D}\bar\psi^\theta \mathcal{D}\psi^\theta e^{iS[A^\theta, \psi^\theta, \bar\psi^\theta]} .$$

After renaming the variables $A^\theta \to A^\mu, \psi^\theta \to \psi, \bar{\psi}^\theta \to \bar{\psi}$, we are left with

$$Z[0] = \Omega_{U(1)} \int [\mathcal{D}A]_{FP} \, e^{i \int d^4z \mathcal{L}} \tag{13.91}$$

where

$$[\mathcal{D}A_\mu]_{F.P.} = \mathcal{D}A_\mu \Delta[A] \prod_x \delta(f(A(x)))$$

and

$$\Omega_{U(1)} = \int \mathcal{D}\theta$$

is the (infinite) "volume" of the gauge-group $U(1)$. We have thus succeeded in factorizing the infinite part of the originally undefined functional integral.

$\Delta[A]$ may be calculated as follows. With $A^\mu(x)$ chosen such that $f(A(x)) = 0$, we have to first order in θ,

$$f(A^\theta(x)) = f(A(x) + \partial\theta(x)) = \int d^4y \left(\frac{\delta f(A(x))}{\delta A^\mu(y)} \right) \partial^\mu \theta(y) + \cdots$$

$$= \int d^4y \mathcal{M}(x,y)\theta(y) + \cdots$$

where

$$\mathcal{M}(x,y) = -\partial_y^\mu \frac{\delta f(A(x))}{\delta A^\mu(y)} \ . \tag{13.92}$$

Hence it follows from (13.90) that

$$1 = \Delta[A] \int \mathcal{D}\theta \prod_x \delta \left(\int d^4y \mathcal{M}(x,y)\theta(y) \right) \ .$$

Using the usual rule for a change of variable in δ-functions, as well as (13.56), we have

$$\Delta[A] = \det \mathcal{M} = \int \mathcal{D}\bar{c}\mathcal{D}c \, e^{-\int \bar{c}\mathcal{M}c} \ , \tag{13.93}$$

with $c, \bar{c}$ Grassmann-valued *spin-zero* (ghost) fields. Notice that unlike Dirac fermions, these ghost fields carry no spin index. $\Delta[A]$ is called the Faddeev–Popov determinant.

A^μ-propagator in α-gauges

The so-called α-gauges were first introduced by 't Hooft.[11] Choose the gauge fixing function $f(x)$ to be some local function of the form

$$f(A(x)) = h(A(x)) - \kappa(x) \ .$$

---

[11]G.'t Hooft, *Nucl. Physics.*, Ser. B (1971) 173.

Now average in (13.91) over all "κ-gauges" with Gaussian weight $e^{-i\frac{\kappa^2}{2\alpha}}$:

$$Z[0] \to Z[0] = \int \mathcal{D}\kappa e^{-\frac{i}{2\alpha}\int d^4z\kappa^2(z)} \int \mathcal{D}A\Delta[A]\delta[h(A) - \kappa] e^{i\int d^4z\mathcal{L}}$$
$$= \int \mathcal{D}A\Delta[A]e^{-\frac{i}{2\alpha}\int d^4zh^2(A)} e^{i\int d^4z\mathcal{L}} .$$

Choose in particular
$$h(A(x)) = \partial_\mu A^\mu(x) .$$

Then we have from (13.92) and (13.93),
$$\Delta[A] = \det\square .$$

Hence
$$Z[0] = \int \mathcal{D}A(\det\square)e^{i\int d^4z[-\frac{1}{4}F_{\mu\nu}F^{\mu\nu} - \frac{(\partial_\mu A^\mu)^2}{2\alpha}]}e^{i\int d^4z\mathcal{L}_F} ,$$

where $\mathcal{L}_F$ stands for the fermionic part of the Lagrangian (13.78). Write

$$\int d^4z \left[-\frac{1}{4}F_{\mu\nu}F^{\mu\nu} - \frac{(\partial_\mu A^\mu)^2}{2\alpha}\right] = \int d^4z \left[-\frac{1}{2}(\partial_\mu A_\nu - \partial_\nu A_\mu)\partial^\mu A^\nu - \frac{1}{2\alpha}(\partial_\mu A^\mu)^2\right]$$
$$= \frac{1}{2}\int d^4z A^\mu \left[g_{\mu\nu}\square - \left(1 - \frac{1}{\alpha}\right)\partial_\mu\partial_\nu\right] A^\nu.$$

Hence we have for the Maxwell part,

$$Z_G[0] = \int \mathcal{D}A(\det\square)e^{\frac{i}{2}\int d^4z A^\mu[g_{\mu\nu}\square - (1-\frac{1}{\alpha})\partial_\mu\partial_\nu]A^\nu} . \tag{13.94}$$

Note that the ghost part completely decouples. This is no longer the case in QCD.

From (13.94) we see that the Feynman propagator of the gauge field is given in compact notation (compare (13.94) with (13.45) and recall that $\Delta_F = 1/\square$),

$$D_F^{\mu\nu} = \left(g_{\mu\nu}\square - \left(1 - \frac{1}{\alpha}\right)\partial_\mu\partial_\nu\right)^{-1}$$
$$= \frac{g^{\mu\nu} - (1-\alpha)\frac{\partial^\mu\partial^\nu}{\square}}{\square} . \tag{13.95}$$

Two choices of α are of particular interest:

$$\alpha = 0: \quad D_F^{\mu\nu} = \frac{g^{\mu\nu} - \frac{\partial^\mu\partial^\nu}{\square}}{\square} \quad \text{(Lorentz gauge)} \tag{13.96}$$

$$\alpha = 1: \quad D_F^{\mu\nu} = \frac{g^{\mu\nu}}{\square} \quad \text{(Feynman gauge)} \tag{13.97}$$

In momentum space, these propagators read, respectively

$$D_F^{\mu\nu}(k) = \begin{cases} -\frac{g^{\mu\nu} - (1-\alpha)\frac{k^\mu k^\nu}{k^2}}{k^2 + i\epsilon} & \text{(α gauges)} \\ -\frac{g^{\mu\nu} - \frac{k^\mu k^\nu}{k^2}}{k^2 + i\epsilon} & \text{(Lorentz gauge)} \\ -\frac{g^{\mu\nu}}{k^2 + i\epsilon} & \text{(Feynman gauge)} \end{cases} . \tag{13.98}$$

Chapter 14

Dyson–Schwinger Equation

The *fully dressed* propagators and vertex-functions form the building blocks of a general diagram in Quantum Field Theory. Their respective perturbation series as obtained from (11.21) can be summed graphically into the form of highly non-linear integral equations, which only admit a perturbative solution, as we show in this chapter.

14.1 Classification of Feynman Diagrams

The following classification of Feynman Diagrams will be of fundamental importance in the subsequent development of the subject:

Tree graphs

Diagrams which do not involve loops have a tree-like structure, and are called *tree-graphs*. Examples are

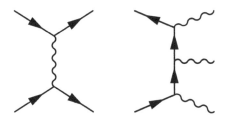

Fig. 14.1. Tree diagrams.

A general Feynman diagram in QED is constructed in terms of the 2- and 3-point functions suitably integrated over the internal space-time variables. The *skeleton* diagram of a general diagram is obtained by stripping it off all corrections to the *bare* vertices, photon and fermion lines. An example is:

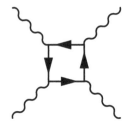

Fig. 14.2. Skeleton diagram.

Conversely we obtain from a given skeleton diagram a general diagram by adding the "meat" to the "bones" via the substitutions.

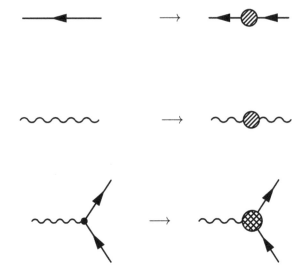

Fig. 14.3. Adding "meat" to the "bones".

Connected diagrams

We have seen in Section 5 of Chapter 11 that disconnected "vacuum" diagrams cancel in the computation of a general Green function. This nevertheless leaves us with disconnected diagrams such as

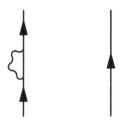

Fig. 14.4. Disconnected diagrams of 2-particle scattering.

contributing to a fermion-4-point function. Such diagrams we call "partially connected". *Connected* Green functions are defined to contain no such disconnected diagrams.

1-particle irreducible (1PI) diagrams

Among the connected diagrams there exist diagrams with the property that they can be separated into two disconnected pieces by cutting a *single* intermediate line. A simple example is provided by

Fig. 14.5. 1-particle reducible diagrams.

Connected diagrams which cannot be separated into two disconnected pieces by cutting a single internal line are called *1-particle irreducible* (1PI). An example of such a diagram is

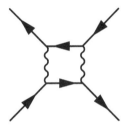

Fig. 14.6. 1-particle irreducible (1PI) diagram.

14.2 Basic building blocks of QED

The classical lagrangian of QED reads

$$\mathcal{L} = -\frac{1}{4}F_{\mu\nu}F^{\mu\nu} + \overline{\psi}(i\partial\!\!\!/ - e_0 A\!\!\!/ - m_0)\psi \ . \tag{14.1}$$

This Lagrange density serves as basis for the perturbative expansion. It contains terms quadratic and cubic in the fields. It is customary to treat the "mass term" $m\overline{\psi}\psi$, being quadratic in the fields, as part of the kinetic terms associated with the fermion field. The absence of an analogous term $\frac{\mu^2}{2}A_\mu^2$ for the gauge field reflects the masslessness of the photon. We claim that the interaction of the fields will not induce such a term either. This is consistent with the experimental fact that the photon is a massless quantum. We similarly claim that the interaction of the fields will not induce other types of interactions on quantum level. It can then be shown that the theory satisfies the basic renormalization criteria to be given later.

The 2-point functions of the photon and fermion,[1]

---

[1] Fully "dressed" 2-point functions are denoted by a "prime". Blobs which are (are not) 1PI are represented by hatched (shaded) circles respectively.

$$\langle \Omega | T A^\mu(x) A^\nu(y) | \Omega \rangle = i D_F'^{\mu\nu}(x-y) = \quad \mu \quad\quad \nu \quad\quad (14.2)$$

$$< \Omega | T \psi_\alpha(x) \bar\psi_\beta(y) | \Omega > = i S_F'(x-y)_{\alpha\beta} = \quad \alpha \quad\quad \beta \quad\quad (14.3)$$

as well as the 3-point function

$$< \Omega | T \psi_\alpha(x) \bar\psi_\beta(y) A^\mu(z) | \Omega >_{amp.} = -i e_0 \Gamma^\mu_{\alpha\beta}(x,y,z) = \quad \mu \quad\quad\quad (14.4)$$

form the basic building block of any Feynman diagram of QED. They are given by the iterative solution of the non-linear diagrammatic equations

Fig. 14.7. Non-linear Dyson–Schwinger equations.

In algebraic form these diagrammatic equations are given in momentum space by

$$i S_F'(p) = i S_F(p) + i S_F(p)(-i\Sigma(p)) i S_F'(p) \quad\quad (14.5)$$

$$i D_F'^{\mu\nu}(k) = i D_F^{\mu\nu}(k) + i D_F^{\mu\lambda}(k)\left(-i\Pi_{\lambda\rho}(k)\right) i D_F'^{\rho\nu}(k) , \quad\quad (14.6)$$

where $\Sigma(p)$ and $\Pi(k)$ are given by the sum of 1-particle irreducible (1PI) diagrams of the type

$$-i\Sigma(p) = \quad\quad + \;' \quad\quad + \quad\quad + \;\cdots \;(14.7)$$

$$-i\Pi_{\lambda\rho}(p) = \text{(diagram)} + \text{(diagram)} + \cdots \tag{14.8}$$

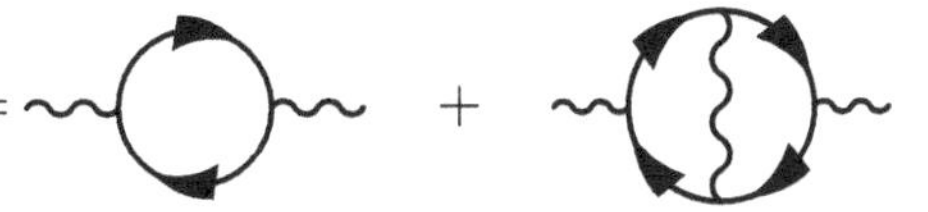

Fig. 14.8. Perturbative expansion of electron self-energy and vacuum polarization.

with

$$S_F(p) = \frac{1}{\not{p} - m_0 + i\epsilon}$$

the Feynman propagator of the fermion, and

$$D_F^{\mu\nu}(k) = -\frac{\mathcal{P}^{\mu\nu}(k)}{k^2 + i\epsilon} + \text{``gauge''} , \qquad \mathcal{P}^{\mu\nu}(k) = g^{\mu\nu} - \frac{k^\mu k^\nu}{k^2} \tag{14.9}$$

the Feynman propagator of the gauge-field in a general gauge. The term abbreviated by "gauge" stands for an expression of the form

$$gauge = k^\mu f^\nu(k) + k^\nu f^\mu(k) \equiv D_F^{\mu\nu}(k)_{long} ,$$

where the function $f^\mu(k)$ depends on the choice of gauge. In the α-gauges (see (13.98))

$$gauge = (1 - \alpha)\frac{k_\mu k_\nu}{(k^2 + i\epsilon)^2} . \tag{14.10}$$

$\mathcal{P}^{\mu\nu}(k)$ is the projector with the property

$$k_\mu \mathcal{P}^{\mu\nu} = 0, \qquad \mathcal{P}^{\mu\lambda}\mathcal{P}_{\lambda\nu} = \mathcal{P}^\mu_\nu . \tag{14.11}$$

$\Sigma(p)$ and $\Pi(k)$ are the full 1PI 2-point functions associated with the fermion and the photon respectively. They are referred to as the electron self-energy and vacuum polarization, respectively. It follows from Lorentz invariance, that they have the following algebraic decomposition

$$\Sigma(p) = \Sigma_1(p^2) + (\not{p} - m)\Sigma_2(p^2) ,$$

$$\Pi^{\mu\nu}(k) = (-k^2 g^{\mu\nu} + k^\mu k^\nu)\pi(k^2) = -\mathcal{P}^{\mu\nu}(k)k^2\pi(k^2) . \tag{14.12}$$

The fact, that the "polarization tensor" $\Pi(k)^{\mu\nu}$ is "transversal", i.e.

$$k_\mu \Pi^{\mu\nu}(k) = 0 \tag{14.13}$$

follows from the gauge invariance of the lagrangian (14.1), as we shall see later on. It is easily seen that the algebraic Eq. (14.5) has the solution

$$S_F'(p) = \frac{1}{\not{p} - m - \Sigma(p)} . \tag{14.14}$$

This expression formally stands for the inverse of the 1PI 2-point function

$$S_F'^{-1}(p) = \not{p} - m_0 - \Sigma(p) \,. \tag{14.15}$$

Substituting (14.12) into (14.6), and using (14.13) we arrive at the algebraic equation

$$iD_F'^{\mu\nu}(k) = iD_F^{\mu\nu}(k) + \mathcal{P}^\mu{}_\rho(k)\pi(k^2)iD_F'^{\rho\nu}(k) \,,$$

or

$$D_F'^{\mu\nu}(k) = -\frac{\mathcal{P}^{\mu\nu}(k)}{k^2 + i\epsilon} + D_F^{\mu\nu}(k)_{long} + \mathcal{P}^\mu{}_\rho(k)\pi(k^2)D_F'^{\rho\nu}(k) \,. \tag{14.16}$$

Make the Ansatz

$$D_F'^{\mu\nu}(k) = -\mathcal{P}^{\mu\nu}(k)D_F'(k^2) + D_F'^{\mu\nu}(k)_{long}$$

and use

$$\mathcal{P}^\mu{}_\rho(k)\mathcal{P}^{\rho\nu}(k) = \mathcal{P}^{\rho\nu}(k), \quad \mathcal{P}^\mu{}_\rho D_F^{\rho\nu}(k)_{long} = 0 \,.$$

We then have from (14.16)

$$-\mathcal{P}^{\mu\nu}(k)D_F'(k^2) + D_F'^{\mu\nu}(k)_{long} = -\frac{\mathcal{P}^{\mu\nu}(k)}{k^2 + i\epsilon} + D_F^{\mu\nu}(k)_{long}$$
$$- \mathcal{P}^{\mu\nu}(k)\pi(k^2)D_F'(k^2) \,.$$

From the tensorial structure we conclude,

$$D_F'^{\mu\nu}(k) = \frac{-\mathcal{P}^{\mu\nu}(k)}{k^2\left(1 - \pi(k^2)\right)} + D_F^{\mu\nu}(k)_{long} \,, \tag{14.17}$$

i.e. the longitudinal part of the propagator is not affected by the interaction. As for the full 1PI 3-point function, it may be written in the form

$$-ie_0\Gamma^\mu(p', p; k) = -ie_0\left(\gamma^\mu + \Lambda^\mu(p', p)\right) \,.$$

The fermion self-energy $\Sigma(p)$, photon polarization tensor $\Pi^{\mu\nu}(k)$, and 1PI irreducible 3-point function $\Gamma^\mu(p', p; k)$ represent the basic building blocks of any n-point function in QED.

14.3 Dyson–Schwinger Equations

We still need to compute $\Sigma(p)$ und $\pi(q)$ appearing in (14.14) and (14.17). They are given by the IPI 2-point functions of the fermion and gauge field, respectively. It is not hard to see that they satisfy the following non-linear integral equations involving the 2- and 3-point functions, which themselves involve $\Sigma_{\alpha\beta}(p)$ and $\pi^{\mu\nu}(k^2)$:

$$-i\Sigma_{\alpha\beta}(p) = \int \frac{d^4k}{(2\pi)^4}\left[(-ie_0)\gamma^\mu iD_F'(k)_{\mu\nu}iS_F'(p - k)(-ie_0)\Gamma^\nu(p - k, p)\right]_{\alpha\beta} \tag{14.18}$$

$$-i\Pi^{\mu\nu}(k) = -\int \frac{d^4p}{(2\pi)^4} tr\left[iS'_F(p)(-ie_0)\gamma^\mu iS'_F(p+k)(-ie_0)\Gamma^\nu(p,p-k)\right] \quad (14.19)$$

and having the diagrammatic representation

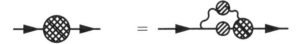

Fig. 14.9. Dyson–Schwinger equation for electron-self-energy.

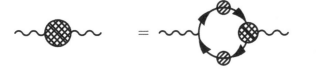

Fig. 14.10. Dyson–Schwinger equations for vacuum polarization.

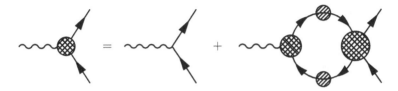

Fig. 14.11. Dyson–Schwinger equation for vertex function.

Notice that one of the vertices on the right-hand side of the first two equations must be a "naked" one, in order to avoid double counting of diagrams such as

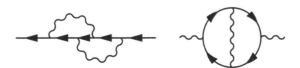

Fig. 14.12. Diagrams included only once in DS equations.

Equations (14.5), (14.6), together with (14.18) and (14.19) are called (unrenormalized) "Dyson–Schwinger equations". They evidently are highly non-linear, and can only be solved perturbatively. The vertex function $\Gamma^\mu(p',p)$ entering in these equations must be computed perturbatively or iteratively from Fig. 14.11 in terms of the 4-point function of the fermion field.

Chapter 15

Regularization of Feynman Diagrams

In the preceding chapter we have obtained the non-linear integral equations satisfied by the (unrenormalized) 2- and 3-point functions of QED. We proceed now to calculate these quantities in second order of perturbation theory. This will involve as a first step the regularization of divergent integrals and the use of some technical tools. For didactic purposes we shall illustrate these tools as we go along. It should also be kept in mind that the results for the 2- and 3-point functions are gauge dependent, since under a gauge transformation the photon and fermion fields transform as in (13.88). We shall use the so-called *Feynmann gauge* (see (13.98)), for which the photon-propagator takes the particularly simple form

$$D_F^{\mu\nu}(k) = \frac{-g_{\mu\nu}}{k^2 + i\epsilon} .$$

In this chapter we shall illustrate two regularization procedures in terms of simple examples: (i) the traditional Pauli–Villars regularization, and (ii) the dimensional regularization. Another procedure based on "Taylor subtraction" will be discussed in Chapter 16, Sections 7 and 9. In the case of QCD the procedure (ii) is particularly suited in order to respect gauge invariance.

15.1 Pauli–Villars and dimensional regularization

Feynman diagrams containing loops involve, in general, divergent integrals. In order to obtain a control of these divergencies we must introduce counter terms involving new parameters turning these integrals finite. Various approaches have been practised, such as *Pauli–Villars*,[1] *Dimensional*,[2] and *heat-kernel*[3]-regularization. Each

---

[1] W. Pauli and F. Villars, *Rev. Mod. Phys.* **21** (1949) 434.

[2] J.J. Giambiagi and C.G. Bollini, *Il Nuovo Cimento B*, **12** (1972) 20; G't Hooft and M. Veltman, *Nucl. Phys.* **B44** (1972) 189.

[3] E. Abdalla, M.C.B. Abdalla and K.D. Rothe, *Non-perturbative Methods in 2-Dimensional Quantum Field Theory* (World Scientific, Second edition, 2001).

one has its merits, the first two being the most popular ones.

Pauli–Villars regularization

By making the replacement

$$\frac{1}{k^2} \rightarrow \frac{1}{k^2} - \frac{1}{k^2 - M^2}$$

we improve the asymptotic behaviour by 2 powers in k. This procedure is referred to as Pauli–Villars regularization, and M as the Pauli–Villars mass. It plays the role of a cutoff giving control of the divergence of a Feynman integral. The simple introduction of a cutoff as upper limit in an integral may have the drawback of destroying Lorentz covariance and gauge invariance of a gauge theory. The latter has in fact been the motivation for introducing the so-called "dimensional regularization" in the context of Quantum Chromodynamics (QCD).

Dimensional regularization

We replace the Pauli–Villars regularization by first Wick rotating p^0 to ip^4, and then continuing in the dimension from space-time $D = 4$ to $D = 4 - \epsilon, \epsilon > 0$ using the following prescription:

$$d^4 p \rightarrow d^4 P = \tilde{\mu}^{4-D} P^{D-1} dP d\Omega_D \tag{15.1}$$

and the rule

$$\int d\Omega_D = \frac{2\pi^{D/2}}{\Gamma(D/2)} , \tag{15.2}$$

where $d^4 P$ is the euclidean integration measure associated with $P = (p^4, \vec{p})$. The parameter $\tilde{\mu}$ with dimensions of a mass (not to be confused with the IR regulating mass μ of the photon in the sequel) has been introduced in order to keep the dimension of the integration measure unchanged. This parameter will play the role of the Pauli–Villars regulator mass in the new scheme. We shall make use of the following result:

$$\int d^D P \frac{1}{(P^2 + a)^\alpha} = \tilde{\mu}^{4-D} \pi^{\frac{D}{2}} \frac{\Gamma\left(\alpha - \frac{D}{2}\right)}{\Gamma(\alpha)} \frac{1}{[a]^{\alpha - \frac{D}{2}}}. \tag{15.3}$$

This result follows by using the above rules and the definition of the Γ-function.[4]

The following calculations are facilitated by making use of the following *Feynman parameter representation* of a product $\prod_j 1/a_j$:

$$\prod_{j=1}^{n} \frac{1}{a_j + i\epsilon} = (n-1)! \int_0^1 d\alpha_1 ... d\alpha_n \frac{\delta(1 - \sum_j \alpha_j)}{(\sum_j \alpha_j a_j + i\epsilon)^n} . \tag{15.4}$$

To prove this formula, we start from the obvious representation

$$\prod_{j=1}^{n} \frac{i}{a_j + i\epsilon} = \int_0^\infty d\beta_1, ... d\beta_n e^{i \sum_j \beta_j (a_j + i\epsilon)} , \tag{15.5}$$

---

[4]We remark that dimensional regularization does not work if γ_5 is involved.

where ϵ is taken to be *positive* in order to ensure convergence of the integral. Introducing in (15.5) the identity

$$1 = \int_0^\infty d\lambda \delta(\lambda - \sum_j \beta_j) = \int_0^\infty \frac{d\lambda}{\lambda} \delta\left(1 - \frac{1}{\lambda}\sum_j \beta_j\right),$$

and making the change of variable

$$\frac{\beta_j}{\lambda} = \alpha_j,$$

Eq. (15.5) takes the form

$$\prod_{j=1}^n \frac{i}{a_j + i\epsilon} = \int_0^\infty d\alpha_1...d\alpha_n \int_0^\infty \frac{d\lambda}{\lambda} \lambda^n \delta(1 - \sum_j \alpha_j) e^{\lambda(i\sum_j \alpha_j a_j - \epsilon \sum_j \alpha_j)}.$$

Recalling

$$\int_0^\infty dx\, x^{n-1} e^{-xA} = \frac{\Gamma(n)}{A^n},$$

and setting

$$A = -i\sum_j \alpha_j a_j + \epsilon \sum_j \alpha_j,$$

the above equation takes the form

$$\prod_{j=1}^n \frac{i}{a_j + i\epsilon} = \Gamma(n) \int_0^1 d\alpha_1...d\alpha_n \frac{\delta(1 - \sum \alpha_j)}{[-i\sum \alpha_j a_j + \epsilon \sum \alpha_j]^n},$$

which proves our assertion (15.4).

15.1.1   Electron self-energy

To second order in the (bare) coupling constant e_0 the electron self-energy in the *Feynman-gauge* is represented by the integral

$$-i\Sigma^{(2)}(p) = (-ie_0)^2 \int \frac{d^4k}{(2\pi)^4} \frac{-ig_{\mu\nu}}{k^2 + i\epsilon} \gamma^\mu \frac{i}{(\not{p} - \not{k}) - m_0 + i\epsilon} \gamma^\nu, \tag{15.6}$$

corresponding to the diagram

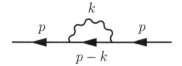

Fig. 15.1. Second order electron self-energy.

It is useful to write (15.6) in the form

$$-i\Sigma^{(2)}(p) = g_{\mu\nu}\gamma^\mu I(p)\gamma^\nu\,, \tag{15.7}$$

where

$$I(p) = (-ie_0)^2 \int \frac{d^4k}{(2\pi)^4} \frac{1}{k^2 - \mu^2 + i\epsilon} \frac{(\slashed{p} - \slashed{k}) + m_0}{(p-k)^2 - m_0^2 + i\epsilon}\,. \tag{15.8}$$

We have introduced a mass μ for the photon, in order to prevent possible infrared divergences. We shall further comment on this point at the end of this section.

Pauli–Villars regularization

The integral (15.8) is superficially *linearly* ultraviolet divergent. We shall first adopt the *Pauli–Villars* regularization by making the substitution

$$\frac{1}{k^2 - \mu^2 + i\epsilon} \longrightarrow \sum_\ell C_\ell \frac{1}{k^2 - M_\ell^2 + i\epsilon}\,.$$

In order for the integral in (15.8) to be finite with this substitution, we must choose

$$C_0 = 1,\ C_1 = -1;\ M_0 = \mu,\ M_1 = \Lambda;\ C_\ell = 0\ \text{for}\ \ell > 1\,,$$

where Λ is a regulating mass which will eventually be taken to infinity. The objective is to have control of the divergence.

Let $\alpha_0 = e_0^2/4\pi$ be the "bare" fine structure constant. Equation (15.8) then takes the form

$$I(p) = -\frac{\alpha_0}{4\pi^3} \int d^4k \sum_{\ell=0}^{1} \frac{C_\ell}{k^2 - M_\ell^2 + i\epsilon} \frac{\slashed{p} - \slashed{k} + m_0}{(p-k)^2 - m_0^2 + i\epsilon}\,. \tag{15.9}$$

For a finite *Pauli–Villars mass* Λ the regularized integrand in (15.9) behaves asymptotically like k^{-5}. This power fall-off is sufficient to render the integral finite. For $\Lambda \to \infty$ we obtain of course a divergent result. The purpose of introducing the *Pauli–Villars regulator* is to be able to control this divergence.

Making use of (15.4) we now rewrite (15.9) in the form

$$I(p) = -\frac{\alpha_0}{4\pi^3} \int d^4k \int_0^1 d\alpha \int_0^1 d\beta\, \delta(1 - \alpha - \beta) \tag{15.10}$$

$$\times \sum_\ell C_\ell \frac{(\slashed{p} - \slashed{k} + m_0)}{\left\{\alpha(k^2 - M_\ell^2) + \beta\left[(p-k)^2 - m_0^2\right] + i\epsilon\right\}^2}\,.$$

Using $\alpha + \beta = 1$ we now rewrite the denominator in the integrand as follows:

$$\alpha(k^2 - M_\ell^2) + \beta\left[(p-k)^2 - m_0^2\right] = k^2 + \beta(p^2 - m_0^2 - 2p\cdot k) - (1-\beta)M_\ell^2\,.$$

Making the change

$$k = k' + \beta p$$

in the integration variable, $I(p)$ takes the form

$$I(p) = -\frac{\alpha_0}{4\pi^3} \int d^4k' \int_0^1 d\beta \sum_\ell C_\ell \frac{(1-\beta)\not{p} - \not{k}' + m_0}{\{k'^2 - \beta^2 p^2 + \beta(p^2 - m_0^2) - (1-\beta)M_\ell^2 + i\epsilon\}^2}.$$

(15.11)

Since the denominator in $I(p)$ only depends quadratically on k', terms in the numerator which are linear in k' will not contribute to the integral, and we are left with $(k' \to k)$

$$I(p) = -\frac{\alpha_0}{4\pi^3} \int d^4k \int_0^1 d\beta \sum_\ell C_\ell \frac{N(\beta;p)}{D_\ell(k^2;\beta;p)}$$

$$N(\beta;p) = (1-\beta)\not{p} + m_0$$

$$D_\ell(k^2;\beta;p) = (k^2 - A_\ell + i\epsilon)^2,$$

(15.12)

where

$$A_\ell = \beta^2 p^2 - \beta(p^2 - m_0^2) + (1-\beta)M_\ell^2.$$

Since the integral in k is absolutely convergent, we may exchange the order of the integrations, and perform first the integration over the momenta. In order to avoid the singularities of the integrand, it is convenient to "rotate" the k^0 integration contour to the positive imaginary axis (*Wick rotation*). This corresponds to the change of integration variables $k^0 \to ik^4$ with the corresponding change in integration measure $(K_\alpha = (k^4, \vec{k}))$:

$$d^4k \to id^4K = \frac{i}{2}K^2 dK^2 d\Omega_4, \quad d\Omega_4 = \sin^2\Theta d\Theta d\Omega, \quad \int d\Omega_4 = 2\pi^2.$$

(15.13)

Note that we cannot do a continuation to the negative imaginary k^0 axis, since the integrand possesses singularities in the lower half of the complex k^0-plane, due to the $i\epsilon$ prescription. We thus obtain with (15.12) $(K^2 = x)$

$$I(p) = \frac{-i\alpha_0}{4\pi} \int_0^1 d\beta \int_0^\infty dx \sum_\ell C_\ell \frac{xN(\beta;p)}{D_\ell^2(-x;\beta;p)}.$$

We may evaluate the integrals by making use of the formula

$$\lim_{X\to\infty} \sum_\ell C_\ell \int_0^X dx \frac{x}{(x+A_\ell)^2} = \lim_{X\to\infty} \sum_\ell C_\ell \left\{ \ln\left(\frac{X+A_\ell}{A_\ell}\right) - 1 \right\}$$

$$= \lim_{X\to\infty} \left\{ \ln\left(\frac{X+A_0}{A_0}\right) - \ln\left(\frac{X+A_1}{A_1}\right) \right\} = -\ln\frac{A_0}{A_1}$$

where we have used $\sum C_\ell = 0$. Making use of this result we have

$$I(p) = \frac{i\alpha_0}{4\pi} \int_0^1 d\beta \left[(1-\beta)\not{p} + m_0\right] \ln\left(\frac{\beta^2 p^2 - \beta(p^2 - m_0^2) + (1-\beta)\mu^2 - i\epsilon}{\beta^2 p^2 - \beta(p^2 - m_0^2) + (1-\beta)\Lambda^2 - i\epsilon}\right)$$

or taking $\Lambda \to \infty$ we obtain, after rescaling,

$$I(p) \to \frac{i\alpha_0}{4\pi} \int_0^1 d\beta \left[(1-\beta)\not{p} + m_0 \right] \ln f(\beta; p^2, m_0^2, \mu^2)$$

$$- i\frac{\alpha_0}{4\pi} (m_0 + \frac{1}{2}\not{p}) \ln \left(\frac{\Lambda^2}{m_0^2} \right) \tag{15.14}$$

where

$$f(\beta; p^2, m_0^2, \mu^2) = \ln \left(\frac{(\beta^2 p^2 - \beta(p^2 - m_0^2) + (1-\beta)\mu^2 - i\epsilon)}{(1-\beta)m_0^2} \right). \tag{15.15}$$

We have chosen the branch of the logarithm such that $iI(p)$ is real for p^2 below the threshold for electron–photon production. Indeed, for $p^2 \leq (m_0 + \mu)^2$ we have in the numerator of the argument of the logarithm,

$$\beta m_0^2 + (1-\beta)\mu^2 - \beta(1-\beta)p^2 \geq \beta m_0^2 + (1-\beta)\mu^2 - \beta(1-\beta)(m_0 + \mu)^2. \tag{15.16}$$

The rhs has an extremum for $\beta = \mu/(m_0 + \mu)$, or $1 - \beta = m_0/(m_0 + \mu)$. At this value, the rhs takes the value zero. It also corresponds to a minimum. Hence $f(\beta; p^2, m_0^2) \geq 0$ is below the electron–photon production threshold. Above this threshold the electron self-energy is seen to become *complex*, corresponding to the fact that for $p^2 > (m_0 + \mu)^2$ electron–photon pairs can be produced. We thus finally have for the second order regularized electron self-energy,[5]

$$\Sigma_{reg}^{(2)}(p; e_0, m_0; M) = \frac{\alpha_0}{4\pi}(4m_0 - \not{p}) \ln \left(\frac{\Lambda^2}{m_0^2} \right) \tag{15.17}$$

$$- \frac{\alpha_0}{2\pi} \int_0^1 d\beta [2m_0 - (1-\beta)\not{p}] \ln f(\beta; p^2, m_0^2, \mu^2).$$

Notice that the superficial linear UV divergence turned out to be a logarithmic one. We shall see in Chapters 16 and 17 the role played by the (for $\Lambda \to \infty$) divergent terms in the renormalization program. This completes the calculation of the electron self-energy to second order in the coupling constant using PV regularization.

Dimensional regularization

We wish now to confront the PV calculation with the dimensional regularization. To this end we return now to (15.12), perform again a Wick rotation, continue in space-time to $D = 4 - \epsilon$ as described at the beginning of this chapter, and set all

---

[5] We recall the following properties of γ-matrices:

$$Sp(\gamma_\mu \gamma_\nu) = 4g_{\mu\nu}$$
$$Sp(\gamma_\mu \gamma_\nu \gamma_\lambda \gamma_\rho) = 4g_{\mu\rho}g_{\nu\lambda} - 4g_{\mu\lambda}g_{\nu\rho} + 4g_{\mu\nu}g_{\lambda\rho}$$
$$\gamma_\lambda \not{p} \gamma^\lambda = \underbrace{-\gamma_\lambda \gamma^\lambda \not{p}}_{-4} + \gamma_\lambda \underbrace{\{\not{p}, \gamma^\lambda\}}_{2p^\lambda} = -2\not{p}.$$

Pauli–Villars coefficients to zero, except for $c_0 = 1$. Integrating over K using (15.3), and making use of the expansion

$$\Gamma\left(2 - \frac{D}{2}\right) \sim \frac{1}{2 - \frac{D}{2}} - \gamma = \frac{2}{\epsilon} - \gamma \qquad (15.18)$$

with

$$\epsilon = 4 - D, \quad \gamma = \text{Euler constant}$$

we then have

$$I(p) = -i\frac{\alpha_0}{4\pi^3}\pi^{\frac{D}{2}}\tilde{\mu}^{4-D}\Gamma\left(2 - \frac{D}{2}\right)\int_0^1 d\beta\frac{(1 - \beta)\slashed{p} + m_0}{A_0^{\alpha - \frac{D}{2}}}$$

$$= -i\frac{\alpha_0}{4\pi}e^{-\frac{\epsilon}{2}\ln\pi}e^{\epsilon\ln\tilde{\mu}}\Gamma\left(\frac{\epsilon}{2}\right)\int_0^1 d\beta[(1 - \beta)\slashed{p} + m_0]e^{-\frac{\epsilon}{2}\ln A_0}$$

$$= -i\frac{\alpha_0}{4\pi^2}\int_0^1 d\beta[(1 - \beta)\slashed{p} + m_0]\left[\left(1 - \frac{\epsilon}{2}\ln A_0\right)\left(\frac{2}{\epsilon} - \gamma\right)(1 + \epsilon\ln\tilde{\mu})\left(1 - \frac{\epsilon}{2}\ln\pi\right)\right]$$

$$= i\frac{\alpha_0}{4\pi}\int_0^1 d\beta[(1 - \beta)\slashed{p} + m_0]\left[\ln\frac{A_0}{\tilde{\mu}^2} - \frac{2}{\epsilon} + \gamma + \ln\pi\right].$$

This can be written in the form

$$I(p) = i\frac{\alpha_0}{4\pi}\int_0^1 d\beta((1 - \beta)\slashed{p} + m_0)f(\beta; p^2, m_0^2, \mu^2) - \frac{2}{\epsilon} + finite\ constants$$

in agreement with (15.14). We see that ϵ plays the role of the Pauli–Villars mass.

15.1.2 Photon vacuum polarization

We now repeat the analogous calculation for the photon 2-point function. This will show that the dimensional regularization simplifies considerably the calculation.

For second order in the coupling constant the photon polarization tensor is represented by the diagram

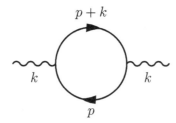

Fig. 15.2. Second order photon vacuum polarization.

with the corresponding 1-loop integral (note the extra minus sign for the fermion loop)

$$-i\Pi^{\mu\nu}(k) = -(-ie_0)^2\int\frac{d^4p}{(2\pi)^4}\,\text{tr}\left(\gamma^\mu\frac{i}{\slashed{p} - m_0 + i\epsilon}\gamma^\nu\frac{i}{\slashed{p} + \slashed{k} - m_0 + i\epsilon}\right). \qquad (15.19)$$

This integral is *superficially quadratically divergent*. On the formal level we observe that

$$-ik^{\mu}\Pi_{\mu\nu}(k) = -(-ie_0)^2 \int \frac{d^4p}{(2\pi)^4} \, \text{tr} \left(\frac{i}{\not{p} - m_0 + i\epsilon} \gamma^{\nu} - \gamma^{\nu} \frac{i}{\not{p} + \not{k} - m_0 + i\epsilon} \right) = 0 \, ,$$

provided that the shift $p + k \rightarrow p$ in the integration variable is performed. Such a shift is however only allowed if the integral is convergent. It nevertheless reflects the underlying *gauge invariance* of the theory. Since we do not want to break this symmetry, our regularization procedure should respect this symmetry.

Pauli–Villars regularization

We first consider again the PV regularization by replacing (15.19) by

$$-i\Pi_{\mu\nu}(k)_{reg} = -e_0^2 \int \frac{d^4p}{(2\pi)^4} \sum_{\ell} C_{\ell} I_{\mu\nu}(p, k; M_{\ell}) \, , \tag{15.20}$$

with[6]

$$I_{\mu\nu}(p; k, M_{\ell}) = \frac{\text{tr}[\gamma^{\mu}(\not{p} + M_{\ell})\gamma^{\nu}(\not{p} + \not{k} + M_{\ell})]}{(p^2 - M_{\ell}^2 + i\epsilon)[(p+k)^2 - M_{\ell}^2 + i\epsilon]} \, , \tag{15.21}$$

where the constants C_{ℓ} and masses M_{ℓ} ($l = 0, 1, 2$) are chosen such that the integrand tends to zero faster than $1/p^4$. This is realized by requiring[7]

$$\sum_{\ell=0}^{2} C_{\ell} = 0 \, , \quad \sum_{\ell=0}^{2} C_{\ell} M_{\ell}^2 = 0 \, , \quad C_0 = 1 \, , \quad M_0 = m_0 \, ; C_{\ell} = 0, \ell > 2 \, . \tag{15.22}$$

These equations have the solution

$$C_1 = 1 \, , \quad C_2 = -2; \quad M_1^2 = m_0^2 + 2\Lambda^2, \quad M_2^2 = m_0^2 + \Lambda^2 \, . \tag{15.23}$$

Now,

$$tr\{\gamma_{\mu}(\not{p} + M_{\ell})\gamma_{\nu}(\not{k} + \not{p} + M_{\ell})\} =: \mathcal{P}_{\mu\nu}(p, k; M_{\ell})$$

$$= 4g_{\mu\nu}\left(M_{\ell}^2 - p \cdot (k+p)\right) + 4[p_{\mu}(k+p)_{\nu} + (k+p)_{\mu}p_{\nu}] \, .$$

---

[6]Note that in contrast to the previous case of the fermion self-energy, we have made the replacement $m_0 \rightarrow M_{\ell}$ everywhere. The reason is that we want to guarantee gauge invariance, i.e. $k^{\mu}\Pi_{\mu\nu}(k) = 0$.

[7]We establish these conditions by making a Taylor expansion of the denominators in powers of p^2/M_{ℓ}^2 and $p.k/M_{\ell}^2$ and requiring that for $p \rightarrow \infty$, $\sum_{\ell} C_{\ell} I_{\mu\nu}(p; k; M_{\ell}) \rightarrow 0 \left(\frac{1}{p^5}\right)$:

$$[p^2 - M_{\ell}^2][(p+k)^2 - M_{\ell}^2] = p^4 + 2p^2(p \cdot k) + p^2 k^2$$
$$- 2p^2 M_{\ell}^2 - 2M_{\ell}^2(p \cdot k) - M_{\ell}^2 k^2 + M_{\ell}^4$$
$$\simeq p^4 \left\{1 + \frac{1}{p^2}\left(k^2 + 2(p \cdot k) - 2M_{\ell}^2\right) + \cdots\right\}$$

$$\frac{1}{[p^2 - M_{\ell}^2][(p+k)^2 - M_{\ell}^2]} \simeq \frac{1}{p^4}\left\{1 - \frac{1}{p^2}(k^2 + 2(p \cdot k) - 2M_{\ell}^2)\right\} + 0\left(\frac{1}{p^6}\right) \, .$$

Making again use of the Feynman parametrization (15.4), expression (15.20) then takes the following form:

$$-i\Pi_{\mu\nu}(k)_{reg} = -e_0^2 \int \frac{d^4p}{(2\pi)^4} \sum_\ell C_\ell \mathcal{P}_{\mu\nu}(p; k, M_\ell) \tag{15.24}$$

$$\times \int_0^1 d\alpha \int_0^1 d\beta \frac{\delta(1 - \alpha - \beta)}{\{\alpha(p^2 - M_\ell^2) + \beta[(k + p)^2 - M_\ell^2] + i\epsilon\}^2} .$$

Now,

$$\int_0^1 d\alpha \int_0^1 d\beta \frac{\delta(1 - \alpha - \beta)}{\{\alpha(p^2 - M_\ell^2) + \beta[(k + p)^2 - M_\ell^2] + i\epsilon\}^2}$$

$$= \int_0^1 d\beta \frac{1}{\{(p + \beta k)^2 + \beta(1 - \beta)k^2 - M_\ell^2 + i\epsilon\}^2} .$$

Setting $p' = p + \beta k$, Eq. (15.24) takes the form

$$-i\Pi_{\mu\nu}(k)_{reg} = -e_0^2 \sum_\ell C_\ell \int \frac{d^4p'}{(2\pi)^4} \int_0^1 d\beta \frac{\left[M_\ell^2 - (p' - \beta k) \cdot (p' + (1 - \beta)k)\right]}{\{p'^2 + \beta(1 - \beta)k^2 - M_\ell^2 + i\epsilon\}^2} g_{\mu\nu}$$

$$- 4e_0^2 \sum_\ell C_\ell \int \frac{d^4p'}{(2\pi)^4} \int_0^1 d\beta \frac{\left[(p' - \beta k)_\mu (p' + (1 - \beta)k)_\nu + (\mu \leftrightarrow \nu)\right]}{\{p'^2 + \beta(1 - \beta)k^2 - M_\ell^2 + i\epsilon\}^2} .$$

After symmetric integration in p' this reduces to

$$-i\Pi_{\mu\nu}(k)_{reg} = -4e_0^2 \sum_\ell C_\ell \int \frac{d^4p'}{(2\pi)^4} \int_0^1 d\beta \Big\{ \Big(M_\ell^2 - p'^2 + \beta(1 - \beta)k^2\Big) g_{\mu\nu}$$

$$+ 2p'_\mu p'_\nu - 2\beta(1 - \beta)k_\mu k_\nu \Big\} \frac{1}{\{p'^2 + \beta(1 - \beta)k^2 - M_\ell^2 + i\epsilon\}^2} .$$

Now, it follows from Lorentz invariance that

$$\int d^4p \, p_\mu p_\nu f(p^2) = \frac{1}{4} g_{\mu\nu} \int d^4p \, p^2 f(p^2) .$$

Hence

$$\Pi_{\mu\nu}(k)_{reg} = (-g_{\mu\nu}k^2 + k_\mu k_\nu)\pi(k^2) + g_{\mu\nu}\hat{\pi}(k^2) ,$$

where $(\alpha_0 = e_0^2/4\pi)$

$$\pi(k^2)_{reg} = i\frac{\alpha_0}{\pi^3} \int d^4p \int_0^1 d\beta \sum_\ell C_\ell \frac{2\beta(1 - \beta)}{\{p^2 + \beta(1 - \beta)k^2 - M_\ell^2 + i\epsilon\}^2}$$

and

$$\hat{\pi}(k^2)_{reg} = -i\frac{\alpha_0}{\pi^3} \int d^4p \int_0^1 d\beta \sum_\ell C_\ell \frac{M_\ell^2 - \frac{1}{2}p^2 - \beta(1 - \beta)k^2}{\{p^2 + \beta(1 - \beta)k^2 - M_\ell^2 + i\epsilon\}^2} .$$

Notice that we have separated the vacuum polarization tensor into a longitudinal (gauge dependent) and a transversal (gauge independent) part.

We perform again a Wick rotation $p^0 \to iP_4$ in the *integration* variable (see (15.13)). This leaves us with

$$\pi(k^2)_{reg} = -2\frac{\alpha_0}{\pi} \int P^2 dP^2 \int_0^1 d\beta \sum_\ell C_\ell \frac{\beta(1-\beta)}{\{P^2 - \beta(1-\beta)k^2 + M_\ell^2 - i\epsilon\}^2} \quad (15.25)$$

and

$$\hat{\pi}(k^2)_{reg} = \frac{\alpha_0}{\pi} \int P^2 dP^2 \int_0^1 d\beta \sum_\ell C_\ell \frac{M_\ell^2 + \frac{1}{2}P^2 - \beta(1-\beta)k^2}{\{P^2 - \beta(1-\beta)k^2 + M_\ell^2 - i\epsilon\}^2} \quad . \quad (15.26)$$

To evaluate the integrals we first introduce a cutoff X in the P-integration. Because of (15.22) the result of the P-integrations will be finite in the limit where $X \to \infty$. Using

$$\int_0^X dx \frac{x^2}{(x+a)^2} = a - 2a\ln\left(\frac{X+a}{a}\right) + X$$

the longitudinal piece (15.26) reduces to

$$\hat{\pi}(k^2)_{reg} = -\frac{\alpha_0}{2\pi} \sum_\ell C_\ell \left\{ 2 \int_0^1 d\beta \left(\beta(1-\beta)k^2 - M_\ell^2\right) \right\} \quad .$$

Because of (15.22) we see that $\hat{\pi}(k^2)_{reg}$ *vanishes*, so that the result for $\Pi_{\mu\nu}(k)_{reg}$ is, in fact, *transversal*:

$$\Pi_{reg}^{\mu\nu}(k) = (-g^{\mu\nu}k^2 + k^\mu k^\nu)\pi(k^2)_{reg} \quad . \quad (15.27)$$

This reflects the gauge-invariance of the polarization tensor already referred to previously.

To compute the transversal piece $\pi(k^2)_{reg}$ we make use of

$$\int_0^X dx \frac{x}{(x+a)^2} = \ln\left(\frac{X+a}{a}\right) - 1 \quad ,$$

and find

$$\pi(k^2)_{reg} = -\frac{2\alpha_0}{\pi} \lim_{X\to\infty} \sum_\ell C_\ell \int_0^1 d\beta\, 2\beta(1-\beta) \left\{ \ln\left(\frac{X + M_\ell^2 - \beta(1-\beta)k^2}{M_\ell^2 - \beta(1-\beta)k^2}\right) - 1 \right\} \quad ,$$

or explicitly

$$\pi(k^2)_{reg} - \frac{2\alpha_0}{\pi} \lim_{X\to\infty} \left\{ \int_0^1 d\beta\, 2\beta(1-\beta)\ln\left(\frac{X + m_0^2 - \beta(1-\beta)k^2}{m_0^2 - \beta(1-\beta)k^2}\right) \right.$$
$$+ \int_0^1 d\beta\, 2\beta(1-\beta)\ln\left(\frac{X + m_0^2 + 2\Lambda^2 - \beta(1-\beta)k^2}{m_0^2 + 2\Lambda^2 - \beta(1-\beta)k^2}\right)$$
$$\left. -2 \int_0^1 d\beta\, 2\beta(1-\beta)\ln\left(\frac{X + m_0^2 + \Lambda^2 - \beta(1-\beta)k^2}{m_0^2 + \Lambda^2 - \beta(1-\beta)k^2}\right) \right\} \quad .$$

Using again (15.22) one finds that the limit $X \to \infty$ is finite. Letting subsequently Λ tend to infinity we are left with

$$\pi(k^2)_{reg}^{PV} = -\frac{\alpha_0}{3\pi}\ln\left(\frac{\Lambda^2}{2m_0^2}\right) + \frac{2\alpha_0}{\pi}\int_0^1 d\beta\beta(1-\beta)\ln\left(\frac{m_0^2 - \beta(1-\beta)k^2 - i\epsilon}{m_0^2}\right) ,$$

(15.28)

where we have used

$$\int_0^1 d\beta\,\beta(1-\beta) = \frac{1}{6} .$$

(15.29)

This concludes the PV calculation for the photon 2-point function.

Dimensional regularization

Keeping only the $\ell = 0$ term in the expansion (15.25) and using (15.3) we obtain

$$\pi(k^2)_{reg} = 2\frac{\alpha_0}{\pi^3}\tilde{\mu}^{4-D}\pi^{D/2}\frac{\Gamma(2-\frac{D}{2})}{\Gamma(2)}\int d\beta\frac{\beta(1-\beta)}{[-\beta(1-\beta)k^2 + m_0^2 - i0)]^{2-D/2}} .$$

Making use of the expansion (15.18)

$$\Gamma\left(2 - \frac{D}{2}\right) \sim \frac{1}{2 - \frac{D}{2}} - \gamma = \frac{2}{\epsilon} - \gamma$$

with

$$\epsilon = 4 - D, \quad \gamma = \text{Euler constant}$$

we obtain for the leading terms in ϵ, which plays here the role of the cutoff Λ,

$$\pi(k^2)_{reg}^{Dim} = 2\frac{\alpha_0}{\pi}(e^{\frac{\epsilon}{2}\ln\tilde{\mu}^2})\pi^{-\frac{\epsilon}{2}}\left(\frac{2}{\epsilon} - \gamma\right)\int_0^1 d\beta\,\beta(1-\beta)e^{-\frac{\epsilon}{2}\ln(-\beta(1-\beta)k^2 + m_0^2 - i0)}$$

$$\simeq 2\frac{\alpha_0}{\pi}\left(1 + \frac{\epsilon}{2}\ln\tilde{\mu}^2\right)\left(1 - \frac{\epsilon}{2}\ln\pi\right)\left(\frac{2}{\epsilon} - \gamma\right)$$

$$\times \int_0^1 d\beta\,\beta(1-\beta)\left[1 - \frac{\epsilon}{2}\ln(-\beta(1-\beta)k^2 + m_0^2 - i0)\right] .$$

Using (15.29) we obtain to order $O(\alpha_0)$,

$$\pi(k^2)_{reg}^{Dim} \simeq \frac{\alpha_0}{3\pi}\left(\frac{2}{\epsilon} - \gamma - \ln\pi + \ln\frac{\tilde{\mu}^2}{m_0^2}\right)$$

(15.30)

$$- 2\frac{\alpha_0}{\pi}\int_0^1 d\beta\beta(1-\beta)\ln\left(\frac{m_0^2 - \beta(1-\beta)k^2 - i0}{m_0^2}\right) ,$$

to be compared with the corresponding PV result (15.28).

15.1.3 The vertex function

To order e_0^2 the unrenormalized *vertex* function is given by the sum of diagrams

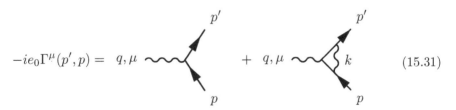

$$-ie_0\Gamma^\mu(p',p) = \qquad + \qquad \qquad (15.31)$$

Fig. 15.3. Second order vertex function.

Writing

$$\Gamma^\mu(p',p) = \gamma^\mu + \Lambda^\mu(p',p)\,,$$

we have, introducing an IR cutoff μ,

$$\Lambda^\mu(p',p) = (-ie_0)^2 \int \frac{d^4k}{(2\pi)^4} \frac{-ig^{\lambda\rho}}{k^2-\mu^2+i\epsilon}\gamma_\lambda \frac{i}{\slashed{p}'-\slashed{k}-m_0+i\epsilon}\gamma^\mu \frac{i}{\slashed{p}-\slashed{k}-m_0+i\epsilon}\gamma_\rho\,.$$

Pauli–Villars regularization

This integral is *logarithmically ultraviolet* divergent. We regularize it by making the following replacement in the photon-propagator:

$$\frac{-g^{\mu\nu}}{k^2-\mu^2+i\epsilon} \longrightarrow \sum_{\ell=0}^{1} C_\ell \frac{-g^{\mu\nu}}{k^2-M_\ell^2+i\epsilon}$$

with

$$C_0 = 1,\ C_1 = -1,\ M_0 = \mu,\ M_1 = \Lambda\,.$$

We have introduced here a mass for the photon, although the integral appears to converge for $\mu = 0$ in the infrared region $k^2 \simeq 0$. It turns out, however, that this is not the case if the electrons are on their mass shell.

The calculation is rather cumbersome. It nevertheless simplifies if one limits oneself to the respective mass shell of the electrons. On shell we have

$$\Lambda^\mu_{reg}(p',p)|_{p'^2=p^2=m_0^2} = -ie_0^2 \int \frac{d^4k}{(2\pi)^4} \sum_\ell \frac{C_\ell}{k^2-M_\ell^2+i\epsilon}$$

$$\times \frac{\gamma_\lambda(\slashed{p}'-\slashed{k}+m)\gamma^\mu(\slashed{p}-\slashed{k}+m)\gamma^\lambda}{(k^2-2p'\cdot k+i\epsilon)(k^2-2p\cdot k+i\epsilon)}\,.$$

Making use of the Feynman parameter representation (12.25) one has

$$\Lambda^\mu_{reg}(p',p) = -ie_0^2 2! \int \frac{d^4k}{(2\pi)^4} \sum_\ell C_\ell \int_0^1 d\alpha_1 \int_0^1 d\alpha_2 \int_0^1 d\alpha_3 \qquad (15.32)$$

$$\times \frac{\delta(1-\alpha_1-\alpha_2-\alpha_3)N^\mu(p,p',k)}{\{\alpha_1(k^2-2p'\cdot k)+\alpha_2(k^2-2p\cdot k)+\alpha_3(k^2-M_\ell^2)+i\epsilon\}^3}\,,$$

where

$$N^\mu(p, p', k) = \gamma_\lambda(\not{p}' - \not{k} + m_0)\gamma^\mu(\not{p} - \not{k} + m_0)\gamma^\lambda .$$

We now make the change of variable (12.28):

$$k = k' + \alpha_1 p + \alpha_2 p' . \tag{15.33}$$

This turns the denominator D_ℓ in the integrand of (15.32) into a form quadratic in k'^2:

$$D_\ell = [q^2\alpha_1\alpha_2 - m^2(\alpha_1 + \alpha_2)^2 - \alpha_3 M_\ell^2 + k'^2]^3 \tag{15.34}$$

where $q = p' - p$. Performing this change of variable also in the numerator, and separating it in its γ-even and odd parts,

$$N^\mu = N^\mu_{even} + N^\mu_{odd} ,$$

we have for N^μ_{even},

$$N^\mu_{even} = m_0\gamma_\lambda \left[\not{p}' - \alpha_1\not{p} - \alpha_2\not{p}' - \not{k}')\gamma^\mu + \gamma^\mu(\not{p} - \alpha_1\not{p} - \alpha_2\not{p}' - \not{k}')\right]\gamma^\lambda , \tag{15.35}$$

and for N^μ_{odd}

$$N^\mu_{odd} = \gamma_\lambda\left\{[(1 - \alpha_2)\not{p}' - \alpha_1\not{p} - \not{k}']\gamma^\mu[(1 - \alpha_1)\not{p} - \alpha_2\not{p}' - \not{k}'] + m_0^2\gamma^\mu\right\}\gamma^\lambda . \tag{15.36}$$

Using the contraction identity

$$\gamma_\lambda\gamma^\mu\gamma^\nu\gamma^\lambda = 4g^{\mu\nu}$$

we obtain for N^μ_{even},

$$N^\mu_{even} = 4m_0[(p + p')^\mu - 2\alpha_1 p^\mu - 2\alpha_2 p'^\mu] ,$$

where we have dropped the term linear in $\not{k}'$, which will eventually drop out after symmetric integration in k'. Since the denominator function is symmetric under the exchange of α_1 and α_2, under the α-integral we can make the replacement

$$N^\mu_{even} \to 4m_0(1 - \alpha_1 - \alpha_2)(p + p')^\mu \to 4m_0\alpha_3(p + p')^\mu . \tag{15.37}$$

The calculation of N^μ_{odd} is a bit more involved. Making use of the contraction identities

$$\gamma_\lambda\gamma^\sigma\gamma^\mu\gamma^\rho\gamma^\lambda = -2\gamma^\rho\gamma^\mu\gamma^\sigma , \quad \gamma_\lambda\gamma^\mu\gamma^\lambda = -2\gamma^\mu$$

we obtain for (15.36),

$$N^\mu_{odd} = -2[(1 - \alpha_1)\not{p} - \alpha_2\not{p}' - \not{k}']\gamma^\mu[(1 - \alpha_2)\not{p}' - \alpha_1\not{p} - \not{k}'] - 2m_0^2\gamma^\mu .$$

Note that the factors now appear inverted!

To implement later the renormalization condition (16.77), we shall need the matrix elements of N^μ taken between the Dirac spinors: $\bar{U}(p')N^\mu U(p)$. This means that we have to move $\not{p}$ and $\not{p}'$ to the right and left respectively, until we can make use of their equations of motion:

$$(\not{p} - m)U(p) = 0, \quad \bar{U}(p')(\not{p}' - m) = 0 .$$

This requires some care and patience. Keeping only terms quadratic in k' (odd terms in k' will cancel in the k'-integral) one obtains after some algebra (the Dirac spinors are understood),

$$N^{\mu}_{odd} = 4\Big\{ \Big(m_0^2 - \frac{q^2}{2} \Big) (1 - \alpha_1)(1 - \alpha_2) - \frac{m_0^2}{2} \tag{15.38}$$
$$+ \frac{m_0^2}{2}[(1 - \alpha_1)(1 - \alpha_2) - \alpha_1(1 - \alpha_1) - \alpha_2(1 - \alpha_2) - \alpha_1\alpha_2]\Big\}\gamma^{\mu}$$
$$- m_0[\alpha_3(1 - \alpha_1)p^{\mu} + \alpha_3(1 - \alpha_2)p'^{\mu}]\Big\} - 2\not{k}'\gamma^{\mu}\not{k}' ,$$

where we have used

$$q^2 = (p' - p)^2 = 2m_0^2 - 2(p \cdot p')^2$$

as well as the δ-function $\delta(1 - \alpha_1 - \alpha_2 - \alpha_3)$.

It is remarkable that the γ-"odd" term N^{μ}_{odd} contains a γ-"even" term proportional to p^{μ} and p'^{μ} when sandwiched between Dirac spinors. Joining N^{μ}_{even} in (15.37) and N^{μ}_{odd} in (15.38), we find

$$N^{\mu} = 4\Big\{ \Big(m_0^2 - \frac{q^2}{2} \Big) (1 - \alpha_1)(1 - \alpha_2) + \frac{m_0^2}{2}[(\alpha_1^2 + \alpha_2^2) - 2(\alpha_1 + \alpha_2)] - 2\frac{k^2}{4} \Big\}\gamma^{\mu}$$
$$+ 4m_0^2\alpha_3(\alpha_1 + \alpha_2)\frac{(p + p')^{\mu}}{2m_0} , \tag{15.39}$$

where we have used (under the k'-integral)

$$\not{k}'\gamma^{\mu}\not{k}' = k'_{\sigma}k'^{\rho}\gamma^{\sigma}\gamma^{\mu}\gamma_{\rho} \to k'^2 g_{\sigma}^{\rho}\gamma^{\sigma}\gamma^{\mu}\gamma_{\rho} = -2\Big(\frac{k'^2}{4} \Big)\gamma^{\mu} ,$$

as well as the symmetry of the α-integral under the substitutions $\alpha_1 \to \alpha_2, \alpha_2 \to \alpha_1$. Between spinors Λ^{μ} (p',p) is thus of the form

$$\bar{U}(p')\Lambda^{\mu}(p', p)U(p) = \bar{U}(p')\Big[\gamma^{\mu}F(q^2) + \frac{(p + p')^{\mu}}{2m_0}G(q^2) \Big] U(p) \tag{15.40}$$

with the form factor $G(q^2)$ given by the *finite* integral (we can set $\sum_{\ell} C_{\ell} \to 1$ and $M_{\ell} \to \mu$ in (15.32))

$$G(q^2)_{reg} = -\frac{\alpha}{\pi} \int_0^1 \prod_{i=1}^3 d\alpha_i \delta\Big(1 - \sum_1^3 \alpha_j \Big) \frac{m_0^2\alpha_3(\alpha_1 + \alpha_2)}{[m_0^2(\alpha_1 + \alpha_2)^2 + \alpha_3\mu^2 - \alpha_1\alpha_2q^2]} . \tag{15.41}$$

The form factor $F(q^2)$ involves a finite piece, plus a divergent piece arising from the k'^2 term in the k'-integration:

$$F(q)_{reg} = \frac{-\alpha}{\pi} \int_0^1 \prod_0^1 d\alpha_i \delta(\alpha_1 + \alpha_2 + \alpha_3 - 1)[I(q^2; \alpha)_{fin} + I_{div}(q^2; \alpha)] \tag{15.42}$$

with the finite part of the k'-integral

$$I(q^2;\alpha)_{fin} = \frac{(1-\alpha_1)(1-\alpha_2)(m_0^2 - \frac{q^2}{2}) + \frac{m_0^2}{2}[(\alpha_1^2 + \alpha_2^2) - 2(\alpha_1 + \alpha_2)]}{[m_0^2(\alpha_1 + \alpha_2)^2 + \alpha_3\mu^2 - q^2\alpha_1\alpha_2]} \quad (15.43)$$

and $I(q^2;\alpha)_{div}$ the divergent part of the integral arising from the k'^2 term in the numerator. Performing a Wick rotation $k^0 \to iK_4$ and integrating, we have

$$I(q^2;\alpha)_{div} = 2\sum_\ell C_\ell \int \frac{dxx\left(\frac{-x}{4}\right)}{[A(q^2;\alpha) + \alpha_3 M_\ell^2 + x]^3} \quad (15.44)$$

where $x = K^2$ and $A(q^2;\alpha)$ stands for

$$A(q^2;\alpha) = m_0^2(\alpha_1 + \alpha_2)^2 - q^2\alpha_1\alpha_2 . \quad (15.45)$$

Explicitly we have,

$$I(q^2;\alpha)_{div} = \int_0^{X\to\infty} dxx\left(\frac{x}{2}\right)\left\{\frac{-1}{[A(q^2;\alpha) + \alpha_3\mu^2 + x]^3} - \frac{-1}{[A(q^2;\alpha) + \alpha_3\Lambda^2 + x]^3}\right\}$$
$$+ \frac{1}{2}\ln[A(q^2;\alpha) + \alpha_3\mu^2] - \frac{1}{2}\ln[A(q^2;\alpha) + \alpha_3\Lambda^2].$$

Putting things together we have for $\Lambda \to \infty$,

$$F(q^2)_{reg} \sim -\frac{\alpha}{\pi}\int_0^1 \prod d\alpha_i\, \delta(1 - \sum\alpha_i)$$
$$\times \left\{\frac{(1-\alpha_1)(1-\alpha_2)(m_0^2 - \frac{q^2}{2}) + \frac{m_0^2}{2}[(\alpha_1^2 + \alpha_2^2) - 2(\alpha_1 + \alpha_2)]}{[m_0^2(\alpha_1 + \alpha_2)^2 + \alpha_3\mu^2 - q^2\alpha_1\alpha_2]}\right.$$
$$\left.+ \frac{1}{2}\ln\left(\frac{m_0^2(\alpha_1 + \alpha_2) + \alpha_3\mu^2 - q^2\alpha_1\alpha_2}{\alpha_3 m_0^2}\right) - \frac{1}{2}\ln\frac{\Lambda^2}{m_0^2}\right\}. \quad (15.46)$$

Dimensional regularization

We return to the computation of the singular term $I(q^2)_{div}$ in (15.42) by dimensional regularization. We wish to use formula (15.3) to compute (15.44). Observing that

$$\int_0^\infty dxx = \int_0^\infty dK^2K^2 \to \frac{1}{\pi^2}\int d^4K$$

we can write

$$I(q^2;\alpha)_{div} = 2\int \frac{d^D K}{\pi^2}\frac{\left(\frac{-K^2}{4}\right)}{(K^2 + a(q^2;\alpha))^3}$$

where

$$a(q^2;\alpha) = A(q^2;\alpha) + \alpha_3\mu^2 .$$

We have, using (15.3) and (15.18),

$$I(q^2;\alpha))_{div} = -\frac{1}{2}\int \frac{d^D K}{\pi^2}\frac{K^2}{(K^2+a(q^2;\alpha))^3}$$

$$= -\frac{1}{2}\int \frac{d^D K}{\pi^2}\frac{1}{(K^2+a(q^2;\alpha))^2} + \frac{1}{2}\int \frac{d^D K}{\pi^2}\frac{a(q^2;\alpha)}{(K^2+a(q^2;\alpha))^3}$$

$$= -\frac{1}{2}\tilde{\mu}^\epsilon \pi^{-\frac{\epsilon}{2}}\frac{\Gamma(\frac{\epsilon}{2})}{\Gamma(2)}\frac{1}{a(q^2;\alpha)^{\frac{\epsilon}{2}}} + \frac{1}{4},$$

or

$$I(q^2;\alpha)_{div} = -\frac{1}{2}(1+\epsilon\ln\tilde{\mu})\left(1-\frac{\epsilon}{2}\ln\pi\right)\left(\frac{2}{\epsilon}-\gamma\right)\left(1-\frac{\epsilon}{2}\ln a(q^2;\alpha)\right) + \frac{1}{4}$$

$$= \frac{1}{2}\ln\frac{a(q^2;\alpha)}{\tilde{\mu}^2} + \left\{-\frac{1}{\epsilon}+\frac{1}{2}\ln\pi+\frac{\gamma}{2}+\frac{1}{4}\right\},$$

where the terms in brackets now play the role of $\ln\frac{\Lambda^2}{m_0^2}$.

The parallel presentation of the Pauli–Villars and dimensional regularization should serve to appreciate the advantage or disadvantage of these two regularization procedures.

Chapter 16

Renormalization

The individual terms contributing to the perturbative expansion of a general Green function are represented by Feynman diagrams. As we have seen in Chapter 15, the integrals representing diagrams involving at least one loop are generally divergent, and are thus *a priori* meaningless. We may nevertheless isolate the *finite* part of a diagram by subtracting a suitable *polynomial* involving infinite constants. Since this procedure does not define *a priori* a unique finite part, there remains an ambiguity in the choice of this polynomial. If we do this for each loop-diagram separately, we appear to generate, in general, an infinite number of undetermined constants as we consider diagrams to all orders of perturbation theory. If this were so, the theory would be meaningless.

There exists, however, a class of theories, for which the number of arbitrary constants is not only finite, but can also be absorbed into the *bare* parameters describing the classical theory, such as *couplings constants, masses* and the *wave function normalization constant Z* appearing in the asymptotic condition (10.16). Such theories are called *renormalizable*. There exist simple criteria to decide whether a theory is renormalizable or not, provided the interaction of the fields does not induce new types of vertices.

The process of renormalization — that is, given a unique meaning to the theory, essentially involves two steps: (i) In the first step one introduces a suitable cutoff in the Feynman integrals, in order to be able to associate with them a finite expression. The result will of course be cutoff dependent. This step is referred to as *regularization* (see previous chapter). (ii) In the second step one eliminates this cutoff dependence by trading the "bare" parameters for physical parameters, which are fixed by comparison with experiment, and cannot be intrinsically calculated from the theory. This "normalization" procedure is referred to as *renormalization*.

In the following sections we shall consider, in particular, the case of electrodynamics. The Feynman rules for this case were given in Section 8 of Chapter 11.

16.1   The principles of renormalization

In this section we shall consider the general principles of *renormalization* leading to unambiguous and finite results for an arbitrary Green function in Quantum Field Theory. We shall illustrate these principles for QED, described by the Lagrange density

$$\mathcal{L} = -\frac{1}{4}F_{\mu\nu}F^{\mu\nu} + \overline{\psi}(i\slashed{\partial} - e_0\slashed{A} - m_0)\psi , \tag{16.1}$$

where e_0 and m_0 are "bare" parameters having the dimensions of charge and mass respectively, and the fields stand for *unrenormalized* fields. The renormalization program consists of essentially three steps:

(i) Regularization

As we have seen, Feynman diagrams involving at least one loop are generally divergent, and hence are *a priori* meaningless and required regularization by effectively introducing a cutoff. As we have exemplified in Chapter 15, there exist many ways of doing this: Pauli–Villars, analytic, dimensional, Taylor subtraction, subtracted dispersion relations and heat-kernel regularization, just to mention the most important ones. This procedure generally leads to the introduction of a parameter carrying the dimensions of a mass. We shall denote this parameter generically by Λ. It is the necessity of introducing some dimensional parameter to regularize the theory, which results in the breaking of scale invariance on quantum level, even if the classical lagrangian is scale invariant, as is the case for QED with massless fermions. Thus after regularization we are left with Green functions which depend on this "cutoff parameter" as well as on the *bare* constants appearing in the defining lagrangian:

$$G_{reg}^{(n,n,\ell)}(x_1...;y_1...;z_1...;m_0,e_0;\Lambda) \tag{16.2}$$
$$= \langle\Omega|T\psi_{\alpha_1}(x_1)...\psi_{\alpha_n}(x_n)\overline{\psi}_{\beta_1}(y_1)...\overline{\psi}_{\beta_n}(y_n)A^{\mu_1}(z_1)...A^{\mu_\ell}(z_\ell)|\Omega\rangle_{reg} .$$

(ii) Renormalization

The result obtained for the regularized Green functions will depend on the *regularization scheme* one employs. Physical results should, of course, not depend on the choice of such a scheme. In a so-called renormalizable theory, we can nevertheless obtain an unambiguous and unique result by imposing suitable *normalization conditions*. The procedure for achieving this is called *renormalization*. In a renormalizable theory the number of normalization conditions to be imposed is at *most* equal to the number of fields (fermions, gauge field) and parameters (electron charge and mass) in the theory. With the aid of these normalization conditions we can eliminate the cutoff dependence, trading the *bare* (meaningless) parameters entering in the classical lagrangian for physically meaningful parameters, which can be determined from experiment.

(iii) Renormalization group equations

It is up to us to choose a suitable set of parametrizations. We thus obtain different results for different parametrizations. All of these results should, however, describe the same physics. The equations relating these different results are called *renormalization group equations.*

Steps (i) and (ii) can be summarized in an algorithm to be described next.

Algorithm

Starting point of our algorithm is the classical lagrangian (16.1). The Feynman diagrams computed from this lagrangian are in general divergent, and thus need to be first defined via some suitable regularization procedure, which will introduce a cutoff Λ into the theory (step (i)). Renormalizability of the theory now means to introduce new parameters via the substitutions[1]

$$m_0 = Z_m^{-1} m \; , e_0 = Z_e^{-1/2} e, \tag{16.3}$$

$$\psi = Z_\psi^{1/2} \tilde{\psi} \; , \; A^\mu = Z_A^{1/2} \tilde{A}^\mu \; , \tag{16.4}$$

and to choose the *renormalization constants* to be suitable functions of the cutoff such as to turn the (still) cutoff dependent expressions for *all* Green functions $\tilde{G}$ of the theory into finite expressions in the limit $\Lambda \to \infty$: $G_{reg} \to \tilde{G} \to G_{ren}$.

Since Z_m, Z_e, Z_ψ and Z_A are dimensionless, they can be chosen to depend on the renormalized parameters of the theory in the form

$$Z = Z\left(e, \frac{\Lambda}{m}\right) \; , \tag{16.5}$$

where e and m are the electric charge and mass as taken from experiment. Renormalizability then means that the regularized Green functions $\tilde{G}$ defined by

$$\tilde{G}^{(n,n,\ell)}(...; e, m; \Lambda) = Z_A^{-\ell/2}\left(e, \frac{\Lambda}{m}\right) Z_\psi^{-n}\left(e, \frac{\Lambda}{m}\right) \tag{16.6}$$

$$\times G_{reg}^{(n,n,\ell)}\left(...; Z_e^{-1/2}\left(e, \frac{\Lambda}{m}\right) e, Z_m^{-1}\left(e, \frac{\Lambda}{m}\right) m; \Lambda\right)$$

has a finite limit for suitably chosen renormalization constants, as we let Λ go to "infinity", where the limit $\Lambda \to \infty$ stands symbolically for "removing the cutoff".[2] If the limit exists, then (16.6) defines in this limit the so-called *renormalized Green functions.*[3] It is very important to realize that this supposes that finiteness is achieved in this way for *all Green functions of the theory.*[4]

This requirement does not fix the constants (16.5) uniquely since different cutoff procedures will lead to different cutoff dependent results. Uniqueness of the Green

---

[1]Our notation for the renormalization constants does not follow convention in order to associate them directly with the fields and parameters involved.

[2]We denote generally by a "tilde" the pre-renormalized Green functions.

[3]Notice that formally the rhs of (16.6) is just the rhs of (16.2) with ψ and A^μ now replaced by the "tilde" fields. We generally denote *pre-renormalized* Green functions by a tilde.

[4]For later notational convenience we shall frequently omit the dependence of $G_{reg}(\cdots; e, m; \Lambda)$ and $\tilde{G}(\cdots; e, m; \Lambda)$ on the parameters., and write $G_{reg}(\cdots)$ and $\tilde{G}(\cdots)$, respectively.

functions is achieved by imposing suitable conditions. To be specific, we take QED as an example.[5]

(a) If m is to have the meaning of a physical mass, we must have for the inverse of the fermion 2-point function: (for notation see table of Section 4 in Chapter 11)

$$\tilde{S}_F^{-1}(p; e, m; \Lambda)|_{\not{p}=m} = 0. \tag{16.7}$$

(b) We require the residue of the pole of the fermion 2-point function to be *one*, in accordance with the *asymptotic condition* (10.16) for the (pre) renormalized field defined in (16.6):

$$(\not{p} - m)\tilde{S}_F(p; e, m; \Lambda)|_{\not{p}=m} = 1. \tag{16.8}$$

(c) For the photon 2-point function it turns out that we need not impose a condition analogous to (a), requiring the photon 2-point function to have the pole at zero-mass. This happens to be guaranteed by the gauge invariance of the theory, as we shall see.

(d) We require the residue at the (zero-mass) pole of the photon 2-point function to be "one", again in accordance with the asymptotic condition:

$$k^2 \tilde{D}_F^{\mu\nu}(k; e, m; \Lambda)_{tr}|_{k^2 \approx 0} \approx -g^{\mu\nu} + \frac{k^\mu k^\nu}{k^2} . \tag{16.9}$$

(e) The Fourier transform of the "amputated" and regularized 3-point function,

$$ie\Gamma_{reg}^\mu(x, y, z)) = < \Omega | T\psi(x)\bar{\psi}(y)A^\mu(z) | \Omega >_{amp, reg}$$

stripped off an overall charge obeys the following relation[6]

$$\tilde{\Gamma}^\mu(p', p; e, m, \Lambda) = Z_1 \Gamma_{reg}^\mu(p', p; Z_e^{-\frac{1}{2}} e, Z_m^{-1} m, \Lambda) , \tag{16.10}$$

with

$$Z_1 = Z_\psi Z_A^{\frac{1}{2}} Z_e^{-\frac{1}{2}} \tag{16.11}$$

and we require that at zero momentum transfer we have for the on-shell 3-point function

$$\tilde{\Gamma}^\mu(p, p) = \gamma^\mu . \tag{16.12}$$

This amounts to requiring e to be the electron charge measured in e^-e^- scattering in the forward direction, a process described exactly by the diagrams of Fig. 11.11.

---

[5]In the literature, the renormalization constants are denoted by Z_1, Z_2, Z_3, the relation to the ones introduced above being

$$Z_A = Z_3 , \quad Z_\psi = Z_2 , \quad Z_e^{-1/2} Z_A^{1/2} Z_\psi = Z_1.$$

[6]Notice that unlike (16.6), the renormalization constants appear with positive powers, since the (fully dressed) external legs have been amputated.

For QED the above normalization conditions fix the renormalization constants as a function of the *physical* parameters e, m and the cutoff. In the limit $\Lambda \to \infty$,

$$\tilde{G}^{(n,n,\ell)}(...; e, m; \Lambda) \to G_{ren}^{(n,n,\ell)}(...; e, m) \qquad (16.13)$$

now defines unique functions called the "renormalized Green functions", in all orders of perturbation theory.

16.2   Renormalizability of QED

The substitutions (16.3) and (16.4) can be viewed as the introduction of counterterms properly chosen to accommodate the renormalization conditions.

The fields and parameters (charge and mass) appearing in the lagrangian (16.1) refer to *unrenormalized* quantities. A perturbative scheme referring *from the start* to renormalized Greens functions in terms of renormalized parameters is obtained by making the substitutions (16.3) and (16.4) in the bare Lagrangian (16.1) and rewriting it in the form

$$\mathcal{L} = \mathcal{L}_0 + \mathcal{L}'_I \, ,$$

where

$$\mathcal{L}_0 = -\frac{1}{4} F_{\mu\nu} F^{\mu\nu} + \overline{\psi}(i\slashed{\partial} - m_0)\psi \, ,$$

and

$$\mathcal{L}'_I = -Z_1 e \overline{\psi} \slashed{A} \psi - \frac{1}{4}\delta Z_A F_{\mu\nu} F^{\mu\nu} + \delta Z_\psi \overline{\psi}(i\slashed{\partial} - m_0)\psi \qquad (16.14)$$

with $A_\mu, \psi, \overline{\psi}$ now *renormalizd* fields, and $\delta Z_A, \delta Z_\psi, Z_1$ given by

$$\delta Z_A = Z_A - 1 \, , \quad \delta Z_\psi = Z_\psi - 1 \quad Z_1 = Z_\psi Z_A^{\frac{1}{2}} Z_e^{-\frac{1}{2}} \, . \qquad (16.15)$$

Note that this amounts to working with Feynman rules given in terms of the Feynman propagators

$$S_F(p) = \frac{1}{\slashed{p} - m_0 + i\epsilon} \, ,$$

$$D_F^{\mu\nu}(k) = \frac{-\mathcal{P}^{\mu\nu}(k)}{k^2 + i\epsilon} + D_F^{\mu\nu}(k)_{long}$$

with the projector (14.9) and a modified interaction $\mathcal{L}'_I$ adjusted to satisfy the normalization conditions (a) through (e).

In a perturbative approach we must think of the renormalization constants as given by a power series in the fine structure constant:

$$\delta Z_i := Z_i\left(e, \frac{\Lambda}{m}\right) - 1 = \sum_{n=1}^{\infty} \alpha^n \delta z_i^{(n)}\left(\frac{\Lambda}{m}\right) \, . \qquad (16.16)$$

The expansion coefficients of these renormalization constants will be determined to every given order in perturbation theory by the normalization conditions (16.7)–(16.12).

The quadratic counterterms

The terms

$$\delta\mathcal{L}_0^{(F)} = \delta Z_\psi \overline{\psi}(i\not{\partial} - m_0)\psi$$

(16.17)

$$\delta\mathcal{L}_0^{(G)} = -\frac{1}{4}\delta Z_A F_{\mu\nu}F^{\mu\nu} \,\hat{=}\, \delta Z_A \frac{1}{2}A_\mu(g^{\mu\nu}\Box - \partial^\mu\partial^\nu)A_\nu$$

in $\mathcal{L}'_I$ play a special role since they are *quadratic* in the fields. Indeed, their role is just to provide the necessary "counterterms" allowing one to implement the normalization conditions (16.8) and (16.9) for the renormalized fields $\tilde{\psi}_\alpha, \tilde{A}^\mu$. Let us examine how this works.

Let us suppose for the moment that the trilinear interaction term in (16.14) is absent. We then obtain for the fermion 2-point function

with

$$\alpha \,\blacktriangleleft\!\!\bullet\!\!\blacktriangleleft\, \beta \;=\; i\delta Z_\psi(\not{p} - m_0)_{\alpha\beta}$$

(16.18)

Hence

$$iS_F(p) \longrightarrow \frac{i}{\not{p}-m_0} + \frac{i}{\not{p}-m_0}i\delta Z_\psi(\not{p}-m_0)\frac{i}{\not{p}-m_0} + ...$$

$$= \frac{i}{\not{p}-m_0}\frac{1}{1+\delta Z_\psi} = \frac{iZ_\psi^{-1}}{\not{p}-m_0} = iZ_\psi^{-1}S_F(p).$$

Similarly we have,

with

$$\lambda \,\sim\!\!\bullet\!\!\sim\, \rho \;=\; i\delta Z_A(-g^{\rho\lambda}k^2 + k^\lambda k^\rho) = -i\delta Z_A k^2\mathcal{P}^{\lambda\rho}(k)$$

(16.19)

Hence

$$D_F^{\mu\nu}(k) \longrightarrow Z_A^{-1/2} D_F^{\mu\nu}(k)_{tr} + D_F^{\mu\nu}(k)_{long} . \qquad (16.20)$$

This means that we may equivalently work with the interaction Lagrangian

$$\mathcal{L} = -\frac{1}{4} F_{\mu\nu} F^{\mu\nu} + \overline{\psi}(i\not{\partial} - m_0)\psi + Z_1 e \overline{\psi} \gamma^\mu \psi A_\nu$$

and new Feynman rules given in terms of the Feynman propagators

$$S_F(p) \longrightarrow \frac{Z_\psi^{-1}}{\not{p} - m_0 + i\epsilon} \qquad (16.21)$$

$$D_F^{\mu\nu}(k)_{tr} \longrightarrow -\frac{Z_A^{-1}}{k^2 + i\epsilon} \mathcal{P}^{\mu\nu}(k) \qquad (16.22)$$

with $\mathcal{P}^{\mu\nu}$ the projector

$$\mathcal{P}^{\mu\nu}(k) = g^{\mu\nu} - \frac{k^\mu k^\nu}{k^2} . \qquad (16.23)$$

We are now ready to implement the rest of the (formal) renormalization program by including the trilinear interaction term.

16.2.1   Fermion 2-point function

The 1PI 2-point function $\Sigma(p)$ (fermion self-energy) is given by an infinite sum of diagrams such as represented for example by the diagrams

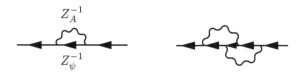

Fig. 16.1. Fermion self-energy diagrams.

With the new propagators (16.21), (16.22), and vertex $Z_1 e = Z_\psi Z_A^{\frac{1}{2}} e_0 \gamma^\mu$ from (16.14), all fermion self-energy contributions factorize, conform to our notation (16.6), as

$$\Sigma(p)_{reg} \longrightarrow Z_\psi \Sigma(p)_{reg} . \qquad (16.24)$$

Correspondingly the Dyson–Schwinger Eq. (14.5) is replaced by the pre-renormalized Dyson–Schwinger equation

$$\tilde{S}_F(p) = \frac{Z_\psi^{-1}}{\not{p} - m_0} + \frac{Z_\psi^{-1}}{\not{p} - m_0} \left(Z_\psi \Sigma(p)_{reg} \right) \tilde{S}_F(p) \qquad (16.25)$$

Eq. (16.25) has the solution[7]

$$\tilde{S}_F(p) = \frac{Z_\psi^{-1}}{\not{p} - m_0 - \Sigma(p)_{reg}} \equiv Z_\psi^{-1} S_F'(p) . \tag{16.26}$$

It is a peculiarity of the proper self-energy, that a multiplicative renormalization (16.24) is not sufficient to render it finite, but requires also a *subtractive* renormalization. To this effect we write $\Sigma(p)_{reg}$ in the form of a Taylor expansion in $\not{p}$ around $\not{p} = m$, where m is the *physical* mass:

$$\Sigma(p)_{reg} = \Sigma(p)_{reg}|_{\not{p}=m} + \left(\frac{\partial \Sigma(p)_{reg}}{\partial \not{p}}\right)_{\not{p}=m} (\not{p} - m) + \Sigma'(p) . \tag{16.27}$$

Note that the calculation of the second term on the rhs is performed after first making the replacement $p^2 = (\not{p})^2$. It contributes to Z_ψ below. Note also that the third term $\Sigma'(p)$ is of order $(\not{p} - m)^2$, which will be important for satisfying the normalization condition. In order for $\tilde{S}(p)$ to have a pole at the physical mass $m = m_0 + \delta m$ we must choose

$$\delta m = \Sigma(p)_{reg}|_{\not{p}=m} . \tag{16.28}$$

Set further

$$B = \left(\frac{\partial \Sigma(p)_{reg}}{\partial \not{p}}\right)_{\not{p}=m} . \tag{16.29}$$

Then we are left with

$$\tilde{S}(p) = \frac{Z_\psi^{-1}}{(\not{p} - m)(1 - B) - \Sigma'(p)} . \tag{16.30}$$

We now require that

$$Z_\psi = \frac{1}{1 - B} . \tag{16.31}$$

We then have

$$\tilde{S}_F(p) = \frac{1}{\not{p} - m - \tilde{\Sigma}(p)}$$

where

$$\tilde{\Sigma}(p) = Z_\psi \Sigma'(p) .$$

Removing the cutoff, we finally have with $\tilde{\Sigma}(p) \to \Sigma(p)_{ren}$,

$$S_F(p)_{ren} = \frac{1}{\not{p} - m - \Sigma(p)_{ren}} . \tag{16.32}$$

Note that since $\Sigma'(p) \approx (\not{p} - m)^2$ for $\not{p} \to m$, the residue at the pole is *one*.

---

[7]Note that the "prime" here, and in general, stands for the subscript "reg".

16.2.2 Photon 2-point function

We proceed as above. The 1PI 2-point function is this time given by the sum of 1PI diagrams such as for example the diagrams

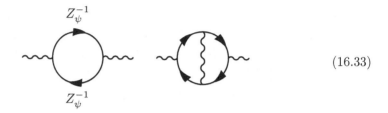

(16.33)

Fig. 16.2. Second order vacuum polarization diagrams.

With the new propagators (16.21), (16.22) and vertex $Z_1 e = Z_\psi Z_A^{\frac{1}{2}} e_0 \gamma^\mu$ from (16.14), all vacuum polarization diagrams factorize, conform to our notation (16.6), as

$$\Pi_{reg}^{\mu\nu}(k) \longrightarrow Z_A \Pi_{reg}^{\mu\nu}(k) \ .$$

Again this generalizes to a diagram of arbitrary order. The pre-renormalized Dyson–Schwinger equation (14.6) is correspondingly replaced by,

$$\tilde{D}_F^{\mu\nu}(k) = Z_A^{-1} D_F^{\mu\nu}(k)_{tr} + D_F^{\mu\nu}(k)_{long} + D_F^{\mu\lambda}(k)\Pi_{\lambda\rho}(k)_{reg}\tilde{D}_F^{\rho\nu}(k) \ .$$

Using (see (15.27)),

$$\Pi_{reg}^{\mu\nu}(k) = (-g^{\mu\nu}k^2 + k^\mu k^\nu)\pi(k^2)_{reg}$$

we obtain from here, in correspondence to (16.26),

$$\tilde{D}_F^{\mu\nu}(k)_{tr} = \frac{-Z_A^{-1}\mathcal{P}^{\mu\nu}(k)}{k^2\left(1 - \pi(k^2)_{reg}\right) + i\epsilon} = Z_A^{-1}D_F'^{\mu\nu}(k)_{tr} \qquad (16.34)$$

and

$$\tilde{D}_F^{\mu\nu}(k)_{long} = D_F^{\mu\nu}(k)_{long} \ , \qquad (16.35)$$

with $\mathcal{P}^{\mu\nu}(k)$ the projector (14.9). Expanding $\pi(k^2)$, in analogy to (16.27), around $k^2 = 0$, the physical photon mass,

$$\pi(k^2)_{reg} = \pi(0)_{reg} + \pi'(k^2) \ , \qquad (16.36)$$

we may rewrite (16.34) as

$$\tilde{D}_F^{\mu\nu}(k)_{tr} = \frac{-Z_A^{-1}\mathcal{P}^{\mu\nu}(k)}{k^2\left(1 - \pi(0)_{reg}\right) - k^2\pi'(k^2) + i\epsilon} \ . \qquad (16.37)$$

Notice that $\pi'(0) = 0$. Hence in order to implement the normalization condition (16.9) we must choose

$$Z_A = \frac{1}{1 - \pi(0)_{reg}} \, . \tag{16.38}$$

We thus have

$$\tilde{D}_F^{\mu\nu}(k)_{tr} = \frac{-\mathcal{P}^{\mu\nu}(k)}{k^2 \left(1 - \tilde{\pi}(k^2)\right) + i\epsilon}$$

where

$$\tilde{\pi}(k^2) = Z_A \pi'(k^2) \, . \tag{16.39}$$

We expect again $\tilde{\pi}(k^2)$ to tend to finite limit $\pi(k)_{ren}$ for suitably chosen renormalization constants, as we remove the cutoff. We thus finally obtain for the *renormalized* transversal photon a 2-point function

$$D_F^{\mu\nu}(k)_{ren} = \frac{-\mathcal{P}^{\mu\nu}(k)}{k^2 \left(1 - \pi(k^2)_{ren}\right) + i\epsilon} \, . \tag{16.40}$$

It is important to note that only the transversal part of the propagator gets renormalized by the interaction! Notice again that $\pi(0)_{ren} = 0$, so that the residue at the pole is *one*.

16.2.3   Vertex function

Similar considerations based on an inspection of the sum of diagrams such as

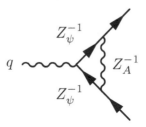

Fig. 16.3. 1-loop QED vertex diagram.

show that with the new set of Feynman rules and vertex $Z_1 e = Z_\psi Z_A^{\frac{1}{2}} e_o$, we have

$$e_0 \Lambda^\mu(p', p)_{reg} \longrightarrow Z_1 e \Lambda^\mu(p', p)_{reg} \tag{16.41}$$

or,

$$\tilde{\Gamma}^\mu(p', p) = Z_1[\gamma^\mu + \Lambda^\mu(p', p)_{reg}]$$

where Z_1 is the renormalization constant defined in (16.11) and *it is conventional to drop an overall charge factor e*. More precisely,

$$\tilde{\Gamma}^\mu(p', p) = Z_1[\gamma^\mu + \Lambda^\mu_{skel}(p', p; S'_F, D'_F, \Gamma; e_0, m_0)] \tag{16.42}$$

where the subscript "skel" is short for "skeleton".

It is by no means trivial to see that for a suitable choice of renormalization constants Z_e, Z_m, Z_1 as a function of the cutoff, (16.42) will define a quantity which is finite in the limit where the cutoff is taken to "infinity".

There persist two problems in this renormalization program:

- *infrared divergences* to cope with, as we have witnessed in the case of $\Sigma(p)_{reg}$ and $\Gamma^\mu(p', p)_{reg}$. The origin of this divergence lies in the fact that we have represented the photon propagator by an isolated pole. This ignores the fact that we can always produce an arbitrary number of *soft* real photons.

- *Overlapping divergences*, such as exemplified by the diagram of Fig. 16.9.

Diagrams exhibiting overlapping divergences pose, in general, a serious problem in the proof of renormalizability of a Quantum Field Theory. In this respect QED is simpler than ϕ^4-theory as J.C. Ward has shown.[8]

16.3   Ward–Takahashi Identity and overlapping divergences

Consider the 1-particle-reducible 3-point function

$$G(x, y, z)^\mu_{\alpha\beta} = \langle \Omega | T \psi_\alpha(x) \bar{\psi}_\beta(y) A^\mu(z) | \Omega \rangle \tag{16.43}$$

having the diagrammatic representation

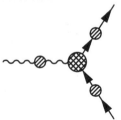

Fig. 16.4. 3-point Green function.

As a first step we wish to express the gauge field in terms of the electromagnetic current $j^\mu(x) = e_0 \bar{\psi}(x) \gamma^\mu \psi(x)$. As is well known this relationship is not unique, and requires a choice of gauge. We shall work in the so-called "alpha-gauges". As we have seen in Section 9 of Chapter 13, these gauges can be implemented by adding the *gauge breaking* term $-\frac{1}{2\alpha}(\partial \cdot A)^2$ to the QED lagrangian. This corresponds to working with the lagrangian

$$\mathcal{L} = -\frac{1}{4} F_{\mu\nu} F^{\mu\nu} + \bar{\psi}(i\partial\!\!\!/ - e_0 A\!\!\!/)\psi - \frac{1}{2\alpha}(\partial \cdot A)^2$$

implying the modified Maxwell equation

$$\left(\Box g^\mu_\lambda - \left(1 - \frac{1}{\alpha}\right)\partial^\mu \partial_\lambda\right) A^\lambda = j^\mu = e_0 \bar{\psi} \gamma^\mu \psi . \tag{16.44}$$

---

[8] J.C. Ward, *Phys. Rev.* **77** (1950) 182.

For $\alpha \neq 0$ the differential operator possesses an inverse obtained by solving the differential equation

$$\left(\Box g_\lambda^\mu - \left(1 - \frac{1}{\alpha}\right)\partial^\mu \partial_\lambda\right)D_F^{\lambda\nu}(x - y;\alpha) = g^{\mu\nu}\delta^4(x - y) .$$

The result is (13.95),

$$D_F^{\mu\nu}(x;\alpha) = \frac{g^{\mu\nu} - \frac{\partial^\mu \partial^\nu}{\Box}}{\Box} + \alpha\frac{\partial^\mu \partial^\nu}{(\Box)^2} .$$

$D_F^{\mu\nu}(x)$ is nothing but the Feynman propagator of the photon field in the "alpha gauges". In momentum space, it reads

$$D_F^{\mu\nu}(k;\alpha) = \frac{-g^{\mu\nu} + \frac{k^\mu k^\nu}{k^2}}{k^2 + i\epsilon} - \alpha\frac{k^\mu k^\nu}{(k^2)^2} .$$

The solution of (16.44) is given by

$$A^\mu(z) = \int d^4z' D_F^{\mu\nu}(z - z';\alpha)j_\nu(z') .$$

We may thus write for the 3-point function (16.43),

$$G(x, y, z)^\mu = \int d^4z' D_F^{\mu\nu}(z - z';\alpha)V_\nu(x, y, z') , \qquad (16.45)$$

where

$$V_\mu(x, y, z)_{\alpha\beta} = \langle\Omega|T\psi_\alpha(x)\bar\psi_\beta(y)j_\mu(z)|\Omega\rangle . \qquad (16.46)$$

Making use of the involution property of Fourier transforms

$$\int d^4z' f(z - z')g(z') = \int \frac{d^4q}{(2\pi)^4}e^{-iq\cdot z}\left[\tilde f(q)\tilde g(q)\right] ,$$

we obtain from (16.45) in momentum space $(k = p' - p)$, after extracting a four-dimensional δ-function,

$$G(p', p, k;\alpha)^\mu = D_F^{\mu\nu}(k;\alpha)V_\nu(p', p) , \qquad (16.47)$$

where $V_\mu(p', p)$ is defined by the three-fold Fourier transform

$$\mathcal{FT}\left[V_\nu(x, y, z)\right] = (2\pi)^4\delta^4(p' - p - k)V_\mu(p', p) .$$

On the other hand, we have from Fig. 16.4,[9]

$$G(p', p, k;\alpha)_{\alpha\beta}^\mu = iD_F'^{\mu\nu}(k;\alpha)iS_F'(p')(-ie_0)\Gamma^\nu(p', p)iS_F'(p) . \qquad (16.48)$$

We have thus two representations of the 3-point function (16.43). Note that contrary to the representation (16.47), the representation (16.48) requires the

---

[9]In this section, $S_F'(p) = S_F(p)_{reg}$.

knowledge of the <u>full</u> photon propagator. We recall, however, that we have for the "divergence" of this propagator, the remarkable property

$$k_\mu D'^{\mu\nu}_F(k;\alpha) = k_\mu D^{\mu\nu}_F(k;\alpha) = -\alpha\frac{k^\nu}{k^2}\,.$$

We thus obtain from (16.47) and (16.48)

$$k_\mu G(p',p,k;\alpha)^\mu = -\alpha\frac{k_\mu}{k^2}V^\mu(p',p) \tag{16.49}$$

and

$$k_\mu G(p',p,k;\alpha)^\mu = e_0\alpha\frac{k_\mu}{k^2}S'_F(p')\Gamma^\mu(p',p)S'_F(p) \tag{16.50}$$

respectively. We conclude that

$$(p'-p)_\mu V^\mu(p',p) = -e_0(p'-p)_\mu S'_F(p')\Gamma^\mu(p',p)S'_F(p). \tag{16.51}$$

We now rewrite the left-hand side of this equation by observing that the current $j^\mu(x)$ is the *Noether current* associated with the *global* $U(1)$ symmetry of the lagrangian (16.1), as represented by the transformation

$$\psi(x) \to e^{i\theta}\psi(x)\,. \tag{16.52}$$

As a result, this current is conserved

$$\partial_\mu j^\mu = 0 \,,\ j^\mu = e_0\bar\psi\gamma^\mu\psi$$

and the corresponding charge

$$Q = \int d^3x\, j^0(x)$$

is the generator of the local symmetry transformation (16.52). The corresponding infinitesimal transformation implies the commutation relations

$$\left[j^0(\vec{x},t),\psi_\alpha(\vec{y},t)\right] = -e_0\delta^3(\vec{x}-\vec{y})\psi_\alpha(\vec{y},t)$$

$$\left[Q,\psi_\alpha(y)\right] = -e_0\psi_\alpha(y)\,, \tag{16.53}$$

$$\left[j^0(\vec{x},t),\bar\psi_\beta(\vec{y},t)\right] = e_0\delta^3(\vec{x}-\vec{y})\bar\psi_\beta(\vec{y},t)\,,$$

which may also be verified explicitly by making use of the canonical commutation relations of the fermion field.

Making use of (16.53) we find from (16.46),

$$\partial^z_\mu V^\mu(x,y,z) = -e_0\left[\delta^3(z-y)\langle\Omega|T\psi(x)\bar\psi(y)|\Omega\rangle - \delta^3(x-z)\langle\Omega|T\psi(x)\bar\psi(y)|\Omega\rangle\right]$$

$$= -e_0\left[\delta^3(z-y)iS'_F(x-y) - \delta^3(x-y)iS'_F(x-y)\right]\,, \tag{16.54}$$

or going to momentum space, this reads

$$(p'-p)_\mu V^\mu(p',p) = -e_0\left[S'_F(p) - S'_F(p')\right]\,. \tag{16.55}$$

Replacing the lhs in (16.51) by this expression, we finally have,[10]

$$(p' - p)_\mu \Gamma^\mu(p', p) = \left[S_F'^{-1}(p') - S_F'^{-1}(p) \right] . \tag{16.56}$$

This is the so-called *Ward–Takahashi identity*. It follows only from the invariance of the QED lagrangian under *global* $U(1)$ transformations. Differentiating (16.56) with respect to p', and setting $p = p'$, one obtains the identity originally obtained by *J.C. Ward*:[11]

$$\Gamma_\mu(p, p) = \frac{\partial}{\partial p^\mu} S_F'^{-1}(p) . \tag{16.57}$$

Recalling (16.27) and (16.28) we can rewrite (16.57) as

$$\frac{\partial \Sigma'(p)}{\partial \not{p}_\mu} = -\Lambda^\mu(p, p)_{reg} . \tag{16.58}$$

The requirement that this identity be satisfied also by the corresponding renormalized quantities leads to the identification $Z_1 = Z_\psi$, as we show next.

Renormalized Ward–Takahashi identity

Recalling from (16.6),

$$Z_\psi^{-1} S_F' = \tilde{S}_F \to S_{ren}, \quad Z_1 \Gamma^\mu = \tilde{\Gamma}^\mu \to \Gamma_{ren}^\mu , \tag{16.59}$$

we have from (16.56),

$$(p' - p)_\mu Z_1^{-1} \tilde{\Gamma}^\mu(p', p) = Z_\psi^{-1} \left[\tilde{S}_F^{-1}(p') - \tilde{S}_F^{-1}(p) \right] . \tag{16.60}$$

It follows from here that the form of the Ward identity is preserved under renormalization, provided that in the limit $\Lambda \to \infty$,

$$Z_1 = Z_\psi . \tag{16.61}$$

Together with (16.11) this implies

$$Z_e = Z_A. \tag{16.62}$$

Taking the limit of infinite cutoff, recalling that

$$S_F^{-1}(p)_{ren} = \not{p} - m - \Sigma(p)_{ren} ,$$

and setting

$$\Gamma_{ren}^\mu(p', p) = \gamma^\mu + \Lambda_{ren}^\mu(p', p)$$

we can equally write for (16.60),

$$(\not{p} - \not{p}) + (p' - p)_\mu \Lambda_{ren}^\mu(p', p) = -\left(\Sigma(p')_{ren} - \Sigma(p)_{ren} \right) + (\not{p}' - \not{p}) ,$$

---

[10]Y. Takahashi, *Il Nuovo Cimento* **6** (1957) 371.
[11]J.C. Ward, *Phys. Rev.* **77** (1950) 293.

or

$$(p' - p)_\mu \Lambda^\mu_{ren}(p', p) = -\left(\Sigma_{ren}(p') - \Sigma_{ren}(p)\right) .$$

From here we obtain, by differentiating with respect to p'^μ and setting $p'^\mu = p^\mu$,

$$\Lambda^\mu_{ren}(p, p) = -\frac{\partial}{\partial p^\mu}\Sigma(p)_{ren} . \tag{16.63}$$

The problem of overlapping divergences in QED

Consider the diagram of Fig. 16.5 with an overlapping, linear divergence.

Fig. 16.5. Diagram with overlapping divergence.

As we see from,

$$\frac{\partial}{\partial p^\mu} \frac{1}{\slashed{k} - \slashed{p} - m} = -\frac{1}{\slashed{k} - \slashed{p} - m}\gamma_\mu \frac{1}{\slashed{k} - \slashed{p} - m} ,$$

differentiation with respect to the *external* momentum p, amounts to inserting a *zero momentum* photon line in every fermion propagator as shown in Fig. 16.6:[12]

$$\frac{\partial \Sigma(p)}{\partial p_\mu} = $$

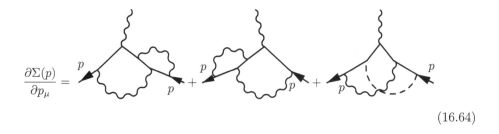

$$\tag{16.64}$$

Fig. 16.6. Ward identity for self-energy diagram of Fig. 16.5.

---

[12] For the sake of visualization, we have replaced one of the photon lines by a dashed line.

This is precisely what is done in the Ward identity (16.63). What has one gained? It turns out that the vertex function $\Lambda^\mu(p',p)$ does NOT involve overlapping divergences, and can thus be renormalized in a conventional way. This was the observation of J.C. Ward. By integration of (16.63), we then obtain the renormalized electron self-energy:

$$\Sigma(p')_{ren} - \Sigma(p)_{ren} = -\int_p^{p'} dq^\mu \Lambda(q,q)^\mu_{ren}.$$

Now, remembering that $\Sigma(p)_{ren}|_{\not{p}=m} = 0$, we finally have

$$\Sigma(p)_{ren} = -\int_{\not{p}=m}^p dq_\mu \Lambda^\mu(q,q)_{ren} .$$

A similar procedure can be followed in the case of the photon vacuum polarization. Let us write in analogy to (16.63),

$$-\frac{\partial}{\partial k_\mu} \pi(k^2)_{ren} = \Delta^\mu(k,k)_{ren} \tag{16.65}$$

where $\pi(k^2)_{ren}$ is given by (16.39), and the right-hand side involves again zero momentum photon insertions in the vacuum polarization diagram in question. Again, Δ^μ does not involve overlapping divergences, and hence can be renormalized in a conventional way. Integrating (16.65), and remembering that $\pi(0)_{ren} = 0$, we obtain,

$$\pi(k^2)_{ren} = \int_0^k dk'_\mu \Delta^\mu(k',k')_{ren} ,$$

or using the parametrization $k_\mu(x) = xk_\mu + (1-x)k'_\mu$, we alternatively have,

$$\pi(k^2)_{ren} = \int_0^1 dx(k-k')_\mu \Delta^\mu(k(x),k(x))_{ren} .$$

For more details see L.H. Ryder.[13]

16.4 1-loop renormalization in QED

Equations (16.31) and (16.38) define the renormalization constants Z_ψ and Z_A to all orders of perturbation theory. As we shall see, these constants play a central role in the Callan–Symanzik equations to be discussed in Chapter 17. We now proceed to calculate the renormalized 2- and 3-point functions, as well as the corresponding renormalization constants, in 1-loop order of perturbation theory.

Computation of Z_ψ and $\Sigma_{ren}(p^2)$

The calculation of the renormalization constant Z_ψ and $\Sigma_{ren}(p^2)$ requires the computation of δm and B from (16.28) and (16.29) respectively. To second order we

---

[13]L.H. Ryder, *Quantum Field Theory* (Cambridge University Press, 1996, second edition).

can set $\alpha_0 = \alpha, m_0 = m$. Replacing p^2 in the argument of $\ln f$ in (15.17) by $(\not p)^2$, and expanding $\Sigma_{reg}(p)$ in the Taylor series (16.27), we then obtain for the electron self-energy,

$$\delta m \simeq \frac{3\alpha}{4\pi} m \ln\left(\frac{\Lambda^2}{m^2}\right) + \delta m' \qquad (16.66)$$

with

$$\delta m' \simeq -\frac{\alpha}{2\pi} \int_0^1 d\beta\, m(1+\beta)\ln\left(\frac{m^2\beta^2 + (1-\beta)\mu^2}{m^2(1-\beta)}\right),$$

and

$$B \simeq -\frac{\alpha}{4\pi} \ln\left(\frac{\Lambda^2}{m^2}\right) + B' \qquad (16.67)$$

with

$$B' \simeq \frac{\alpha}{2\pi} \int_0^1 d\beta \left[(1-\beta)\ln\left(\frac{m^2\beta^2 + (1-\beta)\mu^2}{m^2(1-\beta)}\right) - \frac{2m^2\beta(\beta^2-1)}{\beta^2 m^2 + (1-\beta)\mu^2}\right]. \qquad (16.68)$$

Subtracting from (15.17) $\delta m + B(\not p - m)$, the Λ dependent terms cancel, and we obtain with (16.27) and (16.29) for the renormalized electron self-energy to order $O(\alpha)$ (no wave function renormalization to this order: $\Sigma_{ren}(p) \sim \Sigma'(p)$),

$$\Sigma(p)_{ren} \simeq \frac{\alpha}{2\pi} \int_0^1 d\beta [(1-\beta)\not p - 2m]\ln f(\beta; p^2, m^2, \mu^2) - B'(\not p - m) - \delta m'. \qquad (16.69)$$

One checks explicitly that $\Sigma(p)_{ren}$ is of order $(\not p - m)^2$, as expected.

We notice that the threshold for the cut in the logarithmic function in (16.69) coincides in the limit $\mu \to 0$ with the mass shell $p^2 = m^2$, so that in a physical process, we cannot isolate photons in this limit from physical fermions. We also recall that the above results are gauge dependent, the fermion 2-point function being itself a gauge-dependent quantity. As was shown by Yennie and Fried,[14] this singular divergence can be avoided by choosing the 't Hooft gauge $\alpha = 3$ (see (13.98)). However, this does not solve the problem in general, the proper solution being that of averaging over soft photons taking account of the experimental resolution.[15]

Computation Z_A and $\pi_{ren}(k^2)$

From the result (15.30) for the dimensional regularization,

$$\pi^{dim}(k^2)_{reg} \simeq -\frac{2\alpha_0}{3\pi}\left(\frac{2}{\epsilon} - \gamma - \ln\pi + \ln\frac{\tilde\mu^2}{m_0^2}\right) \qquad (16.70)$$

$$+ 2\frac{\alpha_0}{\pi}\int_0^1 d\beta\,\beta(1-\beta)\ln\left(\frac{m_0^2 - \beta(1-\beta)k^2 - i0}{m_0^2}\right)$$

and (16.38),

$$Z_A = \frac{1}{1 - \pi(0)_{reg}} \approx 1 + \pi(0)_{reg} \qquad (16.71)$$

---

[14] H.M. Fried and D.R. Yennie, *Phys. Rev.* **112** (1958) 1391.
[15] F. Bloch and A. Nordsieck, *Phys. Rev.* **52** (1957) 54.

it follows to second order in dimensional regularization

$$Z_A^{dim}\left(\alpha, \frac{\mu}{m}, \epsilon\right) \simeq 1 - \frac{\alpha}{3\pi}\left(\frac{2}{\epsilon} - \gamma - \ln \pi + \ln \frac{\tilde{\mu}^2}{m^2}\right) . \tag{16.72}$$

On the other hand, from the result (15.28) for the Pauli–Villars regularization, with $M = \Lambda$, we have

$$\pi^{PV}(k^2)_{reg} \approx -\frac{\alpha_0}{3\pi}\left(\ln \frac{\Lambda^2}{2m_0^2}\right) + \frac{2\alpha_0}{\pi}\int_0^1 d\beta\beta(1-\beta)\ln\left(\frac{m^2 - \beta(1-\beta)k^2 - i0}{m^2}\right), \tag{16.73}$$

and

$$Z_A^{PV} \simeq 1 - \frac{\alpha_0}{3\pi}\ln\left(\frac{\Lambda^2}{2m^2}\right) . \tag{16.74}$$

In both cases, we have for the proper renormalized vacuum polarization to order $O(\alpha)$,

$$\pi(k^2)_{ren} \simeq \frac{2\alpha}{\pi}\int_0^1 d\beta\beta(1-\beta)\ln\left(\frac{m^2 - \beta(1-\beta)k^2 - i0}{m^2}\right) . \tag{16.75}$$

Computation of Z_1 and $\Gamma^\mu(p, p')_{ren}$

We first recall (see (15.40)) that *on the fermion mass-shell*, to second order $\Gamma^\mu(p', p)$ had the form

$$\bar{U}(\vec{p}\,', \sigma'))\Gamma^\mu(p', p)U(\vec{p}, \sigma) = \bar{U}(\vec{p}\,', \sigma')[\gamma^\mu + \gamma^\mu F(q^2) + \frac{(p + p')^\mu}{2m}G(q^2)]U(\vec{p}, \sigma) . \tag{16.76}$$

At zero momentum transfer we must require from (16.12) that (we simplify our notation)

$$\bar{U}(\vec{p})\tilde{\Gamma}^\mu(p, p)U(\vec{p}) = Z_1\bar{U}(\vec{p})\Gamma^\mu_{reg}(p, p)U(\vec{p}) = \bar{U}(\vec{p})\gamma^\mu U(\vec{p}) . \tag{16.77}$$

In order to implement the renormalization condition (16.77) we make use of the Gordon relation

$$\bar{U}(\vec{p}\,')\gamma^\mu U(\vec{p}) = \frac{1}{2m}\bar{U}(\vec{p}\,')[(p + p')^\mu + i\sigma^{\mu\nu}(p' - p)_\nu]U(\vec{p}) \tag{16.78}$$

where

$$\sigma^{\mu\nu} = \frac{i}{2}[\gamma^\mu, \gamma^\nu] . \tag{16.79}$$

Hence we have from (16.76) equivalently,

$$\bar{U}(\vec{p}\,')\Gamma^\mu(p', p)U(\vec{p}) = \bar{U}(\vec{p}\,')[(1 + F(q^2) + G(q^2))\gamma^\mu - i\sigma^{\mu\nu}q_\nu G(q^2)]U(\vec{p}) .$$

We define new form factors $F_1(q^2)$ and $F_2(q^2)$,

$$F_1(q^2) = F(q^2) + G(q^2) , \quad F_2(q^2) = G(q^2) \tag{16.80}$$

in terms of which

$$\bar{U}(\vec{p}\,')\tilde{\Gamma}^\mu(p',p)U(\vec{p}) = Z_1\bar{U}(\vec{p}\,')[\gamma^\mu(1+F_1(q^2)_{reg}) - i\sigma^{\mu\nu}q_\nu F_2(q^2)_{reg}]U(\vec{p}) \ . \quad (16.81)$$

Writing Z_1 in the form of the expansion (16.16),

$$Z_1 = 1 + \alpha z_1 + \cdots \ ,$$

we require from (16.77) to order $O(\alpha)$,

$$\alpha z_1 + F_1(0)_{reg} = 0 \ . \quad (16.82)$$

This leads (15.46) and (15.41) to the form factors $(Z_1 - 1 = -F_1(0)_{ren})$

$$F_1(q^2)_{ren} = -\frac{\alpha}{\pi}\int_0^1 \prod d\alpha_i \ \delta\left(1 - \sum \alpha_i\right) \quad (16.83)$$

$$\times \left\{ \frac{(1-\alpha_1)(1-\alpha_2)(m_0^2 - \frac{q^2}{2}) - \frac{m_0^2}{2}[(\alpha_1+\alpha_2)^2 + 2\alpha_1\alpha_2]}{m_0^2(\alpha_1+\alpha_2)^2 + \alpha_3\mu^2 - q^2\alpha_1\alpha_2} \right.$$

$$\left. + \frac{1}{2}\ln\left(m_0^2(\alpha_1+\alpha_2) + \alpha_3\mu^2 - q^2\alpha_1\alpha_2\right) \right\} - \{\mathbf{q^2 = 0}\}$$

$$F_2(q^2)_{ren} = -\frac{\alpha}{\pi}\int_0^1 \prod_i d\alpha_i\delta\left(1 - \sum_1^3 \alpha_j\right)\frac{m^2\alpha_3(\alpha_1+\alpha_2)}{m^2(\alpha_1+\alpha_2)^2 + \alpha_3\mu^2 - \alpha_1\alpha_2q^2} \ . \quad (16.84)$$

Note that this form factor diverges logarithmically as we let the mass of the photon tend to zero. This was the reason for introducing an infrared cutoff to begin with. The origin of this divergence lies in the fact that we have represented the photon propagator by an isolated pole. This ignores the fact that we can always produce an arbitrary number of *soft* real photons, which leads to the replacement of this isolated pole by a *cut* in the complex *k-plane*. These soft photons are excited whenever charged particles suffer a change in velocity, that is, for non-vanishing momentum transfer. As has been shown[16] by *Bloch* and *Nordsieck* this infrared divergence is removed upon averaging over such photons, taking account of the experimental resolution.

Finally, the anomalous magnetic moment is determined by $F_2(0)_{ren}$, whose value is easily calculated. For $q = 0$,

$$F_2(0)_{ren} = -\frac{\alpha}{\pi}\int_0^1 d\alpha_1 \int_0^{(1-\alpha_1)} \frac{(1-\alpha_1-\alpha_2)}{(\alpha_1+\alpha_2)}$$

$$= -\frac{\alpha}{\pi}\left(\alpha_1 + \frac{1}{2}\alpha_1^2\right)\bigg|_0^1 = \frac{\alpha}{2\pi} \ . \quad (16.85)$$

Using the Gordon relation (16.78) we have from (16.81), with $F_1(0)_{ren} = 0$, to lowest order in q^μ,

$$U(\vec{p}\,')[\Gamma^\mu(p',p)U(\vec{p}) \approx \bar{U}(\vec{p}\,')\left[\frac{(p+p')^\mu}{2m} + \left(1 + \frac{\alpha}{2\pi}\right)i\sigma^{\mu\nu}\frac{q_\nu}{2m}\right]U(\vec{p}) \ .$$

---

[16]F. Bloch and A. Nordsieck, *Phys. Rev.* **52** (1957) 54.

Comparing with Section 6.5 of Chapter 6, we identify $F_2(0) = 1 + \frac{\alpha}{2\pi}$ with the anomalous magnetic moment of the electron, first calculated by Schwinger in 1948.[17]

16.5 Composite operators and Wilson expansion

Formally, local composite operators are obtained as the product of local operators in the limit where the arguments tend to the same point. As we know already from the behaviour of 2-point Green functions, these limits are in general singular, so that the corresponding composite operator cannot be defined in this naive sense. By looking at the behaviour of composite operators of local free fields, Wilson proposed for an interacting theory the existence of an expansion for $x \to y$ of the form[18]

$$A(x)B(y) = \sum_{N=0} C_N(x-y)O_N\left(\frac{x+y}{2}\right) \to C_0(x-y)+C_1(x-y)O_1(x)+\cdots \quad (16.86)$$

where O_N are *local* and *finite* operators with $O_0(x) = 1$, and $C_N(x-y)$ are c-number functions, of which some are singular in the limit $x \to y$.

As a simple illustration, we consider a free scalar field theory. Wick's theorem tells us that

$$\varphi(x)\varphi(y) = \langle 0|\varphi(x)\varphi(y)|0\rangle + :\varphi(x)\varphi(y): $$
$$= i\Delta^{(+)}(x-y) + :\varphi(x)\varphi(y): $$
$$= i\Delta^{(+)}(x-y) + :\varphi(x)^2: +O(x-y).$$

This expansion is of the form (16.86) with $A(x) = B(x) = \varphi(x)$, and

$$C_0(z) = i\Delta^{(+)}(z),\; C_1(z) = 1 .$$

$$O_0(z) = 1,\; O_1(z) =: \varphi^2(z): .$$

The fact that the coefficient C_1 is finite reflects that the *scale dimension* of the normal-ordered product $:\varphi(x)\varphi(x):$ is exactly two, and thus identical with its *engineering (canonical)* dimension. Let us denote the engineering dimension of a general operator Q by $[Q]$.[19] Perturbatively we then have in nth order for $z \to 0$, after renormalization,

$$C_N^{(n)}(z) \sim g^n z^{-\delta_{O_N}^{AB}}(\ln\mu|z|)^{\rho_{O_N}^{AB}} , \quad (16.87)$$

where, for a strictly renormalizable theory,

$$\delta_{O_N}^{AB} = [A] + [B] - [O_N] , \text{ if } [g] = 0 .$$

Hence on the perturbative level the coefficients C_N of strictly renormalizable theories tend to zero for terms in the operator expansion involving composite operators O_N with engineering dimension

$$[O_N] > [A] + [B] .$$

[17] J. Schwinger, *Phys. Rev.* **73** (1948) 416.
[18] K. Wilson, *Phys. Rev.* **179** (1969) 1499; and *Phys. Rev. Ser. D*, **3** (1971) 1818.
[19] One measures dimensions in units of "energy" = "mass".

Note that the arguments of the logarithm have to be dimensionless. The parameter μ appearing in (16.87) carries the dimensions of "mass" and is introduced by the need of regularization. It may for instance be the mass parameter introduced in dimensional regularization, or may refer to the point at which the subtraction is performed in a Taylor subtraction scheme (see Section 7).

Non-perturbatively we have seen in Section 2 of Chapter 12, that the logarithms can sum up to a power behaviour, so that after summation to all orders we may have for $z \to 0$,

$$C_N(z) \to z^{-\delta_{O_N}^{AB}} (\mu|z|)^{\gamma_{0_N}^{AB}}$$

with

$$\gamma_{0_N}^{AB} = \gamma_A + \gamma_B - \gamma_{O_N} .$$

Here γ_Q is referred to as the *anomalous dimension* of the operator Q (see Chapter 17).

Following Wilson, the c-cumber functions C_N should be *universal* and characteristic of the operator product in question. According to this hypothesis, we can determine them by considering suitable n-point functions. With the notation

$$G_{AB}(x, y; z_1 \ldots z_n) = <\Omega|TA(x)B(y)\,\varphi(z_1) \cdots \varphi(z_n)|\Omega>$$

we have from (16.86),

$$G_{AB}(x + \frac{\epsilon}{2}, x - \frac{\epsilon}{2}; z_1 \ldots z_n) = \sum_N C_N(\epsilon) G_{O_N}(x; z_1 \ldots z_n) ,$$

where

$$G_{O_N}(x; z_1, \ldots z_n) = <\Omega|TN[A(x)B(x)]\,\varphi(z_1) \cdots \varphi(z_n)|\Omega> ,$$

defines the *normal product* $N[A(x)B(x)]$ of the two (interacting) fields at the same point. For free fields A and B, $N[A(x), B(x)] =: A(x)B(x) :$.

16.6 Criteria for renormalizability

In order to obtain simple criteria for the *renormalizability* of a Quantum Field Theory we need to introduce the notion of the *superficial degree of divergence* of a Feynman diagram.

One obtains a superficial estimate of the degree of divergence of a Feynman diagram G, by simply scaling all *internal* momenta with a common factor λ, and examining the behaviour of the corresponding Feynman integral I_G in the limit of $\lambda \to \infty$:

$$I_G \to^{\lambda^{\omega(G)}} I_G .$$

One calls $\omega(G)$ the *superficial degree of divergence* of the diagram G.

We consider, in general, a theory involving *spin* 0, *spin* 1/2 and *spin* 1 fields. Under a "dilatation", the respective propagators scale as λ^{-2}, λ^{-1} and λ^{-2}. We adopt the following notation:

E_F = Number of external fermion lines
E_B = Number of external boson lines
I_F = Number of internal fermion lines
I_B = Number of internal boson lines

V = Number of vertices
$L = I_F + I_B - (V - 1)$ = Number of independent loops

E_{F_v} = Number of external fermion lines connected to vertex v
E_{B_v} = Number of external boson lines connected to vertex v
I_{F_v} = Number of internal fermion lines connected to vertex v
I_{B_v} = Number of internal boson lines connected to vertex v
δ_v = Number of derivatives contained in vertex v
g_v = Coupling constant associated with the vertex v

We then have (D = Dimension of space-time)

$$\omega(G) = DL + \sum_v \delta_v - 2I_B - I_F \tag{16.88}$$

$$= (D-1)I_F + (D-2)I_B + \sum_v (\delta_v - D) + D \ .$$

We want to express these quantities in terms of the number of *external* fermion and boson lines, and the dimension (in units of mass) of the coupling constants labelling the vertices. Denoting the dimension of a quantity A by $[A]$, we observe that *in 4 space-time dimensions*

$$[\phi] = [A^\mu] = 1 \ , \ [\psi] = \frac{3}{2} \ .$$

Correspondingly we define $\big([v] = $ dimension of vertex, including $[g_v]\big)$

$$[v/g_v] = \frac{3}{2}E_{F_v} + E_{B_v} + \frac{3}{2}I_{F_v} + I_{B_v} + \delta_v \ .$$

Noting that

$$\sum_v I_{F_v} = 2I_F \ , \ \sum_v I_{B_v} = 2I_B \ ,$$

we then have

$$\sum_v \left[\frac{v}{g_v}\right] = \frac{3}{2}E_F + E_B + 3I_F + 2I_B + \sum_v \delta_v \ ,$$

or using (16.88),

$$\sum_v \left([v/g_v] - 4\right) = \frac{3}{2}E_F + E_B + \left(\omega(G) - 4\right) \ .$$

Now, from

$$[d^4x \mathcal{L}_I] = 0$$

we have,

$$[v/g_v] + [g_v] = 4$$

or it follows from above

$$\omega(G) = 4 - \sum_v [g_v] - \frac{3}{2} E_F - E_B \ , \tag{16.89}$$

which is the desired result.

Equation (16.89) shows that if there exist couplings for which $[g_v] < 0$, then the superficial degree of divergence will increase indefinitely with the order of the diagram, independent of the number of (unrenormalized) constants, and the theory is thus called *unrenormalizable*. One is thus led to the following classification of Quantum Field Theories:

(i) *Unrenormalizable QFT*

   According to our above comments these are theories for which

$$\exists v : \ [g_v] < 0$$

   so that $\omega(G) > 0$ for an infinite set of diagrams. Examples are

$$\mathcal{L}_I = g\bar{\psi}\gamma^\rho\psi\partial_\rho\phi, \ g\bar{\psi}\gamma^\rho\gamma^5\psi\partial_\rho\phi, \ \bar{\psi}\gamma_\rho\psi\bar{\psi}\gamma^\rho\psi, \ \phi^{d>4}$$

(ii) *Renormalizable QFT (in the extended sense)*

   These are theories for which $[g_v] \geq 0$, $\forall v$ .. Under these conditions there exists an upper limit of the superficial degree of divergence for any diagram of arbitrary order, so that we can achieve finiteness of the Green functions by adding a *finite* number of counterterms. It may, however, not be possible to absorb all of these counter terms into a redefinition of the unrenormalized parameters of the lagrangian defining the QFT in question. An example is $\mathcal{L}_I = g\bar{\psi}\gamma^5\psi\phi$. The UV divergence of the 4-point function

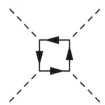

   requires the introduction of a counterterm of the form $\delta\mathcal{L} = \delta Z g^4 \phi^4$ corresponding to a local interaction not contained in the above Lagrangian.

(iii) *Renormalizable QFT (in the strict sense)*

   These are theories for which $[g_v] \geq 0$, $\forall v$ again holds, but for which all the counter terms can be absorbed into a redefinition of the bare parameters and fields of the lagrangian. Examples are

$$\mathcal{L}_I : g_s\bar{\psi}\psi\phi + \lambda\phi^4 \ , g_p\bar{\psi}\gamma^5\psi\phi + \lambda\phi^4 \ , e\bar{\psi}\slashed{A}\,\psi \ , e\bar{\psi}\gamma^5\slashed{A}\,\psi \ , (e\phi^\dagger \overset{\leftrightarrow}{\partial_\mu} A^\mu + e^2\phi^2 A_\mu^2) \ .$$

(iv) *Superrenormalizable QFT*

These are theories for which $[g_v] \geq 1$, $\forall v$. In this case, there exist only a few diagrams which are divergent.

The fact that $\omega(G) < 0$ does not exclude the possibility that G contains a *subdiagram* g with $\omega(g) \geq 0$. Thus consider, for example, the diagram in $g\phi^4$ theory ($[g] = 0$, $E_F = 0$, $E_B = 6$). We have

$$\omega(G) = 4 - \sum [g_v] - \frac{3}{2} E_F - E_B = -2 .$$

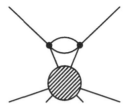

Fig. 16.6. $\omega(G) = -2$ diagram involving logarithmicallly divergent subdiagram.

This diagram nevertheless contains the subdiagram

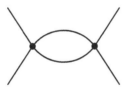

Fig. 16.7. $\omega(g) = 0$ subdiagram of Fig. 16.6.

with $\omega(g) = 0$.

For what follows, we recall the *definition of a subdiagram* (restricted definition): g is a *subdiagram* of $G (g \subset G)$ if *all* lines which connect two vertices in g, also lie in g. We have

Theorem

Let G be a (connected) $1PI$ diagram;
Let g be a (connected) $1PI$ subdiagram of G;
Let $\mathcal{F}_g$ be the family of all (connected) $1PI$ subdiagrams of G.

Then the (euclidean) Feynman integral $I^{(E)}(G)$ is *absolutely* convergent, provided

$$\omega(g) < 0, \ \forall g \in \mathcal{F}_g$$

and

$$\omega(G) < 0 .$$

We have the following two corollaries.

Corollary 1.

In a scalar QFT with derivative couplings, or a theory with non-scalar particles, the presence of polynomials in the numerator of the integrand of a Feynman integral can lead to cancellations reducing the degree of divergence of the integral.

Example: As we have seen in Chapter 15, Figs. 15.1 and 15.2 are only logarithmically divergent.

Corollary 2.

If

$$\omega(g) < 0, \ \forall g \in \mathcal{F}_g$$

but

$$\omega(G) \geq 0 \ ,$$

then, in the limit $\Lambda \to \infty$,

$$I_G^{(E)}(p; m; \Lambda) \to I^{(E)}(p; m) + \mathcal{P}(p; m; \Lambda) \ ,$$

where $\mathcal{P}_\Lambda(p, m; \Lambda)$ is a cutoff dependent polynomial of degree *less or equal* to $\omega(G)$ in the external momenta p, and internal masses m_i.

Proof:

Since $\omega(G)$ measures the degree of homogeneity of I_G in the external momenta and internal masses, we have

$$\frac{\partial^{\omega+1}}{\partial \lambda^{\omega+1}} I_G^{(E)}(\lambda p, \lambda m; \Lambda) \sim \lambda^{\omega-(\omega+1)} = \lambda^{-1} \ ,$$

and hence this expression is absolutely convergent. Corollary 2 then follows upon using the above theorem.

16.7   Taylor subtraction

We have considered so far the renormalization based on two possible regularization procedures — *Pauli–Villars* and *dimensional* regularization. There exist a number of other possibilities for renormalization. In the following, we shall consider that obtained by performing a *Taylor subtraction* of the *integrand* of a Feynman integral. This leads to the so-called *Bogoliubov recursion formula*, whose general solution is given by *Zimmermann's forest formula*.

Renormalization by Taylor subtraction

We shall restrict again our discussion to the case of scalar particles. We suppose the Feynman integral in question to be given in the Feynman-parametric form of Eq. (12.17). Correspondingly we write

$$I_G(p) = \int d\lambda \prod_i d\alpha_i \mathcal{J}_G(p; \alpha, \lambda) \ , \tag{16.90}$$

where p stands for all the external momenta. According to (12.17) the integrand will be a function of the Lorentz invariants $p_i \cdot p_j$, the Feynman parameters α_i and λ. The convergence of the integral is controlled by the behaviour of the integrand at $\lambda = 0$. We observe that repeated differentiation with respect to the external momenta improves this behaviour. The procedure thus consists in subtracting from $\mathcal{J}(p;\alpha)$ a *truncated* Taylor expansion around a conveniently chosen "point" in the space of the Lorentz invariants.

Thus consider for instance the diagram (ϕ^3-theory)

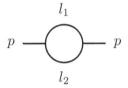

Fig. 16.8. ϕ^3 self-energy.

The corresponding integrand will be a function of the Lorentz invariant p^2. We perform the following Taylor subtraction around $p^2 = 0$:

$$\mathcal{R}_G = (1 - T_G)\mathcal{J}_G(p^2;\alpha,\lambda) , \tag{16.91}$$

with

$$T_G\mathcal{J}_G(p^2;\alpha) = \overbrace{\mathcal{J}_G(0;\alpha)) + p^2 \left(\frac{\partial \mathcal{J}_G(p^2;\alpha,\lambda)}{\partial p^2}\right)_{p^2=0} + \cdots}^{\omega(G)\ terms}. \tag{16.92}$$

The number of terms to be kept in the Taylor series depends on the degree of divergence of the diagram. Since the diagram in question is logarithmically divergent, it suffices to consider the first term in this Taylor series (considering more terms would result in "oversubtraction"). Note that we could have expanded also around another point, other than $p^2 = 0$. The above subtraction for $\omega(G) = 0$ improves the $\lambda \to 0$ behaviour for the example in question sufficiently, to render the λ integration finite. Indeed, making use of the cutting rules of Chapter 12, we have

$$\mathcal{P}(\alpha) = \alpha_1 + \alpha_2$$

corresponding to the tree diagrams

$$\begin{array}{c}\text{(diagram)}\end{array} \tag{16.93}$$

For $Q_G(P,\alpha)$ one correspondingly obtains from (12.14),

$$Q_G(\alpha_1,\alpha_2) = \frac{p^2\alpha_1\alpha_2}{\alpha_1 + \alpha_2} .$$

Now,
$$[g] = 1, \; E_F = 0, \; E_B = 2 \; ,$$

or
$$\omega(G) = 4 - \sum_v [g_v] - \frac{3}{2} E_F - E_B = 0 \; .$$

Hence the integral is logarithmic divergent. From (12.17) we have for $\lambda \to 0$ in the integrand of the λ-integral, after a single Taylor subtraction,

$$\frac{1}{\lambda} \left[e^{i\lambda p^2 \alpha_1 \alpha_2 / (\alpha_1 + \alpha_2)} - 1 \right] e^{-im^2 \lambda (\alpha_1 + \alpha_2)} \to \frac{i\alpha_1 \alpha_2 p^2}{(\alpha_1 + \alpha_2)} \; .$$

The λ-integration is now finite.

Let us exemplify this by calculating the Feynman integral associated with the diagram of Fig. 16.8. Returning to (12.17) we must compute

$$I(p; \alpha) = \int_\varepsilon^\infty \frac{d\lambda}{\lambda} \left[e^{i\lambda(Q(p;\alpha) - m_0^2)} - e^{-i\lambda m_0^2} \right] \; .$$

It is convenient to take $p^2 < 0$ and to deform the integration contour along the real axis to one along the negative imaginary axis, by performing the "rotation"

$$\lambda = -i\rho$$

in the complex λ-plane. The sign in this change of variable reflects the fact that this rotation is to be performed *counterclockwise*. Choosing $p^2 < 0$, that is, below the threshold for particle production, we encounter no singularities of the integrand, so that in this case

$$I(p; \alpha) = \int_\varepsilon^\infty \frac{d\rho}{\rho} \left[e^{-\rho(m_0^2 - Q(p;\alpha))} - e^{-\rho m_0^2} \right] \; , \quad p^2 < 0 \; . \tag{16.94}$$

This leaves us with (12.17) for the second order ϕ^3-selfenergy,

$$-i\Sigma_{reg}(p) = \frac{(-ig_0)^2}{i(4\pi)^2} \int_0^1 \prod_1^2 d\alpha_i \frac{\delta(1 - \Sigma \alpha_i)}{[\mathcal{P}(\alpha)]^2} I(p; \alpha) \; . \tag{16.95}$$

There are several ways of evaluating this integral:

(a) Setting $\epsilon = 0$ and using Pauli–Villars regularization.

(b) We evaluate the integral (16.94) in the limit $\varepsilon \to 0$ by performing the following manipulations:

$$I(p, \alpha) = \int_\varepsilon^\infty d\rho \left(\frac{d\ln \rho}{d\rho} \right) \left[e^{-\rho(m_0^2 - Q(p;\alpha))} - e^{-\rho m_0^2} \right]$$

$$= \int_\varepsilon^\infty d\rho \frac{d}{d\rho} \left\{ \ln \rho \left[e^{-\rho(m_0^2 - Q(p;\alpha))} - e^{-\rho m_0^2} \right] - \ln \rho \frac{d}{d\rho} \left[e^{-\rho(m_0^2 - Q(p;\alpha))} - e^{-\rho m_0^2} \right] \right\}$$

$$= -\ln \varepsilon \left[e^{-\varepsilon(m_0^2 - Q(p;\alpha))} - e^{-\varepsilon m_0^2} \right] + \int_\varepsilon^\infty d\rho \ln \rho \left[(m_0^2 - Q^2) e^{-\rho(m_0^2 - Q^2)} - m_0^2 e^{-\rho m_0^2} \right]$$

$$= -\ln \left(\frac{m_0^2 - Q(p; \alpha)}{m_0^2} \right) \tag{16.96}$$

where we made the change of variable $(m_0 - Q)\rho \to \rho'$, $(m_0\rho) \to \rho'$, respectively. Hence finally we have to second order,

$$\Sigma_{ren}^{(2)}(p^2) = \frac{g_0^2}{16\pi^2} \int_0^1 d\alpha \ln\left(\frac{m^2 - \alpha(1-\alpha)p^2}{m^2}\right) , \qquad (16.97)$$

where we have made the replacements $g_0 \to g$ and $m_0 \to m$.

16.8   Bogoliubov's recursion formula

Differentiation with respect to the *external* momenta of a Feynman graph may not reduce the degree of divergence since the divergent subgraphs may not involve the variable of the differentiation. Bogoliubov, Parasiuk,[20] and Hepp,[21] have shown how to include the divergent subgraphs in the Taylor approach.

Consider a Feynman diagram I_G with $\omega(G) \geq 0$ and let $\mathcal{F}_\gamma$ be the *family of all superficially divergent, connected 1PI subdiagrams γ of G.*[22]

$$\mathcal{F}_\gamma = (\gamma : \omega(\gamma) \geq 0) , (\mathcal{F}_\gamma <_G) .$$

We write (16.90) compactly as $I_G = \int \mathcal{J}_G$. We associate with each such subdiagram γ a counterterm (c.t) rendering it finite. Let us assume that we have done this for *every* subdiagram γ_a. This leaves us with a new, partially regularized integrand $\bar{\mathcal{R}}_G$, $(I_G \to \int \bar{\mathcal{R}}_G)$

$$\bar{\mathcal{R}}_G = \mathcal{J}_G \ + \ \text{Terms resulting from adding c.t. to all } \gamma_a \subset G .$$

We are left with two possibilities:

(i) $\omega(G) < 0$.

   Then
$$\mathcal{R}_G = \bar{\mathcal{R}}_G$$

   and the subsequent integration over α yields a finite result:

$$I_G \to \int \mathcal{R}_G < \infty .$$

(ii) $\omega(G) \geq 0$

   Then we need to perform according to Corollary 2 of Section 6, one final Taylor subtraction,

$$I_G \to \int (1 - T_G)\bar{\mathcal{R}}_G < \infty . \qquad (16.98)$$

[20] N.N. Bogoliubov and O. Parasiuk, *Acta Math.* **97** (1957) 227.
[21] K. Hepp, *Comm. Math. Phys.* **2** (1966) 301.
[22] Taken from C. Itzykson and J.-B. Zuber, *Quantum Field Theory* (McGraw-Hill Inc., 1980).

This leaves us with the question: What is, in general, the connection between $\bar{\mathcal{R}}_G$ and $\mathcal{J}_G$? The difference comes from the counterterms associated with the renormalization of the 1PI subdiagrams $\mathcal{F}_\gamma$ in G.

Claim:[23]

$$\bar{\mathcal{R}}_G = \mathcal{J}_G + \sum_{\{\gamma_1 \ldots \gamma_s\}} \mathcal{J}_G / \{\gamma_1 \ldots \gamma_s\} \prod_{a=1}^{s} (-T_{\gamma_a} \bar{\mathcal{R}}_{\gamma_a}) , \qquad (16.99)$$

where $\bar{\mathcal{R}}_{\gamma_a}$ are <u>determined recursively</u>. Our notation is as follows:

1. $\{\gamma_1 \ldots \gamma_s\}$ stands for a *partition* of G in superficially divergent disjoint subgraphs
$$\gamma_i \cap \gamma_j = \phi , \ i \neq j = 1, \ldots, s$$

2. $T_{\gamma_a} \bar{\mathcal{R}}_{\gamma_a}$ denotes the Taylor expansion of the modified integrand $\bar{\mathcal{R}}_{\gamma_a}$ in the independent external momenta of γ_a up to order $\omega(\gamma_a)$.

3. $\mathcal{J}_G / \{\}$ stands for the contribution to the integrand $\mathcal{J}_G$ of the *lines and vertices external* to $\{\}$, as well as the *propagators* pertaining to the lines that join $\{\}$ to the rest of G.

4. The sum extends only over <u>disjoint</u> proper subdiagrams; this allows for an independent renormalization of the subgraphs in question.

Notice that (16.99) represents an iterative formula, since the renormalized subgraphs are to be computed from an analogous formula. We shall now illustrate these rather formal developments in terms of a concrete example.

16.9 Overlapping divergences

Consider the 2-loop diagram in QED:

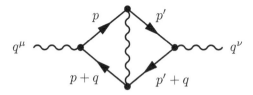

Fig. 16.9. Diagram with an overlapping divergence.

with $\omega(G) = 2$. This diagram is superficially quadratically divergent. We have two (left-right) disjoint subdiagrams $\omega(\gamma_1) = 0$ and $\omega(\gamma_2) = 0$, respectively. Note that $\{\gamma_1, \gamma_2\}$ is not an allowed partition, since the two subdiagrams share a common internal line. One speaks of *overlapping* divergences. Correspondingly

$$\bar{\mathcal{R}}_G = \mathcal{J}_G - \mathcal{J}_{G/\gamma_1} T_{\gamma_1} \bar{\mathcal{R}}_{\gamma_1} - \mathcal{J}_{G/\gamma_2} T_{\gamma_2} \bar{\mathcal{R}}_{\gamma_2} .$$

---

[23]N.N. Bogoliubov and O.S. Shirkov, *Introduction to the Theory of Qantized Fields* (Interscience, New York, 1959); We follow the discussion of Itzikson and Zuber.

Now, γ_1 and γ_2 have a superficial degree of divergence $\omega = 0$ and contain no divergent subgraphs, so that

$$\bar{\mathcal{R}}_{\gamma_1} = \mathcal{J}_{\gamma_1} \, , \quad \bar{\mathcal{R}}_{\gamma_2} = \mathcal{J}_{\gamma_2} \, .$$

Hence

$$\mathcal{J}_{G/\gamma_1} T_{\gamma_1} \bar{\mathcal{R}}_{\gamma_1} = T_{\gamma_1} \mathcal{J}_{G/\gamma_1} \mathcal{J}_{\gamma_1} = T_{\gamma_1} \mathcal{J}_G$$

$$\mathcal{J}_{G/\gamma_2} T_{\gamma_2} \bar{\mathcal{R}}_{\gamma_2} = T_{\gamma_2} \mathcal{J}_G \, ,$$

or

$$\bar{\mathcal{R}}_G = (1 - T_{\gamma_1} - T_{\gamma_2}) \mathcal{J}_G \, . \tag{16.100}$$

Since $\omega(G) = 2$, we conclude that

$$\mathcal{R}_G = (1 - T_G) \bar{\mathcal{R}}_G = (1 - T_G)(1 - T_{\gamma_1} - T_{\gamma_2}) \mathcal{J}_G \, . \tag{16.101}$$

Note that

$$\mathcal{R}_G \neq (1 - T_G)(1 - T_{\gamma_1})(1 - T_{\gamma_2}) \mathcal{J}_G \, .$$

The r.h.s. would be valid, if the subdiagrams γ_1 and γ_2 were not overlapping. Bogoliubov's recursion formula gives the correct treatment of the renormalization of overlapping divergences!

Notice that (16.99) is a recursive formula. The operations are understood to be performed in momentum space as Taylor subtractions around momentum zero in the external variables *of the subdiagrams*.

Let us make all this more concrete. Consider the left partition γ_1 of the diagram represented in Fig. 16.9. It is represented by the *amputated* vertex diagram of Fig. 16.10.

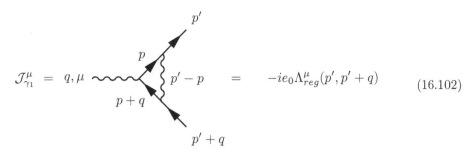

Fig. 16.10. 1-loop QED vertex diagram.

From (16.81) and (16.82) we see that

$$\Lambda^\mu_{ren}(p', p' + q) = \Lambda^\mu_{reg}(p', p' + q) + (Z_1 - 1)\gamma^\mu$$

with $(Z_1 - 1) = -\alpha_0 F_1(0)_{reg}$. This choice of Z_1 enforces the renormalization condition (16.77), that is

$$\Lambda^\mu_{ren}(p, p) = 0 \, .$$

Hence

$$(Z_1 - 1)\gamma^\mu = -\Lambda^\mu_{reg}(p, p) \tag{16.103}$$

or in the Feynman gauge,

$$(Z_1 - 1)\gamma^\mu = -\int \frac{d^4p}{(2\pi)^4} \frac{-ig_{\rho\lambda}}{p^2 - i\epsilon} e_0 \gamma^\rho S_F(p) \gamma^\mu S_F(p) e_0 \gamma_\lambda . \tag{16.104}$$

Notice that the integral does not involve a trace, but must be proportional to γ^μ because of Lorentz covariance.

Since the vertex diagram is logarithmically divergent we only need the first leading term of the Taylor expansion for γ_1 and γ_2. From (16.103) this means

$$T_{\gamma_1} \mathcal{J}^\rho_{\gamma_1} = T_{\gamma_1} \mathcal{J}^\rho_{\gamma_2} = -(Z_1 - 1)\gamma^\rho . \tag{16.105}$$

In the Feynman gauge the integral representing Fig. 16.9 is

$$\mathcal{J}_G = (-ie_0)^4 \int \frac{d^4p}{(2\pi)^4} \int \frac{d^4p'}{(2\pi)^4} \frac{-ig_{\rho\lambda}}{(p - p')^2 - i\epsilon} \tag{16.106}$$
$$\times tr \left[\gamma^\lambda i S_F(p+q) \gamma^\mu i S_F(p) \gamma^\rho i S_F(p') \gamma^\nu i S_F(p' + q) \right] .$$

Putting things together by adjoining the remaining part of the eight-dimensional integral to (16.105), we have with (16.100),

$$\bar{\Pi}^{\mu\nu} = e_0^4 \int \frac{d^4p}{(2\pi)^4} \int \frac{d^4p'}{(2\pi)^4} \frac{-i}{(p - p')^2 - i\epsilon} \tag{16.107}$$
$$\times tr[\gamma_\rho S_F(p+q)\gamma^\mu S_F(p)\gamma^\rho S_F(p')\gamma^\nu S_F(p' + q)]$$
$$+ \int \frac{d^3p'}{(2\pi)^4} tr\{(Z_1 - 1)(-ie_0\gamma^\mu)[iS_F(p')(-ie_0\gamma^\nu)iS_F(p' + q)]\}$$
$$+ \int \frac{d^3p}{(2\pi)^4} tr\{[(iS_F(p+q)(-ie_0\gamma^\mu)iS_F(p)](Z_1 - 1)(-ie_0\gamma^\nu)\}.$$

This takes care of the subdiagrams. The integral is still quadratically divergent, so a final operator $(1 - T_G)$ still has to be applied to the integrand according to (16.98).

Let us check that the divergences of the subgraphs have been taken care of. The following argument has been given by S. Weinberg:[24] Replace Z_1 in (16.107) by (16.104). We may then rearrange the terms as follows:

$$\bar{\Pi}^{\mu\nu} = -ie_0^4 \int \frac{d^4p}{(2\pi)^4} \int \frac{d^4p'}{(2\pi)^4}$$
$$\times \left[\frac{1}{(p - p')^2 - i\epsilon} tr[\gamma_\rho S_F(p+q)\gamma^\mu S_F(p)\gamma^\rho S_F(p')\gamma^\nu S_F(p' + q)] \right.$$
$$- tr \left[\frac{\gamma^\rho S_F(p)\gamma^\mu S_F(p)\gamma_\rho}{p^2 - i\epsilon} S_F(p')\gamma^\nu S_F(p' + q) \right]$$
$$\left. - tr \left[S_F(p+q)\gamma^\mu S_F(p) \frac{\gamma^\rho S_F(p')\gamma^\nu S_F(p')\gamma_\rho}{p'^2 - i\epsilon} \right] \right].$$

---

[24]S. Weinberg, *The Quantum Theory of Fields* (Cambridge University Press, 1996).

To show that this expression takes care of the divergences of the subgraphs, we first consider p' fixed, and let p go to infinity. We see that the difference between the second and third terms tends to zero. The same applies to the second and fourth terms, when p' is held fixed and p goes to infinity. There remains an overall quadratic infinity in form of a second order polynomial $\frac{1}{4}\delta Z_A(-g^{\mu\nu}q^\nu q^2 + q^\mu q^\nu)$ (see (16.17)), from which Z_A is obtained. The divergent part is finally removed by the operator $(1 - T_G)$ in (16.101).

We have presented various schemes of renormalization. Some, such as the Taylor subtraction and Bogoliubov recursion procedure did not make reference to the "normalization conditions" to be imposed on the Green functions. Thus, instead of the Taylor expansion in (16.91) around $p^2 = 0$, we could have made an expansion around some other point, as long as we take proper care of the normalization conditions (16.7) to (16.12). We shall return to this matter in the chapter on the "Renormalization group".

16.10 Dispersion relations: a brief view

There exist essentially two "dispersive" approaches to the approximative calculation of transition amplitudes. Both of them are related in that the absorptive part is obtained as the sum over the intermediate physical states participating in the interaction.

Källen–Lehmann representation

Consider the commutator of two general scalar field operators $\phi(x)$. Using translational invariance and the completeness of states $|\alpha>$, we have

$$< \Omega|[\phi(x),\phi(y)]|\Omega >= \sum_\alpha [< \Omega|\phi(0)|\alpha >< \alpha|\phi(0)|\Omega > e^{ip_\alpha \cdot (x-y)} - (x \to y)]$$

with the sum running over a complete set of positive energy states, which can be rewritten as

$$< \Omega|[\phi(x),\phi(y)]|\Omega >= \int \frac{d^4q}{(2\pi)^3} \rho(q)(e^{-iq\cdot(x-y)} - e^{iq\cdot(x-y)})$$

where $\rho(q^2)$ is the density

$$\rho(q^2) = (2\pi)^3 \sum_\alpha \delta^4(q - p_\alpha)|< \Omega|\phi(0)|\alpha > |^2.$$

This density is a positive Lorentz invariant and vanishes outside the forward light cone. It can thus be written as

$$\rho(q^2) = \theta(q^0)\sigma(q^2), \quad \text{with} \quad \sigma(q^2) = 0 \quad \text{if} \quad q^2 < 0 .$$

We have then

$$< \Omega|[\phi(x),\phi(y)]|\Omega >= i \int_0^\infty dm'^2 \; \sigma(m'^2)\Delta(x - y; m'^2). \tag{16.108}$$

The LSZ asymptotic condition (10.16) implies

$$< \Omega | [\phi(x), \phi(y)] | \Omega > = i Z_\phi \Delta(x - y; m^2) + i \int_{m'^2 > m^2}^{\infty} dm'^2 \, \sigma(m'^2) \Delta(x - y; m'^2) \, .$$

Taking the time-derivative on both sides, we have with (7.36)

$$1 = Z + \int_{m'^2 > m^2} dm'^2 \sigma(m'^2)$$

or positivity of $\sigma(m^2)$ implies

$$Z \leq 1 \, ,$$

with $Z_\phi = 0$ being reached only for a free field. In momentum space, Eq. (16.108) takes the form of a dispersion relation:[25]

$$\Sigma(p^2) = \int_{s > m^2}^{\infty} ds \frac{\sigma(s)}{s - p^2 + i\epsilon} \, . \tag{16.109}$$

The analytic approach

The analyticity properties of Feynman amplitudes have been the subject of extensive studies documented in a number of books. A nice account has been given by Bjorken and Drell.[26] These studies have played a central role in the so-called "S-matrix theory" of the 1960's.[27]

The "dispersive" approach was initiated in 1954 with the work of M. Gell-Mann, M.L. Goldberger and W. Thirring.[28] The analytical approach based on the study of perturbative expansions has received most of the attention. In this section we only briefly touch on this subject, by providing a representation of the renormalized self-energy in ϕ^3 theory as a dispersion relation. The analytic structure of the corresponding Feynman diagrams in lowest non-trivial order of perturbation theory is particularly simple, but serves to illustrate some points, such as the Taylor expansion in (16.91).

Consider a general self-energy diagram of ϕ^3-theory as shown in Fig. 16.11:

$$p \; -- \; \text{(diagram)} \; --- \; p$$

Fig. 16.11. General self-energy diagram.

Its analyticity properties on the physical sheet in the external momentum are determined by the vanishing of the argument of the exponential in (12.17), or the

[25] G. Källan, *Helv. Phys. Acta* **25** (1952) 417; H. Lehmann, *Nuovo Cimento* **11** (1954) 342.
[26] J.D. Bjorken and S.D. Drell, *Relativistic Quantum Fields* (McGraw-Hill, 1965).
[27] G.F. Chew, *S-Matrix Theory of Strong Interactions* (W.A. Benjamin Inc., New York, 1962); S. Mandelstam, *Rept. Progr. Phys.* **25**.
[28] M. Gell-Mann, M.L. Goldberger and W. Thirring, *Phys. Rev.* **95** (1954) 1612.

denominator in (12.18) ($\mathcal{P}_G(\alpha)$ plays no role) in the multidimensional parameter space (p, α):

$$J \equiv Q_G(p; \alpha) - \sum_i \alpha_i m_i^2 = 0 ,$$

supplemented by the condition that the solution $\alpha^{(0)}$ satisfies

$$\textit{either} \quad \left.\frac{\partial J}{\partial \alpha_i}\right|_{\alpha_i^{(0)}} = 0 , \quad \textit{or} \quad \alpha_i^{(0)} = 0 .$$

These are the *Landau conditions*.

As an example we have for the diagram of Fig. 16.8, taking account of the δ-function in α,

$$J(p^2; \alpha) = p^2 \alpha(1 - \alpha) - m^2 ,$$

$$\frac{\partial}{\partial \alpha} J = p^2 (1 - 2\alpha) = 0$$

implying $\alpha^{(0)} = \frac{1}{2}$ and a singularity at $p^2 - 4m^2 = 0$.

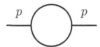

Fig. 16.12. Self-energy diagram of ϕ^3 theory.

Indeed, the diagram is represented by the integral (see Eq. (16.97)),

$$\Sigma_{ren}^{(2)}(p^2) - \Sigma_{ren}^{(2)}(-\mu^2) = \frac{g^2}{16\pi^2} \int_0^1 d\beta \ln \left(\frac{m^2 - \beta(1 - \beta)p^2 - i\epsilon}{m^2 + \beta(1 - \beta)\mu^2}\right) . \tag{16.110}$$

We see that the integrand exhibits a logarithmic cut in p^2 extending to the right in the complex p^2 plane from $p^2 = m^2/\beta(1 - \beta)$ to $p^2 = \infty$. The discontinuity across the cut is given by

$$\ln(m^2 - \beta(1 - \beta)(p^2 + i\epsilon)) - \ln(m^2 - \beta(1 - \beta)(p^2 - i\epsilon)) = -2\pi i\theta(-m^2 + \beta(1 - \beta)p^2)$$

or we obtain for the discontinuity of $\Sigma(p^2)$

$$disc\Sigma^{(2)}(p^2) = (-2\pi i)\frac{g^2}{16\pi^2} \int_0^1 d\beta \theta(-m^2 + \beta(1 - \beta)p^2)$$

$$= (-2\pi i)\frac{g^2}{16\pi^2} \int_0^1 d\beta \left(-(\beta - \frac{1}{2})^2 + \left(\frac{1}{4} - \frac{m^2}{p^2}\right)\right) ,$$

$$= -i\frac{g^2}{8\pi} \sqrt{1 - \frac{4m^2}{p^2}} \theta(p^2 - 4m^2) . \tag{16.111}$$

Let us repeat this calculation starting from the basic integral

$$-i\Sigma^{(2)}(p^2) = (-ig_0)^2 \int \frac{d^4q}{(2\pi)^4} \frac{i^2}{[q^2 - m_0^2 + i0][(q-p)^2 - m_0^2 + i0]} .$$

This integral is ultraviolet divergent, but its regularized expression only differs from the renormalized expression by a cutoff dependent additive constant which does not contribute to the discontinuity to be calculated.

The calculation is simplified by going into the frame $\vec{p} = 0$:

$$-i\Sigma(p_0^2) = g_0^2 \int \frac{dq^0}{2\pi} \int \frac{d^3q}{(2\pi)^3}$$

$$\times \frac{1}{[q^0 - \omega_q + i0][q^0 + \omega_q - i0][(q^0 - p^0) - \omega_q + i0][(q^0 - p^0) + \omega_q - i0]} ,$$

where $\omega_q = \sqrt{\vec{q}^2 + m_0^2}$. The location of the poles in the complex q^0-plane and the contour of integration are depicted in the figure below.

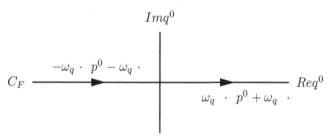

Fig. 16.13. q^0-integration contour and singularities for $\Sigma^{(2)}(q_0^2)$.

We have

- $p^0 < 2m_0$

  The q^0-contour is never "pinched" by the singularities in q^0 as $\vec{q}$ varies, so that there is no singularity in p^0 for $p^0 < 2m_0$.

- $p^0 \geq 2m_0$

For $p_0 = 2m_0$ the q_0-contour is pinched at $q^0 = m$ as $\vec{q} \to 0$, and there will be a cut from $p^0 = 2m_0$ to ∞. In order to compute the discontinuity across this cut we distort the contour across the singularity at $q^0 = \omega_q$, keeping track of this pole, which contributes with an infinitesimal clockwise contour C_ϵ around it:

$$\Sigma_{C_\epsilon}^{(2)}(p_0^2) = g_0^2 \int \frac{d^3q}{(2\pi)^3} \int_{-\infty}^{\infty} \frac{dq^0}{2\pi} \frac{(-2\pi i)}{2p^0 \omega_q} \delta(q^0 - \omega_q) \left\{ \frac{1}{[p^0 - 2\omega_q + i0]} \right\} . \quad (16.112)$$

The discontinuity of $\Sigma(p)$ arises from here, and is

$$disc\Sigma(p^0) = \Sigma_{C_\epsilon}(p^0 + i0) - \Sigma_{C_\epsilon}(p^0 - i0)$$

$$= g_0^2 \int \frac{d^3q}{(2\pi)^3} \int_{-\infty}^{\infty} \frac{dq^0}{2\pi}$$

$$\times \frac{(-2\pi i)}{2p^0 \omega_q} \delta(q^0 - \omega_q) \left\{ \frac{1}{[p^0 - 2\omega_q + i0]} - \frac{1}{[p^0 - 2\omega_q - i0]} \right\} . \tag{16.113}$$

Recalling that

$$\frac{1}{x + i0} - \frac{1}{x - i0} = -2i\pi\delta(x)$$

the remaining integral is easily done:

$$disc\Sigma(p^0) = g_0^2(-2\pi i) \int \frac{d^3q}{(2\pi)^3} \frac{1}{2p^0 \omega_q} \delta(p^0 - 2\omega_q)$$

$$= \frac{-ig_0^2}{(2\pi)^2} \int d\Omega \int_0^{\infty} d\omega \, \omega (\omega^2 - m_0^2)^{\frac{1}{2}} \delta\left(\frac{\omega - \frac{1}{2}p^0}{2p_0^2}\right)$$

$$= -i \frac{g_0^2}{8\pi} \sqrt{1 - \frac{4m_0^2}{p_0^2}} \theta(p^0 - 2m_0). \tag{16.114}$$

Using relativistic invariance, and setting now $m_0 = m$ and $g_0 = g$, this result agrees with (16.111).

Starting from the usual Cauchy integral

$$\Sigma(p^2) = \frac{1}{2\pi i} \int_{\mathcal{C}} ds \frac{\Sigma(s)}{s - p^2}$$

with $\mathcal{C}$ given by the contour closed at infinity, enclosing the cut, we cannot discard the contribution at ∞. Hence we write the Cauchy integral for $\Sigma(p^2)/(p^2 - \mu^2)$. Taking account of the pole at $p^2 = \mu^2$ we obtain this way the subtracted dispersion relation

$$\Sigma(p^2) = \Sigma(\mu^2) + \frac{(p^2 - \mu^2)}{2\pi i} \int_{4m^2}^{\infty} \frac{disc\Sigma(s)}{(s - p^2 + i0)(s - \mu^2 + i0)} \tag{16.115}$$

$$= \Sigma(\mu^2) - \frac{p^2 - \mu^2}{8\pi} \int_{4m^2}^{\infty} ds \frac{g^2 \sqrt{1 - \frac{4m^2}{s}}}{(s - p^2 + i0)(s - \mu^2 + i0)} . \tag{16.116}$$

This corresponds to a single, final Taylor subtraction in (16.91) for $\omega(G) = 0$. Bogoliubov's recursion formula provides the generalization thereof to arbitrary diagrams with divergent subgraphs.

Cutcosky rules

Let us rewrite the result (16.113) as follows:

$$-idisc\Sigma(p^2) = g_0^2 \int \frac{d^4q}{(2\pi)^4} \frac{(-2\pi i)\delta(q^0 - \omega_q)\theta(q^0)(-2\pi i)\delta(p^0 - q^0 - \omega_q)\theta(p^0 - q^0)}{2p^0 \omega_q}$$

$$= g_0^2 \int \frac{d^4q}{(2\pi)^4} (-2\pi i)^2 \delta(q^2 - m_0^2)\theta(q^0)\delta((p - q)^2 - m_0^2)\theta(p^0 - q^0).$$

This result provides an example of the so-called *Cutkoski rules* which state, that the discontinuity of $\Sigma(p^2)$ across the cut in the physical region $p^2 \geq m_0^2$ is obtained in the present case by putting the "particles" in the intermediate state on their respective mass shell, with the rule[29]

$$\frac{1}{q^2 - m_0^2 + i0} \longrightarrow (-2\pi i)\theta(q^0)\delta(q^2 - m_0^2)$$

$$\frac{1}{(q-p)^2 - m_0^2 + i0} \longrightarrow (-2\pi i)\theta(q^0 - p^0)\delta((q-p)^2 - m_0^2) ,$$

as represented in the following diagram.

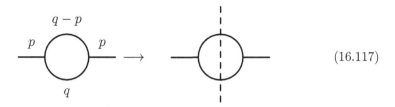

$$(16.117)$$

Fig. 16.14. Cutkosky rule discontinuity.

This observation establishes the link with the Källan–Lehmannn representation.

[29]R.E. Cutkosky, *J. Math. Phys.* **1** (1960) 429.

Chapter 17

Broken Scale Invariance and Callan–Symanzik Equation

If there were no dimensional parameters in the theory, then dimensional arguments alone (plus translational invariance) would uniquely fix, up to a normalization constant, the form for the 2-point function of a scalar field to be

$$\langle\Omega|\phi(x)\phi(y)|\Omega\rangle = \frac{1}{2\pi^2}\frac{1}{(x-y)^2} \ . \tag{17.1}$$

The function (17.1) evidently scales with the *canonical* dimension of the scalar field under the *dilation* $x \to \lambda x$ of the coordinates,

$$\langle\Omega|\phi(\lambda x)\phi(\lambda y)|\Omega\rangle = \lambda^{-2d_\phi}\langle\Omega|\phi(x)\phi(y)|\Omega\rangle, \quad d_\phi = 1 \ . \tag{17.2}$$

If we take the vacuum to be invariant under such a scale transformation, then we deduce from (17.2) the transformation law

$$\phi(x) \to U[\lambda]\phi(x)U^{-1}[\lambda] = \phi'(x) = \lambda\phi(\lambda x) \ . \tag{17.3}$$

The corresponding Lagrange density describing the dynamics of the scalar field would thus scale as follows under a dilatation

$$\mathcal{L}(x) \to \mathcal{L}'(x) = \lambda^4\mathcal{L}(\lambda x) \ .$$

Hence in this case the corresponding *action* is left invariant by the unitary operator inducing the dilatation (17.3). According to the theorem of Noether, we expect a conserved current corresponding to this continuous symmetry, called the *dilatation current S^μ*. In analogy to electrodynamics we further expect a Ward identity to be associated with the conservation of this current.

De factum there exists no non-trivial Quantum Field Theory exhibiting such dilatation invariance. Even if the defining classical lagrangian does not involve any dimensioned parameter (such as would be the case for electrodynamics of massless fermions), the need of renormalization will require the introduction of dimensioned parameters, which break the dilatation invariance of the classical theory. One speaks of *broken scale invariance*, and the modified Ward Identities reflecting this fact are called the *Callan–Symanzik equations*.

One would nevertheless be tempted to argue that despite the appearance of dimensioned "mass" parameters in the theory, these parameters should not play a role at high energies. This "intuitive feeling" turns out to be incorrect. It turns out to be however almost correct in the case of so-called *asymptotically free* theories, where the violation of the "canonical" scaling law is *soft* in the sense that it is only broken by logarithmic factors. An important example of such theories is Quantum Chromodynamics (QCD). In general, however, the Green functions of the theory will exhibit a powerlike (or logarithmic) deviation from the canonical behaviour at large (euclidean) momenta (short distances). Thus for the 2-point function of a scalar field one would have for example for $(x - y)^2 \to 0$,

$$\langle \Omega | \phi(x)\phi(y) | \Omega \rangle \approx \frac{1}{[(x-y)^2]} \frac{1}{[\mu^2(x-y)^2]^{\gamma_\phi}} \tag{17.4}$$

with μ a parameter with the dimension of a mass introduced by the renormalization procedure. Here γ_ϕ is called *anomalous* dimension of the scalar field $\phi(x)$. As we shall see, such anomalous dimensions can be calculated in perturbation theory from the knowledge of the wave function renormalization constants. They are the analogue of the critical exponents of systems at criticality in solid state physics. This is the reason, why solid state physicists have made extensive use of field-theoretic methods in order to compute such critical exponents.

The anomalous scaling behaviour also holds if the lagrangian contains massive particles. In order to see this, we shall have to examine more closely the behaviour of Feynman diagrams in the so-called *deep euclidean region*, where all the external momenta in the diagram, continued analytically to the euclidean domain, tend to infinity. The general result runs under the name of *Weinbergs's theorem*. It will eventually imply that the Callan–Symanzik equation, which is an inhomogeneous equation, reduces to an homogeneous one in the deep-euclidean region.

17.1   Scale transformations

We begin by recalling *Noether's Theorem*. (See Section 1 of Chapter 9.) Let $\phi_\alpha(x), \alpha = 1, ..., n$ stand for n complex fields whose dynamics is described by a Lagrange density

$$\mathcal{L}(x) = \mathcal{L}\big(\{\phi_\alpha(x), \partial_\mu \phi_\alpha(x)\}\big)$$

with

$$S[\phi] = \int d^4x \mathcal{L}(x) \tag{17.5}$$

the corresponding action. Consider a *linear* infinitesimal transformation

$$\phi_\alpha(x) \to \phi_\alpha(x) + \delta_\epsilon \phi_\alpha(x) \tag{17.6}$$

with

$$\delta_\epsilon \phi_\alpha(x) = \epsilon_a \mathcal{D}^a_{\alpha\beta} \phi_\beta(x) \ ,$$

where ϵ^a are A infinitesimal parameters characterizing the transformation. If it is a symmetry transformation, then it leaves the action (17.5) invariant, implying that the corresponding change in the Lagrange density must be representable as the divergence of a 4-vector field:

$$\delta_\epsilon \mathcal{L}(x) = \epsilon_a \partial_\mu \mathcal{F}^{\mu,a}(x) \ .$$

Noether's theorem then states that the following A "currents" $\mathcal{J}^{\mu,a}$ $(a = 1 \cdots, A)$ are conserved:

$$\mathcal{J}^{\mu,a} = \sum_{\alpha,\beta} \left(\frac{\partial \mathcal{L}}{\partial \partial_\mu \phi_\alpha} \mathcal{D}^a_{\alpha\beta} \phi_\beta + h.c. \right) - \mathcal{F}^{\mu,a}, \quad \partial_\mu \mathcal{J}^{\mu,a} = 0 \ . \qquad (17.7)$$

If, on the other hand, the transformation (17.6) does not represent a symmetry transformation of the action (17.5), then these conservation laws will be replaced by

$$\partial_\mu \mathcal{J}^{\mu,a} = \Delta^a \ , \qquad (17.8)$$

where the rhs is so far unspecified.

We shall be interested in particular in the Noether current S^μ associated with *dilatations* or *scale transformations* $x^\mu \longrightarrow \lambda x^\mu$.

Like all other transformations of physical interest, scale transformations can be implemented in a linear or non-linear way. We shall be interested in a linear implementation on the fields:

$$\vec{\phi}(x) \underset{U}{\longrightarrow} U[\lambda]\vec{\phi}(x)U^{-1}[\lambda] = \vec{\phi}'(x) = \lambda^{\mathbf{d}} \vec{\phi}(\lambda x),$$

or in infinitesimal form $(\lambda = 1 + \epsilon)$

$$\delta_\epsilon \vec{\phi}(x) = \epsilon(\mathbf{d} + \mathbf{1}x^\mu \partial_\mu)\vec{\phi}(x) \ . \qquad (17.9)$$

Here $\mathbf{d}$ is in general a matrix acting on the vector field $\vec{\phi}(x)$. In a *classical* field theory involving no dimensioned parameters, $\mathbf{d}$ would be a diagonal matrix multiplying all *Bose* fields by *one* and all *Fermi* fields by 3/2, according to their *canonical* (engineering) dimensions. In models involving a mass term, this symmetry will be explicitly broken. In a QFT this symmetry will further be broken by renormalization, since it requires the introduction of new dimensioned scale parameters. To give an example consider the theory of interacting bosons and fermions,

$$\mathcal{L} = \mathcal{L}_S + \mathcal{L}_{SB} \qquad (17.10)$$

where

$$\mathcal{L}_S = \overline{\psi} i \partial\!\!\!/ \psi + \frac{1}{2} \partial_\mu \phi \partial^\mu \phi + g_0 \overline{\psi}\gamma_5 \psi \phi - \frac{\lambda_0}{4!} \phi^4$$

and $\mathcal{L}_{SB}$ is the symmetry breaking term

$$\mathcal{L}_{SB} = -m_0 \overline{\psi}\psi - \frac{1}{2}\mu_0^2 \phi^2 \ .$$

In the absence of $\mathcal{L}_{SB}$, the transformations

$$\phi(x) \to \phi'(x) = \lambda\phi(\lambda x) , \quad \psi(x) \to \psi'(x) = \lambda^{3/2}\psi(\lambda x) \qquad (17.11)$$

would result in

$$\mathcal{L}_S(x) \to \mathcal{L}'_S(x) = \lambda^4 \mathcal{L}(\lambda x) ,$$

thus implying the invariance of the action (17.5) under the transformation (17.11). For $\mathcal{L}_{SB}$ we however have

$$\mathcal{L}_{SB}(x) \longrightarrow \mathcal{L}'_{SB}(x) = -m_0\lambda^3\overline{\psi}\psi(\lambda x) - \frac{1}{2}\mu_0^2\lambda^2\phi^2(\lambda x) ,$$

so that scale invariance will be broken by this term. Infinitesimally we have from (17.11) (we omit the ϵ-parameter)

$$\delta\psi = \left(\frac{3}{2} + x^\mu\partial_\mu\right)\psi, \quad \delta\phi = (1 + x^\mu\partial_\mu)\phi ,$$

or

$$\delta\mathcal{L}_S = (4 + x^\mu\partial_\mu)\mathcal{L}_S = \partial_\mu(x^\mu\mathcal{L}_S) \qquad (17.12)$$

$$\delta\mathcal{L}_{SB} = -(3 + x^\mu\partial_\mu)m_0\overline{\psi}\psi - (2 + x^\mu\partial_\mu)\frac{\mu_0^2}{2}\phi^2$$

$$= \partial_\mu(x^\mu\mathcal{L}_{SB}) + m_0\overline{\psi}\psi + \mu_0^2\phi^2 . \qquad (17.13)$$

We see that (17.12) gives a vanishing contribution to the action, while

$$\delta S_{SB}[\phi] = \int d^4x\Delta , \quad \Delta = m_0\overline{\psi}\psi + \mu_0^2\phi^2 .$$

Retracing the derivation of the Noether theorem, we see that this means that the corresponding Noether current is no longer conserved,

$$\partial_\mu \mathcal{S}^\mu = \Delta , \qquad (17.14)$$

which is the local version of the statement that masses break scale invariance.

It is important to realize that what we have done is *not* "dimensional analysis". Indeed, a dimensional analysis requires to also scale the masses as $m \to \lambda^{-1}m$, and is *always* an *exact symmetry* of the action.

Example of a scale current

Consider the Lagrange density

$$\mathcal{L} = \frac{1}{2}\partial_\mu\phi\partial^\mu\phi - \frac{\lambda}{4!}\phi^4. \qquad (17.15)$$

We have (see (17.12))

$$\mathcal{F}^\mu = x^\mu\mathcal{L}_S ,$$

so that according to (17.7) the corresponding scale (dilatation) current, now denoted by S^μ, is given by

$$S^\mu = \partial^\mu \phi (1 + x_\nu \partial^\nu)\phi - x^\mu \mathcal{L} = \phi \partial^\mu \phi + x_\nu \mathcal{J}^{\mu,\nu} , \qquad (17.16)$$

where $\mathcal{J}^{\mu,\nu}$ is the canonically constructed energy momentum tensor

$$\mathcal{J}^{\mu,\nu} = \frac{\partial \mathcal{L}}{\partial(\partial_\mu \phi)}\partial^\nu \phi - g^{\mu\nu}\mathcal{L} = \partial^\mu \phi \partial^\nu \phi - g^{\mu\nu}\mathcal{L} .$$

Note that this energy-momentum tensor is in general not symmetric, and is conserved only with respect to the first index:

$$\partial_\mu \mathcal{J}^{\mu,\nu} = 0 .$$

Using the equations of motion

$$\Box \phi = \frac{\partial \mathcal{L}}{\partial \phi}$$

one easily checks that the current (17.16) is conserved, as expected.

In QFT the need for renormalization will force us to introduce dimensioned parameters, even if such parameters were absent in the classical action. Accordingly the fields $\phi(x)$ will transform with *anomalous* scale dimensions. These anomalous dimensions turn out to be computable in perturbation theory. The basic idea has its roots in Eq. (17.14), and the *Ward identities* for renormalized n-point functions implied by it.

17.2   Unrenormalized Ward identities of broken scale invariance

We now consider the Ward identities for the *unrenormalized* n-point functions following from broken scale invariance. For the sake of concreteness the reader may think of the theory of a massive, real scalar field described by a Lagrange density

$$\mathcal{L} = \frac{1}{2}(\partial_\mu \phi)^2 - \frac{1}{2}m_0^2 \phi^2 - \frac{g_0}{4!}\phi^4 . \qquad (17.17)$$

The divergence of the dilatation current is in this case given by (17.14) with $\Delta = m_0^2 \phi^2$, and

$$S^\mu = \partial^\mu \phi (1 + x_\nu \partial^\nu)\phi - x^\mu \mathcal{L} - m_0^2 \phi^2 .$$

Although the corresponding charge is not conserved, we still have the equal time commutation relations

$$\left[S^0(y), \phi(x) \right]_{ET} = -i\delta^3(\vec{x} - \vec{y})\delta\phi(x) \qquad (17.18)$$

for the unrenormalized fields. We define as usual the time-ordered product of n fields $A(x)$,

$$T\left[A(x_1)...A(x_n)\right] = \sum_P \delta_P \theta(x_{i_1}^0 - x_{i_2}^0)...\theta(x_{i_{n-1}}^0 - x_{i_n}^0)A(x_{i_1})...A(x_{i_n})$$

where the sum runs over all permutations in time of the fields, and δ_p is the "signature" of the permutation; i.e. $1(-1)$ for even (odd) permutation of commuting (anticommuting) fields. We then have from (17.14) and (17.18) the *Ward Identity*

$$\frac{\partial}{\partial y^\mu}\langle\Omega|T\big(\mathcal{S}^\mu(y)\phi(x_1)...\phi(x_n)\big)|\Omega\rangle = \langle\Omega|T\Delta(y)\phi(x_1)...\phi(x_n)|\Omega\rangle$$

$$-i\sum_{i=1}^{n}\delta^4(y-x_i)\langle\Omega|T\phi(x_1)...\phi(x_{i-1})\delta\phi(x_i)\phi(x_{i+1})...\phi(x_n)|\Omega\rangle .$$

Integrating both sides of the equation, and dropping the surface term, we arrive at

$$\int d^4y\langle\Omega|T\big(\Delta(y)\phi(x_1)...\phi(x_n)\big)|\Omega\rangle = \qquad\qquad (17.19)$$

$$i\sum_{i=1}^{n}\langle\Omega|T\phi(x_1)...\phi(x_{i-1})\delta\phi(x_i)\phi(x_{i+1})...\phi(x_n)|\Omega\rangle .$$

These equations are called *zero-energy theorems*, since the integral over Δ carries zero energy and momentum in Fourier space. (The *soft-pion theorems* of current algebra are special cases of (17.19).) Identifying $\delta\phi$ with

$$\delta\phi(x) = (d_\phi + x^\mu\partial_\mu)\phi(x) , \quad d_\phi = 1 , \qquad\qquad (17.20)$$

and pulling the derivative in (17.20) past the time-ordering operation in (17.19) (since the scalar fields commute at equal times, this does not give rise to "contact" or "seagull" terms), we can write the Ward identity in the form

$$\sum_{i=1}^{n}\left(d_\phi + x_i^\mu\frac{\partial}{\partial x_i^\mu}\right)\langle\Omega|T\phi(x_1)...\phi(x_i)...\phi(x_n)|\Omega = -i\int d^4y\langle\Omega|\Delta(y)\phi(x_1)...\phi(x_n)|\Omega\rangle.$$

$$(17.21)$$

Taking the Fourier transform of both of these equations, and making use of translational invariance, we obtain from here for the lhs,[1]

$$\int d^4x_1...d^4x_n e^{ip_1\cdot x_1}...e^{ip_n\cdot x_n}\left(nd_\phi + \sum_{i=1}^{n}x_i^\mu\frac{\partial}{\partial x_i^\mu}\right)\langle\Omega|T\phi(x_1-x_n)...\phi(0)|\Omega\rangle$$

$$= \int dy e^{iy\cdot(\sum_1^n p_i)}\int\prod_1^{n-1}dy_i e^{ip_1\cdot y_1}...e^{ip_{n-1}\cdot y_{n-1}}\left(nd_\phi + \sum_{i=1}^{n-1}y_i^\mu\frac{\partial}{\partial y_i^\mu}\right)$$

$$\times\langle\Omega|T\phi(y_1)...\phi(y_{n-1})\phi(0)|\Omega\rangle .$$

Making use of

$$\int d^4y e^{ip\cdot y}y^\mu\partial_\mu f = -\int d^4y(4+iy^\mu p_\mu)e^{ip\cdot y}f(y) = -\left(4+p_\mu\frac{\partial}{\partial p_\mu}\right)\int d^4y e^{ip\cdot y}f(y)$$

---

[1]Make change of variable $x_i = y_i + y , i = 1\cdots n-1 , x_n = y$, so that

$$\sum_{i=1}^{n}x_i\frac{\partial}{\partial x_i} \to \sum_{i=1}^{n-1}(y_i+y)\frac{\partial}{\partial y_i} - \sum_{i=1}^{n-1}y\frac{\partial}{\partial y_i} = \sum_1^{n-1}y_i\frac{\partial}{\partial y_i} .$$

we arrive at the momentum-space version of the Ward identity (17.21):

$$\left[nd_\phi - 4(n-1) - \sum_{r=1}^{n-1} p_r^\mu \frac{\partial}{\partial p_r^\mu} \right] G^{(n)}(p_1...p_n) = -iG_\Delta^{(n)}(0;p_1...p_n) \,. \qquad (17.22)$$

Here $G^{(n)}(p_1 \cdots p_n)$ stands for the n-point Green function after separation of the overall momentum-conservation delta function:

$$\int \prod_{r=1}^n dx_r e^{i\sum_1^n p_r \cdot x_r} \langle \Omega|T\phi(x_1)...\phi(x_n)|\Omega\rangle = (2\pi)^4 \delta(\Sigma p_i) G^{(n)}(p_1...p_n) \,,$$

and $G_\Delta^{(n)}$ is the corresponding function with a "mass insertion":

$$\int \prod_1^n dx_r e^{i\sum p_r \cdot x_r} \langle \Omega|T\tilde\Delta(0)\phi(x_1)...\phi(x_n)|\Omega\rangle = (2\pi)^4 \delta(\Sigma p_r) G_\Delta^{(n)}(0;p_1...p_n) \,,$$

with $\tilde\Delta(0)$ the Fourier transform of $\Delta(x)$ evaluated at zero momentum:

$$\tilde\Delta(0) = \int d^4y \Delta(y) \,.$$

The first two terms on the lhs of (17.22) just reflect the engineering dimension $(nd_\phi - 4(n-1))$ of the n-point function $G^{(n)}$. Making use of ordinary dimensional analysis, we may write $G^{(n)}$ in the form

$$G^{(n)}(p_1...p_n) = \mathcal{S}^{\frac{nd_\phi - 4(n-1)}{2}} F^{(n)} \left(\frac{\mathcal{S}}{m_0^2}, \frac{p_i \cdot p_j}{\mathcal{S}}, g_0 \right) \qquad (17.23)$$

where we have introduced the Lorentz invariant variable

$$\mathcal{S} = \sum_{r=1}^n p_r^2 = \sum_{r=1}^{n-1} p_r^2 + \left(\sum_1^{n-1} p_r \right)^2 . \qquad (17.24)$$

Note that we have explained the dependence of $F^{(n)}$ on the unrenormalized mass m_0 and (dimensionless) coupling constant g_0. Returning to (17.23) and (17.22), we note that

$$\sum_1^{n-1} p_r^\mu \frac{\partial}{\partial p_r^\mu} \mathcal{S} = 2\mathcal{S} \,, \qquad \sum_1^{n-1} p_r^\mu \frac{\partial}{\partial p_r^\mu} (p_i \cdot p_j) = 2(p_i \cdot p_j) \,.$$

We see from here that

$$\sum_1^{n-1} p_r^\mu \frac{\partial}{\partial p_r^\mu} \frac{(p_i \cdot p_j)}{\mathcal{S}} = 0 \,.$$

Hence we can write

$$\sum_1^{n-1} p_r^\mu \frac{\partial}{\partial p_r^\mu} F^{(n)} = -m_0 \frac{\partial}{\partial m_0} F^{(n)} \left(\frac{\mathcal{S}}{m_0^2}, \frac{p_i \cdot p_j}{\mathcal{S}}, g_0 \right) \,, \qquad (17.25)$$

or we obtain from (17.23),

$$\sum_{1}^{n-1} p_r^\mu \frac{\partial}{\partial p_r^\mu} G^{(n)} = \left(nd_\phi - 4(n-1) - m_0 \left(\frac{\partial}{\partial m_0} \right)_{g_0,\lambda} \right) G^{(n)} .$$

We conclude that on the formal level we can write the Ward identity (17.22) in the form[2]

$$m_0 \left(\frac{\partial}{\partial m_0} \right)_{g_0,\Lambda} G^{(n)}(p_1...p_n) = -iG_\Delta^{(n)}(0; p_1...p_n) . \tag{17.26}$$

After all this work, this is a surprisingly simple, but not unexpected result. Indeed, in this form we recognize that the unrenormalized Ward identity (17.21) essentially states that differentiation of a general Green function with respect to the bare mass m_0, and subsequent multiplication with m_0, simply amounts to a zero-momentum "mass insertion" in each internal boson line. Thus we have for a ϕ-field propagator

$$m_0 \frac{\partial}{\partial m_0} \frac{i}{p^2 - m_0^2} = \frac{i}{p^2 - m_0^2} (-i2m_0^2) \frac{i}{p^2 - m_0^2}$$

with the corresponding diagrammatic representation

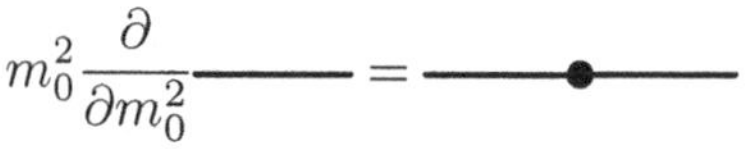

$$m_0^2 \frac{\partial}{\partial m_0^2} \underline{\qquad\qquad} = \underline{\qquad\bullet\qquad} \tag{17.27}$$

Fig. 17.1. Dot represents mass insertion $-i2m_0^2$.

17.3 Broken scale invariance and renormalized Ward identities

In the case of the Ward identities associated with gauge invariance of QED or QCD, renormalization can be performed consistent with this gauge invariance, and the corresponding Ward identity remains form-invariant, as we have seen. This is quite different from the case of the Ward identities associated with scale transformations. Even if we start from a scale invariant lagrangian on classical level, there exists no regularization procedure which will respect this invariance. The need for introducing dimensioned parameters into the theory will destroy this symmetry already on the 1-loop level. As a result, the Ward identity for the unrenormalized proper functions will change in form under renormalization. The result is known as the *Callan–Symanzik equation*, to be discussed next.

In order to conform to usage, we first define the 1PI (amputated) Green functions

$$i\Gamma_{reg}^{(n)}(\{p_i\}) = \prod_{j=1}^{n} \left[G_{reg}^{(2)}(p_j) \right]^{-1} G_{reg}^{(n)}(\{p_i\}) . \tag{17.28}$$

---

[2]S. Coleman, *Selected Erice Lectures* (Cambridge University Press).

The factor of "i" has been introduced to conform to convention (see Chapter 20, Eq. (20.12)). From Fig. 17.1 we conclude that in terms of the proper Green functions, Eq. (17.26) reads

$$m_0 \left(\frac{\partial}{\partial m_0} \right)_{g_0, \Lambda} \Gamma^{(n)}(p_1 \ldots p_n)_{reg} = \Gamma_\Delta^{(n)}(0; p_1 \ldots p_n)_{reg} . \tag{17.29}$$

We now derive from (17.29) the corresponding Ward identities for the renormalized proper functions. To this end, we review briefly the basic principles of renormalization.

Starting point is the *unrenormalized* Lagrange density (17.17). The Green functions calculated in terms of this lagrangian are given by divergent integrals, and thus need to be regularized by a suitable cutoff procedure. This leads to cutoff dependent Green functions. In a renormalizable theory, the cutoff dependence of an n-point unrenormalized Green function can be eliminated by (i) a judicious choice of the bare parameters as functions of the cutoff, accompanied by (ii) a multiplication of the regularized Green function with a suitably chosen cutoff dependent constant $Z_\phi^{\frac{1}{2}}$ for each of the n fields in the corresponding vacuum expectation value of the time-ordered product. It is convenient to write

$$g_0 = Z_g^{-\frac{1}{2}} \left(g, \frac{\Lambda}{m} \right) g, \quad m_0 = Z_m^{-1} \left(g, \frac{\Lambda}{m} \right) m .$$

Renormalizability then means that for a proper choice of the renormalization constants, the function

$$\widetilde{\Gamma}^{(n)}(p_1 \ldots p_n; g; m; \Lambda) = Z_\phi^{n/2} \Gamma_{reg}^{(n)}(p_1 \ldots p_n; Z_g^{-\frac{1}{2}} g, Z_m^{-1} m; \Lambda) \tag{17.30}$$

possesses a finite limit as we remove the cutoff ($\Lambda \to \infty$), keeping g and m fixed. Note that for the proper Green functions the wave function renormalization constant Z_ϕ appears with inverse powers as compared to (16.6). The multiplicative renormalization amounts to the substitution

$$\phi = Z_\phi^{1/2} \phi_{ren}$$

where ϕ_{ren} is referred to as the renormalized field.

In a practical calculation the renormalization constants will naturally emerge as functions of the cutoff Λ, *bare* coupling constant g_0 and the renormalized or bare mass. In line with (16.3) and (16.4) we define them by

$$g = Z_g^{\frac{1}{2}} \left(g_0, \frac{\Lambda}{m_0} \right) g_0 , \quad \phi_{ren} = Z_\phi^{-\frac{1}{2}} \left(g_0, \frac{\Lambda}{m_0} \right) \phi . \tag{17.31}$$

Unless required, we shall not distinguish between the different functional dependencies, and denote all of these functions by one and the same symbol, Z_σ.

We now apply the "bare mass insertion operator" $\left(m_0 \frac{\partial}{\partial m_0} \right)_{g_0, \Lambda}$ to

$$\Gamma_{reg}^{(n)}(\{p\}; g_0, m_0; \Lambda) = Z_\phi^{-\frac{n}{2}} \widetilde{\Gamma}^{(n)}(\{p\}; g, m; \Lambda) , \tag{17.32}$$

or (17.29) becomes,

$$\left(m_0 \frac{\partial}{\partial m_0}\right)_{g_0,\Lambda} \left(Z_\phi^{-\frac{n}{2}} \tilde{\Gamma}^{(n)}(\{p\}; g, m; \Lambda)\right) = Z_\Delta^{-1} Z_\phi^{-\frac{n}{2}} \tilde{\Gamma}_\Delta^{(n)}(\{p\}; g, m; \Lambda).$$

Note that the partial differentiation with respect to the *bare* mass is to be performed, holding the cutoff and *bare* coupling constant fixed. Making use of the chain rule of differentiation, we obtain for the lhs, upon multiplication by Z_Δ,

$$Z_\Delta \left[\left(m_0 \frac{\partial m}{\partial m_0}\right)_{g_0,\Lambda} \frac{\partial}{\partial m} + \left(m_0 \frac{\partial g}{\partial m_0}\right)_{g_0,\Lambda} \frac{\partial}{\partial g} - \frac{n}{2} \left(m_0 \frac{\partial \ln Z_\phi}{\partial m_0}\right)_{g_0,\Lambda}\right] \tilde{\Gamma}^{(n)}.$$

$$(17.33)$$

In perturbation theory, Z_Δ is given by the expansion

$$Z_\Delta = 1 + \sum_\ell \mathcal{Z}_\Delta^{(\ell)} \left(\frac{\Lambda}{m_0}\right) g_0^\ell.$$

The cutoff dependent terms are defined only up to finite additive constants. We can make use of this freedom to fix these finite parts by requiring for Z_Δ the normalization

$$Z_\Delta m_0 \left(\frac{\partial m}{\partial m_0}\right)_{g_0,\Lambda} = m.$$

Note that this normalization is consistent with lowest order perturbation theory, where $Z_\Delta = 1$ and $m = m_0$. Having thus fixed Z_Δ we further define, together with (17.33),

$$\tilde{\beta}\left(g, \frac{\Lambda}{m}\right) = m \frac{\partial}{\partial m} g\left(g_0, \frac{\Lambda}{m}\right) = -\Lambda \frac{\partial}{\partial \Lambda} g\left(g_0, \frac{\Lambda}{m}\right) \tag{17.34}$$

and

$$\tilde{\gamma}\left(g, \frac{\Lambda}{m}\right) = \frac{1}{2} m \frac{\partial}{\partial m} \ln Z_\phi\left(g_0, \frac{\Lambda}{m}\right) = -\frac{1}{2} \Lambda \frac{\partial}{\partial \Lambda} \ln Z_\phi\left(g_0, \frac{\Lambda}{m}\right). \tag{17.35}$$

Note that we have used dimensional analysis to replace the derivative with respect to the mass by a derivative with respect to the cutoff. This shows that *we only need the knowledge of the (divergent) cutoff dependent terms in order to calculate the β-function and anomalous dimension.*

We now suppose that (17.33) tends to a finite limit as we let the cutoff go to infinity. Hence each of the three terms in (17.33) must also have this property. Accordingly, as a function of g and m, the cutoff dependent constants $\tilde{\beta}$ and $\tilde{\gamma}$ must have a finite limit as we let the cutoff tend to infinity:

$$\beta(g) = \lim_{\Lambda \to \infty} \tilde{\beta}\left(g, \frac{\Lambda}{m}\right), \quad \gamma(g) = \lim_{\Lambda \to \infty} \tilde{\gamma}\left(g, \frac{\Lambda}{m}\right).$$

In the limit $\Lambda \to \infty$ we thus finally obtain from (17.33) the differential equation for the renormalized proper n-point functions

$$\left[m \frac{\partial}{\partial m} + \beta(g) \frac{\partial}{\partial g} - n \gamma_\phi(g)\right] \Gamma^{(n)}(\{p\}) = \Gamma_\Delta^{(n)}(\{p\}). \tag{17.36}$$

This is the *Callan–Symanzik* equation replacing the naive Ward identity (17.29). Several remarks are in order:[3]

1. The generalization to more complicated field theories involving different types of interactions, masses and coupling constants is obvious, with every type of field being associated its own anomalous dimension:

$$\left[\sum_\ell m_\ell \frac{\partial}{\partial m_\ell} + \sum_r \beta_r \frac{\partial}{\partial g_r} - \sum_{k=1}^N n_k \gamma_{\phi_k}\right] \Gamma^{(n_1 \cdots n_N)}(\{p\}) = \Gamma_\Delta^{(n_1 \cdots n_N)}(\{p\}) .$$

2. The *Callan–Symanzik* equation for Green functions involving other types of operators such as partially conserved currents, for instance, will involve different anomalous dimensions, *but the β-functions, which make only reference to the underlying dynamics will remain unchanged.*

3. From a practical point of view these equations will only be useful, provided we are allowed to calculate the beta function and anomalous dimension perturbatively. We shall see in Section 8 that this will be the case for QED in the *low* energy domain (and QCD in the high energy domain). However, before we can demonstrate this, we make an excursion to the deep euclidean behaviour of Feynman graphs.

17.4   Weinberg's Theorem

The Callan–Symanzik equation (17.36) can straightforwardly be continued to the euclidean region, where scalar products will be given in terms of the (euclidean) metric of R^4. We shall argue in the following section that in the *deep euclidean* region, where all the components of the momenta tend to infinity, the right-hand side of the *CS* equation can be neglected, thus turning this inhomogeneous differential equation into a homogeneous one.

Let us consider a scalar field theory without derivative couplings. At first we assume that the Feynman integral is convergent. As we have seen in Section 1 of Chapter 12, such an integral can be put into the form ($\omega(G) < 0$),

$$I_G(p) = i^{1-V-\omega(G)/2} \frac{\Gamma(-\frac{\omega(G)}{2})}{[(4\pi)^2]^L} \times \tag{17.37}$$

$$\int \prod_{\ell=1}^I d\alpha_\ell \delta\left(1 - \sum_1^I \alpha_\ell\right) \frac{[Q_G(p,\alpha) - \sum_1^I \alpha_\ell m_\ell]^{\frac{\omega(G)}{2}}}{\mathcal{P}_G(\alpha)^2} ,$$

where $\omega(G)$ is the *superficial degree of divergence* of the diagram G (see (16.88)),

$$\omega(G) = 4L - 2I = 2\big(I - 2(V-1)\big) ,$$

---

[3]K. Symanzik, *Comm. Math. Phys.* **18** (1970) 227; C.G. Callan, *Phys. Rev. Ser. D* **2** (1970) 1541; R.J. Crewther, *Asymptotic Behaviour in Quantum Field Theory* (Coference Cargèse, 1975) p. 345.

and where $\mathcal{P}_G(\alpha)$ and $Q_G(p,\alpha)$ are the homogeneous polynomials of the Feynman parameters defined in (12.13) and (12.14), respectively.

The theorem to be stated below applies to Feynman amplitudes continued to euclidean momenta. Performing the continuation $p = (p^0, \vec{p}) \to P = (ip^4, \vec{p})$ in (17.37) with

$$Q_G(p; \alpha) \to Q_G^{(E)}(P; \alpha) \, ,$$

and scaling all external momenta as $P \longrightarrow \lambda P$, $Q_G^{(E)}(P; \alpha)$ scales as

$$Q_G^{(E)}(\lambda P; \alpha) = \lambda^2 Q_G^{(E)}(P; \alpha) \, .$$

Correspondingly we have,

$$I_G^{(E)}(\lambda P) = \frac{i^{1-V-\omega(G)/2} \Gamma\left(-\frac{\omega(G)}{2}\right)}{[(4\pi)^2]^L} \lambda^{\omega(G)} \qquad (17.38)$$
$$\times \int \prod_{\ell=1}^{I} d\alpha_\ell \delta\left(1 - \sum \alpha_\ell\right) \frac{[Q_G^{(E)}(P,\alpha) - \frac{1}{\lambda^2} \sum \alpha_\ell m_\ell^2]^{\frac{\omega(G)}{2}}}{[\mathcal{P}_G(\alpha)]^2} \, .$$

Definition

Euclidean momenta tending to infinity in all directions we call *nonexeptional.*

Statement 1

Provided that the zero mass limit of (17.38) exists for nonexceptional momenta, i.e.

$$\int \prod_{\ell=1}^{I} d\alpha_\ell \delta\left(1 - \sum \alpha_\ell\right) \frac{[Q^{(E)}(P; \alpha)]^{\frac{\omega(G)}{2}}}{[\mathcal{P}(\alpha)]^2} < \infty \, ,$$

it follows that

$$I_G^{(E)}(\{\lambda P\}) \sim \lambda^{\omega(G)} \, . \qquad (17.39)$$

Note that this is a non-trivial assumption. Indeed, infrared (IR) divergences can invalidate this result.

In the case where the Feynman diagram is not UV convergent, the above result is modified by logarithmic powers, as the following examples illustrate.

Some Examples:

Consider the following diagram:

Fig. 17.2. Second order 4-point function in ϕ^4-theory.

This diagram has superficial degree of divergence $\omega(G) = 0$, and thus requires regularization. The result after renormalization can be read off from Eq. (16.97):

$$I^{(E)}(-P^2) = \frac{g^2}{2(4\pi)^2} \int_0^1 d\alpha \ln\left(1 + \alpha(1-\alpha)\frac{P^2}{m^2}\right).$$

The asymptotic behaviour is given by

$$I^{(E)}(P) \to \frac{g^2}{2(4\pi)^2} \ln\frac{P^2}{m^2} \tag{17.40}$$

and corresponds to a behaviour (17.39) with logarithmic corrections.

Consider now the above diagram with n *mass insertions* $-im_0^2 \int d^4y : \phi^2(y)$:

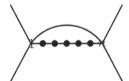

Fig. 17.3. Diagram with mass insertions.

This diagram can be viewed as an $4+n$-point function with n (zero-momentum) external lines, with a factor $-im_0^2$ for each proper of a mass insertion. It is represented by the (Minkowski) Feynman integral

$$I_{n\Delta}(p^2) = (-im_0^2)^n \frac{(-ig_0)^2}{2} \int \frac{d^4k}{(2\pi)^4} \frac{(i)^{n+2}}{(k^2 - m_0^2 + i\epsilon)^{n+1}[(p+k)^2 - m_0^2 + i\epsilon]}.$$

For $n \geq 1$ this integral is UV convergent. Doing a Wick rotation, $k^0 \to ik^4$, and $p^0 \to ip^4$, we have

$$I_{n\Delta}(-P^2) = \frac{i}{2}(m_0^2)^n g_0^2 \int \frac{d^4K}{(2\pi)^4} \frac{(-1)^{n+2}}{(K^2 + m_0^2)^{n+1}[(P+K)^2 + m_0^2]}$$

where $P = (p^4, \vec{p})$, etc. For $n > 1$ the asymptotic behaviour of $I_{n\Delta}^{(E)}$ is easily read off:

$$I_{n\Delta}(-P^2) \simeq \frac{i}{2}\frac{g_0^2}{(4\pi)^2}\frac{m_0^{2n}}{P^2}(-1)^{n+2}\int_0^\infty dx\frac{\pi^2 x}{(x + m_0^2)^{n+1}} = \frac{i}{n(n-1)}\frac{g_0^2}{(2\pi)^4}\frac{m_0^2}{P^2},$$

where we have used (15.1) and (15.2) with $D = 4$, and have set $K^2 = x$.

For $n = 1$ we have, using (15.4) and setting $K = K' - \alpha P$, $K'^2 = x$,

$$I_\Delta(-P^2) \simeq \frac{i}{2}\frac{(m_0^2 g_0^2)}{(4\pi)^2} \int_0^1 d\alpha_1 \int_0^1 d\alpha_2 \int_0^1 d\alpha \int_0^\infty x dx \frac{\delta(\alpha_1 + \alpha_2 + \alpha - 1)}{[x + \alpha(1-\alpha)P^2 + m_0^2]^3}$$

$$\simeq \frac{i}{2}\frac{(m_0^2 g_0^2)}{(4\pi)^2} \int_0^1 d\alpha_1 \int_0^1 d\alpha_2 \int_0^1 d\alpha \frac{1}{2}\frac{\delta(\alpha_1 + \alpha_2 + \alpha - 1)}{\alpha(1-\alpha)P^2 + m_0^2}$$

$$\simeq \frac{i}{2}\frac{(m_0^2 g_0^2)}{(4\pi)^2} \int_0^1 d\alpha \frac{\alpha}{\alpha(1-\alpha)P^2 + m_0^2}.$$

The asymptotic behaviour of this integral is determined by the integration region around $\beta = 1$. We have for $n = 1$, that is, for *one* mass insertion,

$$I_\Delta(-P^2) = \frac{ig_0^2}{2(4\pi)^2} \frac{m_0^2}{P^2} \int_0^1 d\alpha \frac{\alpha}{\alpha(1-\alpha) + \frac{m_0^2}{P^2}}$$

$$\approx \frac{ig_0^2}{2(4\pi)^2} \frac{m_0^2}{P^2} \ln \frac{P^2}{m_0^2} + O\left(\frac{1}{(P^2)^2}\right) .$$

In particular, we see that $(m_0 \to m)$

$$\frac{I_\Delta(-P^2)}{I(-P^2)} \simeq \frac{im^2}{P^2} . \tag{17.41}$$

The reason for the power behaviour for $n > 1$ is that in Fig. 17.3 the inclusion of mass insertions amounts to the introduction of *superrenormalizable* quadratic interactions in the lagrangian, whereas in the case of Fig. 17.2, one is dealing with a *renormalizable* interaction, characterized by one dimensionless coupling constant. The picture behind this is that for $n = 0$ (Fig. 17.2) momentum flows equally through both branches connecting the vertices with the non-exceptional external lines, whereas for $n \geq 1$ the momentum prefers to flow along the "upper branch" in Fig. 17.3.

This observation is at the heart of

Weinberg's Theorem:

Renormalized Feynman diagrams, in which only strictly renormalizable couplings are involved, scale in the case of *all momenta being non-exceptional* is as follows,

$$I_G^{(E)}(\lambda P) \approx \lambda^{\omega(G)} (\ln \lambda)^{n(G)} ,$$

where $n(G)$ is some integer.[4]

17.5   Solution of CS equation in the deep euclidean region

The CS equation may evidently be continued analytically in all the momenta to the euclidean domain, by simply performing the substitutions $p^0 \to ip^4$, the scalar product now being positive, semi-definite. Having done this, we now consider the solution of the CS equation in the *deep euclidean region* for *non-exceptional* momenta that is, in the limit where all components of the momenta grow large. Since the mass operator ϕ^2 carries the dimension $[\phi^2] = 2$, we have for a mass insertion in the proper functions, in the deep euclidean region (see (17.41)),[5]

$$\frac{\Gamma_\Delta(\{\lambda p\}; ...)}{\Gamma(\{\lambda p\}; ...)} \sim \lambda^{-2} \mathcal{P}\left(\ln \frac{\lambda}{m}\right) \tag{17.42}$$

---

[4]S. Weinberg, *Phys. Rev.* **118** (1960) 838.
[5]From here on, Γ denotes the *renormalized* proper Green function as defined in (20.12).

in every order of perturbation theory. Correspondingly the euclidean asymptotic n-point proper function satisfies the *homogeneous* CS equation

$$\left[m\frac{\partial}{\partial m} + \beta(g)\frac{\partial}{\partial g} - n\gamma_\phi(g) \right] \Gamma^{(as)}(\{p\};g,m) = 0 \ . \tag{17.43}$$

Although Weinberg's theorem only holds in every given order of perturbation theory, we now assume that

(i) the CS equation also holds for the complete proper function, and

(ii) Weinberg's theorem continues to hold for non-exceptional momenta in the deep Minkowskian region.

We now seek the solution of the asymptotic CS equation (17.43). To this end it is convenient to rewrite the CS equation in a slightly different form.

The asymptotic CS equation states that a small change in the mass-parameter m can be compensated by a corresponding change in the coupling constant g and a suitable rescaling of the fields associated with the wave function renormalization constant.

Now, a change in the mass parameter corresponds to a change of the scale in which we measure momenta and other dimensional quantities. Hence let us consider what happens under a rescaling of the mass. Let D be the engineering dimension of the proper function in question. Ordinary dimensional analysis then allows us to write (g assumed dimensionless)

$$\Gamma(\{\lambda p\};g,m) = \lambda^D \Gamma\left(\{p\};g,\frac{m}{\lambda}\right) . \tag{17.44}$$

Hence, we can view a scaling $m \to m/\lambda$ of the mass as a scaling $p \to \lambda p$ in the momentum. Noting that

$$m\frac{\partial}{\partial m}f\left(\frac{m}{\lambda}\right) = \frac{m}{\lambda}f'\left(\frac{m}{\lambda}\right); \quad \lambda\frac{\partial}{\partial\lambda}f\left(\frac{m}{\lambda}\right) = -\frac{m}{\lambda}f'\left(\frac{m}{\lambda}\right) , \tag{17.45}$$

we thus have

$$m\frac{\partial}{\partial m}\Gamma(\{\lambda p\}...) = \left(-\lambda\frac{\partial}{\partial\lambda} + D\right)\Gamma(\{\lambda p\}, ...) . \tag{17.46}$$

It is convenient to define a new variable t by

$$t = \ln\lambda . \tag{17.47}$$

We may then rewrite the asymptotic CS equation in the form

$$\left(-\frac{\partial}{\partial t} + \beta(g)\frac{\partial}{\partial g} + D - \gamma(g)\right)\Gamma^{(as)}(\{e^t p\};g,m) = 0 , \tag{17.48}$$

with $\gamma(g) = n\gamma_\phi(g)$ and $t \to \infty$.

Generally, we must keep in mind, that ordinary dimensional analysis tells us that

$$\Gamma(\{e^t p_i\},g,m) = e^{Dt}\Gamma(\{p_i\},g,e^{-t}m) .$$

The behaviour of the proper function in the deep euclidean limit is thus controlled by the limit $m \to 0$. The first and third terms in the CS equation take care of dimensional analysis. The other two terms take care of anomalous behaviour in m introduced by the mass dependence of the renormalization constants. We distinguish three cases:

(i) $\beta(g) = 0,$ $\gamma(g) = 0$

This is the case, if the theory is finite to begin with, so that in the limit $\Lambda \to \infty$, the Z_i do not depend on m_0 (or m):,

$$Z_i\left(g_0, \frac{\Lambda}{m}\right) \to Z_i(g_0) \ .$$

In that case the asymptotic CS equation reads

$$\left(-\frac{\partial}{\partial t} + D\right)\Gamma^{(as)}(\{e^t p\}; g, m) = 0\,,$$

implying for $t \to \infty$ the asymptotic behaviour

$$\Gamma(\{e^t p\}; g, m) \to e^{Dt}\Phi(\{p\}; g, e^{-t}m).$$

The CS equation thus tells us that the proper function scales in the deep euclidean region with the *canonical* dimension, including the possibility of a broken scale invariance such as (17.4).

(ii) $\beta(g) = 0,$ $\gamma(g) \neq 0$

There is no coupling constant renormalization; the mass dependence only enters via the cutoff dependent logarithms $\ln(\Lambda^2/m^2)$ appearing in the wave function renormalization constant in perturbation theory (and the possible failure of the $m \to 0$ limit). We have in this case, for $t \to \infty$,

$$\Gamma(\{e^t p\}; g, m) \to e^{t(D-\gamma(g))}\Phi(\{p\}; g, e^{-t}m)\,,$$

and one says that the proper function scales with an anomalous dimension $D - \gamma(g)$.

(iii) $\beta(g) \neq 0,$ $\gamma(g) \neq 0$

In this case the solution of Eq. (17.48) is more involved. Consider the auxiliary equation

$$\left(-\frac{\partial}{\partial t} + \beta(g)\frac{\partial}{\partial g}\right)\bar{g}(g, t) = 0 \tag{17.49}$$

satisfying the "initial" condition

$$\bar{g}(g, 0) = g\,. \tag{17.50}$$

Making use of (17.49), we now observe that

$$\left(-\frac{\partial}{\partial t} + \beta(g)\frac{\partial}{\partial g}\right)e^{-\int_g^{\bar{g}(g,t)} dg' \frac{\gamma(g')}{\beta(g')}} = \gamma(g)e^{-\int_g^{\bar{g}(g,t)} dg' \frac{\gamma(g')}{\beta(g')}}\,.$$

Hence we conclude that in this case,

$$\Gamma(\{e^t p\}, g, m) \to e^{tD} \Phi(\{p_i\}; \bar{g}(g,t), e^{-t} m) e^{-\int_g^{\bar{g}(g,t)} dg' \frac{\gamma(g')}{\beta(g')}} \tag{17.51}$$

where the additional dependence on m in the deep euclidean region (see (17.58)) now arises from the dependence of t on the running coupling constant. The function $\bar{g}(g,t)$ is referred to as the *running coupling constant*, which together with $\gamma(\bar{g}(g,t))$ now controls the asymptotic behaviour.

Alternative form of the solution

It is convenient to rewrite Eq. (17.49) for $\bar{g}(g,t)$ in a different form. To this end, consider the function f defined by

$$\beta(g) \frac{\partial}{\partial g} \bar{g}(g,t) =: f(g,t) . \tag{17.52}$$

From (17.50) we see that

$$f(g,0) = \beta(g) . \tag{17.53}$$

One readily checks that $f(g,t)$ is itself again a solution of (17.49), though with a different boundary condition. Indeed, using Eq. (17.52),

$$-\frac{\partial}{\partial t} f(g,t) = -\beta(g) \frac{\partial}{\partial g} \left(\frac{\partial}{\partial t} \bar{g}(g,t) \right) = -\beta(g) \frac{\partial}{\partial g} f(g,t)$$

so that we have

$$\left(-\frac{\partial}{\partial t} + \beta(g) \frac{\partial}{\partial g} \right) f(g,t) = 0 .$$

From here and (17.53) we infer

$$f(g,t) = \beta(\bar{g}(g,t)) .$$

Hence we may write (17.52) also in the form

$$\frac{\partial}{\partial t} \bar{g}(g,t) = \beta(\bar{g}(g,t)) . \tag{17.54}$$

We now use (17.54) in order to put the solution (17.51) into an alternative form. To this end we make the change of variable $g' \to t'$ in the exponential of (17.51), leaving us with

$$\int_g^{\bar{g}(g,t)} dg' \frac{\gamma(g')}{\beta(g')} = \int_0^t dt' \left(\frac{\partial \bar{g}(g,t')}{\partial t'} \right) \frac{\gamma(\bar{g}(g,t'))}{\beta(\bar{g}(g,t'))} = \int_0^t dt' \gamma(\bar{g}(g,t')) . \tag{17.55}$$

We may thus write (17.51) in the alternative form

$$\Gamma^{(as)}(\{e^t p\}; g, m) = e^{tD} \Phi(\{p\}; \bar{g}(g,t), e^{-t} m) e^{-\int_0^t dt' \gamma(\bar{g}(g,t'))} . \tag{17.56}$$

17.6   Asymptotic behaviour of Γ and zeros of the β-function

From perturbation theory we know that $\beta(0) = 0$. We also know the behaviour of the beta function $\beta(g)$ in the neighbourhood of $g = 0$. The behaviour of the beta-function "far away" from $g = 0$ is in general not known. For the sake of discussion let us nevertheless assume that the beta-function vanishes at a discrete set of values g_i^* of g. Depending on the sign of β near $g = 0$ we distinguish two situations, as depicted in the figures below.

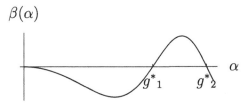

Fig. 17.4(a). β-function for case (a).

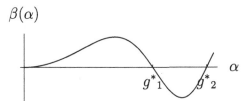

Fig. 17.4(b). β-function for case (b).

It then follows from (17.49) and the figures above that

Case (a);

$$\bar{g}(g,t) \xrightarrow[t \to \infty]{} \begin{cases} 0 \,, & g \in (0, g_1^*) \\ g_2^* \,, & g \in (g_1^*, g_2^*) \end{cases}$$

$$\bar{g}(g,t) \xrightarrow[t \to -\infty]{} g_1^* \,, \quad g \in (0, g_2^*)$$

Case (b);

$$\bar{g}(g,t) \xrightarrow[t \to \infty]{} g_1^* \,, \quad g \in (0, g_2^*)$$

$$\bar{g}(g,t) \xrightarrow[t \to -\infty]{} \begin{cases} 0 \,, & g \in (0, g_1^*) \\ g_2^* \,, & g \in (g_1^*, g_2^*) \end{cases}$$

This means:

In *case (a)* are the "even" ("odd") zeros of β "UV(IR) attractors" (fix points).

In *case (b)* are the "odd" ("even") zeros of β "UV (IR) attractors" (fix points).

Suppose that g^* is such a relevant fix point. In that case we have from (17.56)

$$\Gamma(\{e^t p_i\}, g, m) \longrightarrow e^{t\left(D - \gamma(g^*)\right)} \Phi\left(\{p_i\}, \bar{g}(g, t), e^{-t} m)\right) e^{-\epsilon(t)} ,$$

where

$$\epsilon(t) = \int_0^t dt' \left[\gamma(\bar{g}(g, t')) - \gamma(g^*)\right] . \tag{17.57}$$

Introduce

$$t = \frac{1}{2} \ln \frac{S}{m^2}, \quad with \quad S = \sum_1^n p_r^2 .$$

We distinguish different possibilities:

(1) $\epsilon(t)$ *is finite in the limit $t \to \infty$.*

In this case we have

$$\Gamma(\{e^t p_i\}, g, m) \longrightarrow (\sqrt{S})^D \left(\frac{S}{m^2}\right)^{-\frac{1}{2}\gamma(g^*)} \Phi\left(\{p_i\}; g^*, \frac{m}{\sqrt{S}}\right) e^{-\epsilon(\infty)} . \tag{17.58}$$

Let us examine under what condition this is realized. We consider two examples:

(a) Assume that the beta-function behaves like $\beta(g) \simeq b(g_1^* - g)$ for $0 < g < g_1^*$ with $b > 0$ (case (b)). In that case we have from (17.54)

$$\frac{\partial}{\partial t} \bar{g} = \beta(\bar{g}) \simeq -b(\bar{g} - g_1^*)$$

near the fix-point. We may solve this differential equation by separation of variables:

$$t = -\int_g^{\bar{g}} \frac{dg'}{b(g' - g_1^*)} , \quad g \text{ close to } g_1^*$$

or

$$\bar{g} - g_1^* \simeq (g - g_1^*)e^{-bt} = ce^{-bt} .$$

Hence $\bar{g}$ is driven toward g_1^* as $t \to \infty$:

$$\gamma(\bar{g}(g, t)) \approx \gamma(g_1^* + ce^{-bt}) \simeq \gamma(g_1^*) + \gamma'(g_1^*)ce^{-bt}$$

where 'prime' denotes 'derivative'. Substitution into (17.57) thus yields $\epsilon(\infty) < \infty$. Hence $\Gamma^{(n)}$ scales with anomalous dimension $\gamma(g_1^*)$.

(b) An example of a relevant IR fixed point at $g = 0$ is provided by a ϕ^4 theory (see (17.69)):

$$\beta(g) \simeq \frac{3g^2}{(4\pi)^2} = bg^2 . \tag{17.59}$$

Hence

$$\frac{\partial \bar{g}}{\partial t} = b\bar{g}^2$$

has the solution

$$\bar{g}(g, t) = \frac{g}{1 - bgt} \; .$$

With $\gamma = cg^2$ one finds in the IR limit $t \to -\infty$,

$$\epsilon(t = -\infty) = -\frac{cg}{b} \; .$$

A particularly interesting case is, where

(2) $\epsilon(t)$ *does not have a finite limit for* $|t| \to \infty$.

An example of immediate interest is provided by QCD,[6] where

$$\beta(g) \simeq -bg^3 \; , \; b > 0 \, .$$

Hence $g* = 0$ is an UV fixed point. We may then solve (17.54) near the fixed point, i.e.

$$\frac{\partial}{\partial t} \bar{g} = -b\bar{g}^3$$

to give

$$t = \frac{1}{2b} \left(\frac{1}{\bar{g}^2} - \frac{1}{g^2} \right) \qquad \to \qquad \bar{g}^2(g, t) \simeq \frac{g^2}{1 + 2bt} \; . \tag{17.60}$$

In order to compute $\epsilon(t)$ for $t \to \infty$ we again need only the (known) leading term in the perturbative expansion of the anomalous dimension:

$$\gamma(\bar{g}(g, t)) \sim c\bar{g}(g, t)^2 \; .$$

Noting that $\gamma(g^*) = 0$, we have from (17.57),

$$\epsilon(t) \sim \int_0^t dt' \gamma(\bar{g}(g, t'))$$

$$\sim cg^2 \int_0^t dt' \frac{1}{1 + 2bg^2 t'} = \frac{c}{2b} \ln (1 + 2bg^2 t)$$

or

$$e^{\epsilon(t)} \sim \left[\ln \frac{S}{m^2} \right]^{\frac{c}{4b}} \; .$$

We thus obtain in this case logarithmic corrections to the naive asymptotic behaviour:

$$\Gamma(\{e^t p\}; g, m) \to (\sqrt{S})^D \left[\ln \left(\frac{S}{\Lambda_s^2} \right) \right]^{-\frac{c}{4b}} \Phi \left(\{p_i\}, g*, \frac{m}{\sqrt{S}} \right) \tag{17.61}$$

with Λ_s being some strong interaction scale.

---

[6]D.J. Gross and F. Wilczek, *Phys. Rev. Lett.* **30** (1973) 1343; H.D. Politzer, *Phys. Rev. Lett.* **30** (1973) 1346.

17.7 Perturbative calculation of $\beta(g)$ and $\gamma(g)$ in ϕ^4 theory

Before engaging in an explicit perturbative calculation of the anomalous dimension and beta-function in ϕ^4-theory, it is convenient to rewrite the normalization conditions in a slightly different form. The wave function renormalization constant and coupling constant renormalization constant are calculated from the knowledge of $\Gamma^{(2)}_{reg}(p)$ (see (20.11)) and $\Gamma^{(4)}_{reg}(p)$, respectively. To second order we have

$$-i\Sigma(p^2)_{reg} \simeq \!\!\!\bigcirc\!\!\!\qquad\qquad i\Gamma^{(4)}_{reg}(p^2) \simeq -ig_0 + \;\;\Join \tag{17.62}$$

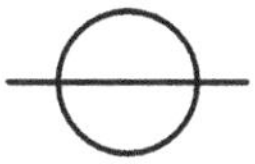

Fig. 17.5. Proper diagrams contributing to $\beta(g)$ and $\gamma(g)$.

In terms of $\Sigma(p^2)_{reg}$ the normalization conditions state (compare with (16.29) and (16.31))

$$\left(m^2 - m_0^2 - \Sigma_{reg}(m^2)\right) = 0$$

$$Z_\phi \left(1 - \frac{\partial \Sigma_{reg}(p^2)}{\partial p^2}\Big|_{p^2 = m^2}\right) = 1 \; . \tag{17.63}$$

Furthermore (see (20.12) for convention)

$$(Z_\phi^{\frac{1}{2}})^4 \Gamma^{(4)}_{reg}(0) = -g \; . \tag{17.64}$$

ϕ^4 beta-function

The β-function is calculated from (17.64). There exist *three* topologically distinct 1-loop graphs contributing to the 4-point function in order g^2:

$$\begin{matrix} 1 & & 3 \\ & \Join & \\ 2 & & 4 \end{matrix}\qquad\qquad \begin{matrix} 1 & & 2 \\ & \Join & \\ 3 & & 4 \end{matrix}\qquad\qquad \begin{matrix} 1 & & 2 \\ & \Join & \\ 4 & & 3 \end{matrix}$$

Fig. 17.6. Topologically distinct diagrams.

(Note that the relabelling of the integration variables $z_1 \leftrightarrow z_2$ is taken care of by the factor $1/2!$ arising from the expansion of the exponential in the formula (11.21). Consider the sum of (amputated) 1PI diagrams in momentum space:

$$I(p^2) = 3\frac{(4!)^2}{2!}\left(-\frac{ig_0}{4!}\right)^2 \int \frac{d^4k}{(2\pi)^4} \frac{i}{[k^2 - m_0^2 - i\epsilon]} \frac{i}{[(p-k)^2 - m_0^2 + i\epsilon]} \tag{17.65}$$

where the factor $\frac{(4!)^2}{2!}$ is a combinatorial factor arising from the $(4!^2/2)$ possible contractions with $\mathcal{H}_I = \frac{g_0}{4!}\phi^4$.

Regularizing a la *Pauli–Villars* we have for the sum of the three graphs

$$I_{reg}(p^2) = \frac{3g_0^2}{2} \int \frac{d^4k}{(2\pi)^4} \sum_\ell C_\ell \frac{1}{[k^2 - M_\ell^2 + i\epsilon]} \frac{1}{[(p-k)^2 - m_0^2 + i\epsilon]} , \qquad (17.66)$$

where

$$C_0 = 1, \ M_0 = m_0$$
$$C_1 = -1, \ M_1 = \Lambda.$$

Making use of the Feynman parameter representation (15.4), we may write this in the form

$$I_{reg}(p^2) = \frac{3g_0^2}{2} \int \frac{d^4k}{(2\pi)^4} \sum_\ell C_\ell$$

$$\times \int_0^1 d\alpha \frac{1}{\left\{ (k - \alpha p)^2 + \alpha(\alpha - 1)p^2 - (1 - \alpha)M_\ell^2 - \alpha m_0^2 + i\epsilon \right\}^2} .$$

Making the change of variable $k' = k - \alpha p$, and performing the *Wick rotation* $k^0 = ik^4, p^0 = ip^4$, we obtain (see (15.13))

$$I_{reg}(-P^2) = \frac{3ig_0^2/2}{(4\pi)^2} \int_0^1 d\alpha \sum_\ell C_\ell \int_0^\infty dx \frac{x}{\left\{ x + \alpha(1 - \alpha)P^2 + \alpha m_0^2 + (1 - \alpha)M_\ell^2 \right\}^2} ,$$

where we have set $K^2 = x$. Making use of the above expressions for C_0 and C_1, we obtain from here

$$I_{reg}(-P^2) = \frac{3ig_0^2/2}{(4\pi)^2} \int_0^1 d\alpha \int_0^{X \to \infty} dx \, x \cdot$$

$$\left\{ \frac{1}{[x + \alpha(1 - \alpha)P^2 + m_0^2]^2} - \frac{1}{[x + \alpha(1 - \alpha)P^2 + \alpha m_0^2 + (1 - \alpha)\Lambda^2]^2} \right\} ,$$

where we have introduced a cutoff X for the x-integration, since each term, taken separately, diverges logarithmically. Noting that

$$\int_0^X dx \frac{x}{(x + B)^2} = \ln \left(\frac{x + B}{B} \right) + \frac{B}{x + B} \Big|_0^X$$

$$\longrightarrow \ln \frac{X}{B} - 1 + \frac{B}{X} ,$$

we obtain from above

$$I_{reg}(-P^2) = \frac{-3ig_0^2/2}{(4\pi)^2} \int_0^1 d\alpha \ln \left(\frac{\alpha(1 - \alpha)\frac{P^2}{m_0^2} + 1}{\alpha(1 - \alpha)\frac{P^2}{m_0^2} + \alpha + (1 - \alpha)\frac{\Lambda^2}{m_0^2}} \right) ,$$

where we have taken the limit $X \to \infty$. Retaining only the terms which do not vanish as we let the cutoff $\Lambda \to \infty$ go to infinity, we have for the complete second order *Minkowski* 1-particle irreducible 4-point function, now including also the point-like interaction (see (20.12) for definition of $\Gamma^{(n)}$),

$$\Gamma_{reg}^{(4)}(p^2) = -g_0 - \frac{3g_0^2/2}{(4\pi)^2} \left[\int_0^1 d\alpha \ln \left(\frac{1 - \alpha(1-\alpha)\frac{p^2}{m_0^2}}{1-\alpha} \right) - \ln \frac{\Lambda^2}{m_0^2} \right] . \quad (17.67)$$

Setting $p^\mu = 0$, we have with (17.64) for the renormalized coupling,

$$g = g_0 + \frac{3}{2} \frac{g_0^2}{(4\pi)^2} \left[\int_0^1 d\alpha \ln(1-\alpha) - \ln \frac{\Lambda^2}{m_0^2} \right] . \quad (17.68)$$

From here we obtain for the β-function in 1-loop order,

$$\beta^{(2)}(g) = \frac{3g^2}{(4\pi)^2} . \quad (17.69)$$

In terms of g the renormalized 4-point function takes the form

$$\Gamma^{(4)}(p^2) \simeq -g - \frac{3g^2/2}{(4\pi)^2} \int_0^1 d\alpha \ln \left(1 - \alpha(1-\alpha)\frac{p^2}{m^2} \right) . \quad (17.70)$$

Note that to order g^2 there is no contribution from the wave function renormalization constant.

ϕ^4 Anomalous dimension

As we have seen in Section 16.2, the calculation of Z_ϕ, and consequently of γ_ϕ to order g^2 requires the calculation of the proper self-energy diagram of Fig. 17.7. This 2-loop diagram with superficial degree of divergence $\omega = 2$ involves an overlapping divergence. Following closely the reasoning taking one from the expansion (16.27) to (16.31) in the case of QED, we see that to second order g^2,

$$Z_\phi \approx 1 + \left(\frac{\partial \Sigma(q^2)_{reg}}{\partial q^2} \right)_{q^2=0} . \quad (17.71)$$

From here and (17.35) we then calculate the anomalous dimension $\gamma_\phi(g)$.

Since we are dealing with a diagram with overlapping divergences, we shall use the BPH prescription (16.101) already applied in the context of the diagram of Fig. 16.9. We label the self-energy diagram as shown in Fig. 17.7. The diagram involves two overlapping subgraphs γ_1 (upper half) and γ_2 (lower half) sharing one common propagator. We have for the full Feynman diagram (we suppress the $i\epsilon$ prescriptions),

$$-i\Sigma(q^2) = (4!)^2 \left(\frac{g_0}{4!} \right)^2 \int \frac{d^4 p'}{(2\pi)^4} \int \frac{d^4 p}{(2\pi)^4} \frac{i}{p^2 - m_0^2} \frac{i}{(p+q-p')^2 - m_0^2} \frac{i}{p'^2 - m_0^2} \quad (17.72)$$

where $(4!)^2$ is a combinatorial factor.

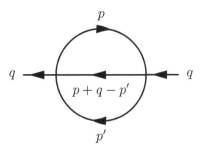

Fig. 17.7. ϕ^4 self-energy diagram with overlapping divergence.

We shall follow the notation of (16.100) and define $\mathcal{J} \equiv \Sigma(q^2)$. According to the BPH prescription we expand the integrand of the upper half (lower half) subgraphs γ_1 (γ_2) in a Taylor series around $p' = q = 0$ ($p = q = 0$), keeping only the first leading term:

$$T_{\gamma_1} \mathcal{J}_{\gamma_1} = g_0^2 \int \frac{d^4p}{(2\pi)^4} \frac{1}{(p^2 - m_0^2)^2}$$

and

$$T_{\gamma_2} \mathcal{J}_{\gamma_2} = g_0^2 \int \frac{d^4p'}{(2\pi)^4} \frac{1}{(p'^2 - m_0^2)^2}.$$

Following the recipe (16.101) we then have

$$\Sigma(q^2) = g_0^2 \left\{ \int \frac{d^4p'}{(2\pi)^4} \int \frac{d^4p}{(2\pi)^4} \frac{1}{p^2 - m_0^2} \frac{1}{(p+q-p')^2 - m_0^2} \frac{1}{p'^2 - m_0^2} \right.$$
$$- \int \frac{d^4p'}{(2\pi)^4} \frac{1}{(p'^2 - m_0^2)} \int \frac{d^4p}{(2\pi)^4} \frac{1}{(p^2 - m_0^2)^2}$$
$$\left. - \int \frac{d^4p}{(2\pi)^4} \frac{1}{(p^2 - m_0^2)} \int \frac{d^4p'}{(2\pi)^4} \frac{1}{(p'^2 - m_0^2)^2} \right\}.$$

The same reasoning as in the case of Fig. 16.9 shows that the divergences of the subgraphs have been taken care of. These cancellations are realized by the second integral in the second and third lines at fixed p' and p respectively. The remaining integration implies an overall quadratic divergence, resulting in a second order cutoff dependent polynomial $\delta\mathcal{L}_\phi$ of the form $\delta Z_\phi(q^2 - m_0^2)$ (compare with (16.17)) from where the wave function renormalization constant can be read off. More concretely: Doing a Wick rotation, continuing q^0 to iQ^4, and making use of the Feynman parameter formula (12.25), we obtain, using dimensional regularization,

$$\mathcal{J} - T_{\gamma_1}\mathcal{J} = i^2 \frac{g_0^2}{(4\pi)^2} \int \frac{d^4P'}{(2\pi)^4} \frac{-1}{P'^2 + m_0^2} \int_0^1 d\alpha \frac{\ln\left[\alpha(1-\alpha)(P'-Q)^2 + m_0^2\right]}{m_0^2}$$

where, as expected, the singularity in $\epsilon = D - 4$ was cancelled by the counterterm $T_{\gamma_1}\mathcal{J}$, and the limit $\epsilon \to 0$ has been taken. Comparing with (17.70) we see that this is nothing but the corresponding figure in Fig. 17.5 with a pair of external legs closed to a loop. Taking the derivative with respect to Q^μ, the remaining counterterm $T_{\gamma_2}\mathcal{J}$ gives a vanishing contribution and we obtain,

$$\frac{\partial\Sigma(-Q^2)}{\partial Q^\mu} = -\Sigma'(-Q^2)2Q^\mu = \frac{g_0^2}{(4\pi)^2}\int\frac{d^4P'}{(2\pi)^4}\frac{1}{(P'^2+m_0^2)}\frac{-2\alpha(1-\alpha)(P'-Q)^\mu}{[\alpha(1-\alpha)(P'-Q)^2+m_0^2]}.$$

Using Lorentz covariance,

$$\Sigma'(-Q^2) = -\frac{g_0^2}{(4\pi)^4}\int_0^{\Lambda^2}P'^2dP'^2\frac{1}{(P'^2+m_0^2)}\frac{\alpha(1-\alpha)}{[\alpha(1-\alpha)(P'-Q)^2+m_0^2]},$$

or setting Q=0,

$$\Sigma'(0) = -\frac{g_0^2}{(4\pi)^4}\int_0^{\Lambda^2}P'^2dP'^2\frac{1}{(P'^2+m_0^2)}\frac{\alpha(1-\alpha)}{[\alpha(1-\alpha)P'^2+m_0^2]}.$$

From (17.71) we thus obtain for $\Lambda\to\infty$,

$$Z_\phi \simeq 1 - \frac{g_0^2}{(4\pi)^4}\ln\frac{\Lambda^2}{m_0^2} + const.$$

and

$$\gamma_\phi(g_0) \simeq \frac{g_0^2}{(4\pi)^4}.$$

Our result for the renormalization constant and anomalous dimension differs from the one quoted by Itzykson and Zuber.[7]

Checking the results

Let us check our results. From (17.70) we obtain for the leading asymptotic behaviour for euclidean momentum,

$$\Gamma^{(4)}(-P^2) \simeq -g - \frac{3g^2}{2(4\pi)^2}\ln\left(\frac{P^2}{m^2}\right).$$

Differentiating this expression with respect to the mass and (renormalized) coupling constant we have with (17.69),

$$m\frac{\partial}{\partial m}\Gamma^{(4)}(-P^2) \simeq \frac{3g^2}{(4\pi)^2}\ ,\ \ \beta(g)\frac{\partial}{\partial g}\Gamma^{(4)}(-P^2) \simeq -\frac{3g^2}{(4\pi)^2}\ ,\ \ \gamma_\phi(g)\Gamma^{(4)} = 0(g^3)$$

or

$$\left(m\frac{\partial}{\partial m} + \beta(g)\frac{\partial}{\partial g} - \gamma_\phi(g)\right)\Gamma_{as}^{(4)}(-P^2) = O(g^3)\ .$$

Since the β-function (17.69) is positive, $g = 0$ is an IR fixed point, and we cannot draw any conclusion concerning the high-energy behaviour of the ϕ^4-theory. Notice that this holds for all Green functions. The same is true for QED as we show next.

---

[7]C. Itzykson and J.-B. Zuber, *Quantum Field Theory* (McGraw-Hill, 1980), p. 652.

17.8   QED β-function and anomalous dimension

The success of QED at low energies (Lambshift, etc.) can be understood in terms of the behaviour of the β-function near zero coupling and the smallness of the electromagnetic coupling itself ($\alpha = 1/137$). Let us examine therefore the behaviour of the QED β-function near zero coupling.

The β-function and anomalous dimension of QED requires the computation of

$$\beta(\alpha) = \alpha_0 \left(m \frac{\partial}{\partial m} \ln Z_e \right)_{m_0, \alpha_0, \Lambda} = -\alpha_0 \left(\Lambda \frac{\partial}{\partial \Lambda} \ln Z_e \right)_{m_0, \alpha_0, \Lambda}$$

and

$$\gamma_A(\alpha) = \frac{1}{2} \left(m \frac{\partial}{\partial m} \ln Z_A \right)_{m_0, \alpha_0, \Lambda} = - \left(\frac{\Lambda}{2} \frac{\partial}{\partial \Lambda} \ln Z_A \right)_{m_0, \alpha_0, \Lambda} .$$

Because of the QED-Ward identity (16.61), $Z_1 = Z_\psi$. Hence $Z_e = Z_A$, or we have to arbitrary order

$$\beta(\alpha) = 2\gamma_A(\alpha). \tag{17.73}$$

From the result (15.30) for the *dimensional regularization*

$$\pi_{reg}^{dim}(k^2) \simeq -\frac{\alpha_0}{3\pi} \left(\frac{2}{\epsilon} - \gamma - \ln \pi + \ln \frac{\mu^2}{m^2} \right) + 2\frac{\alpha_0}{\pi} \int d\beta \, \beta(1-\beta) \ln \left(\frac{m^2 - \beta(1-\beta)k^2}{m^2} \right) \tag{17.74}$$

we have from (16.38) to order $O(\alpha^2)$,

$$Z_A^{dim} \left(\alpha, \frac{\mu}{m}, \epsilon \right) \simeq 1 - \frac{\alpha}{3\pi} \left(\frac{2}{\epsilon} - \gamma - \ln \pi + \ln \frac{\mu^2}{m^2} \right) \tag{17.75}$$

and from the result (15.28) for the Pauli–Villars regularization

$$\pi_{reg}^{PV}(k^2) \simeq -\frac{\alpha_0}{3\pi} \ln \left(\frac{\Lambda^2}{2m^2} \right) + \frac{2\alpha_0}{\pi} \int_0^1 d\beta \beta(1-\beta) \ln \left(\frac{m^2 - \beta(1-\beta)k^2}{m^2} \right) , \tag{17.76}$$

we have correspondingly,

$$Z_A^{PV} \left(\alpha, \frac{\Lambda}{m} \right) \simeq 1 - \frac{\alpha}{3\pi} \ln \left(\frac{\Lambda^2}{2m^2} \right) . \tag{17.77}$$

Hence we obtain for *both* regularizations

$$\beta(\alpha) \simeq \frac{2\alpha^2}{3\pi} , \quad \gamma_A(\alpha) \simeq \frac{\alpha}{3\pi} . \tag{17.78}$$

Notice that in lowest order of perturbation theory, $\beta(\alpha)$ is *positive*. Hence unlike QCD, $\alpha = 0$ is *not* a "relevant UV fix-point" for QED, but rather an infrared (IR) fix-point. This explains the success of QED at low energies.

17.9   QED β-function and leading log summation

The asymptotic behaviour (17.56) of a general proper function shows that it deviates, in general, from that expected from dimensional analysis. Let us examine in some detail what this precisely means from the point of view of perturbation theory. We shall illustrate the ideas for QED.

Consider the renormalized photon 2-point function in the Lorentz gauge:

$$D_F^{\mu\nu}(k;\alpha,m) = \frac{-\mathcal{P}^{\mu\nu}(k)}{k^2\big(1 - \pi_{ren}(k^2;\alpha,m)\big)} \,,$$

where $\mathcal{P}^{\mu\nu}$ is the usual projector

$$\mathcal{P}^{\mu\nu}(k) = g^{\mu\nu} - \frac{k^\mu k^\nu}{k^2} \,.$$

The corresponding 1PI function, that is, $1 - \pi_{ren}(k^2)$ of (16.75) satisfies asymptotically the Callan–Symanzik equation

$$\left(m\frac{\partial}{\partial m} + \beta(\alpha)\frac{\partial}{\partial\alpha} - 2\gamma_A(\alpha)\right)(1 - \pi_{ren}(k^2)) \approx 0 \,. \tag{17.79}$$

Let us check this in 1-loop order. From (16.75) we have for $\pi(k^2)_{ren}$, and K euclidean,

$$\pi_{ren}^{(1-loop)}(-K^2) = \frac{2\alpha}{\pi}\int_0^1 d\beta\,\beta(1-\beta)\ln\left(\frac{m^2 + \beta(1-\beta)K^2}{m^2}\right)$$

$$\simeq \frac{\alpha}{3\pi}\ln\left(\frac{K^2}{m^2}\right) \,.$$

Recalling that

$$\frac{\beta(\alpha)}{\alpha} = 2\gamma_A(\alpha) = \frac{2\alpha}{3\pi} \,, \tag{17.80}$$

and performing the differentiations, one verifies that (17.79) is satisfied asymptotically to order $O(\alpha^2)$. We now take the converse point of view.

Instead of computing the beta-function from the coupling constant renormalization constant via (17.34), we can compute it from the knowledge of the asymptotic behaviour of $\pi(k^2)$ using the asymptotic CS equation (17.79). To this end it is useful to rewrite this equation in terms of the variable introduced in (17.47), and using the Ward identity (17.80):

$$\left(-\frac{\partial}{\partial t} + \beta(\alpha)\frac{\partial}{\partial\alpha}\right)\frac{\pi(e^{2t}k^2)}{\alpha} = \frac{\beta(\alpha)}{\alpha^2} \,. \tag{17.81}$$

The merit of writing the CS equation in this form is that it no longer involves the anomalous dimension. We can thus use the knowledge of an asymptotic expansion of

$\pi(e^{2t}k^2)_{ren}$ in powers of $\ln t$ (leading log summation) in order to obtain conversely a perturbative expansion for the β-function. Introducing the ansatz

$$\pi(e^{2t}k^2) = \sum_{n=1} \alpha^n (a_n + b_n t + c_n t^2 + d_n t^3 + \cdots)$$

$$\beta(\alpha) = \alpha^2 \sum_{n=1} (\beta_1 + \beta_2 \alpha + \beta_3 \alpha^2 + \cdots)$$

and equating separately terms in powers of α and $t = \ln u$, with $u = \frac{k^2}{m^2}$, we find

$$\pi(e^{2t}k^2) = \alpha(b_1 \ln u + a_1) + O\left(\frac{1}{u}\ln u\right)$$

$$+\alpha^2(b_2 \ln u + a_2)$$
$$+\alpha^3(c_3 \ln^2 u + b_3 \ln u + a_3)$$
$$+\alpha^4(d_4 \ln^3 u + c_4 \ln^2 u + b_4 \ln u + a_4)$$
$$+\cdots$$

with the coefficients constrained by the relations

$$c_1 = d_1 = \cdots = 0$$
$$c_2 = d_2 = \cdots = 0$$
$$2c_3 + b_1 b_2 = 0, \quad d_3 = 0, \quad etc.$$

and

$$\beta_1 = -b_1$$
$$\beta_2 = -b_2$$
$$\beta_3 = -b_3 - b_1 a_2, \quad etc.$$

This result is in agreement with Weinberg's theorem. Notice that in order α^n there are no higher terms than $\ln^{n-1} u$. By performing a 3-loop calculation of $\pi(k)_{ren}$, de Rafäel and Rosner obtained in this way[8] the following result for the QED-β function up to fourth order in α:

$$\beta(\alpha) = \frac{2\alpha^2}{3\pi} + \frac{\alpha^3}{2\pi^2} - \frac{121\alpha^4}{144\pi^3} + O(\alpha^5). \qquad (17.82)$$

The result is represented in Fig. 17.8 below.[9]

It is interesting that there has been discussion in the literature about the uniqueness of the beta function. Noting that the renormalized proper functions are unique under renormalization, the above procedure should also guarantee the uniqueness of the beta function. Notice that this beta-function is "universal" in the sense, that it applies to all Green functions of the theory.

---

[8] Eduardo de Rafäel and Jonathan L. Rosner, *Ann. of Phys.* **82**(2), (1974) pp. 301–608.

[9] The third term in this expression disagrees in sign with the one quoted in N.N. Bogoliubov and N.N. Shirkov, *Introduction to the Theory of Quantized Fields*, (1959).

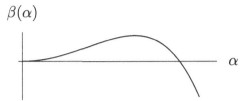

$\beta(\alpha)$

α

Fig. 17.8. E. de Rafäel and J.L. Rosner.

17.10 Infrared fix point of QED and screening of charge

We have seen that the β-function of QED in 1-loop approximation is given by

$$\beta(\alpha) = \frac{2\alpha^2}{3\pi} + 0(\alpha^3) \ , \ \alpha = \frac{e^2}{4\pi} \ .$$

Hence QED is not asymptotically free, and as a consequence its behaviour in the deep euclidean region is unknown. Nevertheless, QED has been remarkably successful in the low energy region. This success, usually attributed to the smallness of the fine structure constant ($\alpha = 1/137$) is at least consistent with the fact that $\alpha = 0$ is an infrared (IR) fixed point of QED. Let us therefore assume that the value $\alpha = 1/137$ lies to the left of the first non-trivial zero of the QED β-function. In view of its smallness the running coupling constant may be taken to satisfy the equation

$$\frac{\partial \bar{\alpha}}{\partial t} \simeq \frac{2\bar{\alpha}^2}{3\pi} \ .$$

Separating variables, integration of this equation yields $\bar{\alpha}(\alpha, t)$

$$\int_0^t dt' = \int_\alpha^{\bar{\alpha}(\alpha,t)} d\alpha' \frac{1}{\frac{2}{3\pi}\alpha'^2} = -\frac{3\pi}{2}\left(\frac{1}{\bar{\alpha}} - \frac{1}{\alpha}\right) \ ,$$

or

$$\bar{\alpha}(\alpha, t) \simeq \frac{\alpha}{1 - \frac{2\alpha}{3\pi}t} \ . \tag{17.83}$$

This is the case considered in (17.60). The pole on the r.h.s. is referred to as the "Landau pole". The IR limit corresponds to $t \to -\infty$. The result shows that significant deviations from $\alpha = 1/137$ will occur only for (t is negative!) "microscopic" energy values of $\sqrt{\mathcal{S}}$:

$$\frac{\sqrt{\mathcal{S}}}{m_e} = O\left(e^{-\frac{3\pi}{2\alpha}}\right) \ .$$

Rigorously speaking, this result applies only to the *deep infrared* region. From a phenomenological point of view one may nevertheless think of the running coupling constant (17.83) as replacing α in Thomson-scattering at (very) small momentum transfer q^2, with $\mathcal{S}$ replaced by q^2. Note however that in this picture we are actually far away from the non-exceptional region since the external momenta are restricted in this case to the electron mass shell $p_i^2 = m_e^2$.

Since a small momentum transfer corresponds in Thompson scattering to a large separation R of the charges, we may equally well think of the running coupling constant as a function of R. From the phenomenological point of view we could thus be led to replace the classical Coulomb potential by

$$V(R) = \frac{\bar{\alpha}(R)}{R} = \frac{\alpha}{R(1 + \frac{2\alpha}{3\pi}\ln(m_e R))} .$$

Note that expansion of the denominator in powers of α yields

$$V(R) \simeq \frac{\alpha}{R}\left(1 - \frac{2\alpha}{3\pi}\ln(m_e R)\right), \tag{17.84}$$

which is just the famous *Uhlenbeck formula*.

The potential (17.84) can be viewed as expressing the effect of screening of the 2-point charges by virtual electron-positron pairs. The creation of such pairs should set in at separations R of the order of the inverse electron mass, as brought also in evidence by the argument of the logarithmic factor. Notice that this "vacuum polarization" leads to an enhancement (reduction) of the effective charge inside (outside) the radius of the electrons.

Chapter 18

Renormalization Group

In Section 1 of Chapter 16 we imposed suitable conditions in order to define the wave function renormalization constant, as well as the renormalized mass and coupling as a function of the cutoff and the bare parameters. We have done this *without* introducing further dimensioned parameters into the theory. This is an option we are free to take, but need not be the most adequate one from the point of view of perturbation theory. Thus from the point of view of the convergence of perturbation theory it could turn out to be advantageous to define the coupling constant in ϕ^4-theory, instead of (17.64), by

$$\Gamma_{ren}^{(4)}(p_1 \cdots p_4)\Big|_{p_i \cdot p_j = -\mu^2(\delta_{ij} - \frac{1}{4})} = -\bar{g}(\mu) \ , \tag{18.1}$$

where μ is a new parameter with the dimensions of a mass, and $\Gamma^{(n)}(p_1, \cdots)$ stands for the 1PI renormalized proper n-point Green function. Similarly, we could choose for the mass *parameter* a value not being identical with the *physical* mass associated with the field $\phi(x)$. To be concrete let us take the coupling constant parameter to be parametrized by μ.[1] The physical consequences should of course be independent of how we choose our parametrization. This means that the vertex functions computed for different choices of parametrizations must be related in a definite way. The *Renormalization Group Equation (RGE) expresses in differential form this relationship as a function of μ.*[2]

18.1 The Renormalization Group equation

The renormalization group equation expressing the independence of the *renormalized* 1PI functions of the normalization point μ is a homogeneous equation taking the form

$$\left[\mu \frac{\partial}{\partial \mu} + \bar{\beta}(\bar{g}, \frac{\mu}{m})\frac{\partial}{\partial \bar{g}} - \bar{\gamma}(\bar{g}, \frac{\mu}{m})\right] \Gamma^{(n)}(p; \bar{g}, m, \mu) = 0 \tag{18.2}$$

---

[1] We shall not parametrize the mass to simplify the discussion.

[2] E.C.G. Stueckelberg and A. Peterman, *Helv. Phys. Acta* **26** (1953) 499; M. Gell-Mann and F. Low, *Phys. Rev.* **95** (1954), 1300; N.N. Bogoliubov and D.V. Shirkov, *Introduction to the Theory of Quantized Fields* (Interscience, New York, 1959).

reminiscent of the *asymptotic* Callan–Symanzik equation (17.43), where the nomenclature is the same except for the overhead "bar" distinguishing the beta-function and anomalous-dimension from those of the CS equation.

We begin by considering some examples.

Example 1: Coupling constant in ϕ^4-theory

The coupling constant g need not be defined at $p^2 = 0$ as in (17.64), but can be chosen to be defined at some euclidean point $p^2 = -\mu^2$. From (17.70) we have for the renormalized 4-point 1PI function,

$$\Gamma^{(4)}(p^2) = -g - \frac{3g^2/2}{(4\pi)^2} \int_0^1 d\beta \ln\left(1 - \beta(1-\beta)\frac{p^2}{m^2}\right) + O(g^4) \qquad (18.3)$$

where g is the usual renormalized coupling constant, normalized at zero momentum. Defining a new coupling constant by

$$\bar{g}(\mu) = -\Gamma^{(4)}(-\mu^2)$$

we have

$$\bar{g}(\mu) \simeq g + \frac{3g^2/2}{(4\pi)^2} \int_0^1 d\beta \ln\left(1 + \beta(1-\beta)\frac{\mu^2}{m^2}\right) + O(g^4) . \qquad (18.4)$$

In terms of this *parametrized coupling constant* the renormalized 4-point function (18.3) now takes the form

$$\bar{\Gamma}^{(4)}(p^2; \bar{g}, m, \mu) = -\bar{g} - \frac{3\bar{g}^2}{2(4\pi)^2} \int_0^1 d\beta \ln\left(\frac{m^2 - \beta(1-\beta)p^2}{m^2 + \beta(1-\beta)\mu^2}\right) + O(\bar{g}^3)$$

where the possibility of inverting the relation between g and $\bar{g}$ has been assumed in the second term on the right.

Viewed as a function of $\bar{g}(\mu)$ and μ, the independence of μ of (18.3) is thus expressed by the differential equation

$$\mu \frac{d}{d\mu}\bar{\Gamma}^{(4)}(p; \bar{g}, m, \mu) = \left(\mu \frac{\partial}{\partial\mu} + \bar{\beta}(g, \frac{\mu}{m})\frac{\partial}{\partial\bar{g}}\right)\bar{\Gamma}^{(4)}(p^2; \bar{g}, m, \mu) + O(\bar{g}^3) , \qquad (18.5)$$

where $\bar{\beta}$ is the function[3]

$$\bar{\beta}\left(g, \frac{\mu}{m}\right) \equiv \mu \frac{\partial}{\partial\mu}\bar{g}\left(g, \frac{\mu}{m}\right) , \qquad (18.6)$$

and we have made explicit the dependence of $\bar{g}$ on g, m and μ. Notice that here the right-hand side *defines* the left-hand side. Instead of considering the coupling $g = \bar{g}(0)$ and $\bar{g} = \bar{g}(\mu)$ it is useful to consider them at neighbouring points μ_1 and μ_2. According to (18.4) we then have

$$\bar{g}_2 = \bar{g}_1 + \frac{3\bar{g}_1^2}{32\pi^2} \int_0^1 d\beta \ln\left(\frac{m^2 + \beta(1-\beta)\mu_2^2}{m^2 + \beta(1-\beta)\mu_1^2}\right) + O(\bar{g}_1^3) . \qquad (18.7)$$

---

[3]This $\bar{\beta}$ function is not to be confused with the β function of the Callan–Symanzik equation, except for the observation made later on.

Notice that in going from (18.4) to this expression we have to order $O(\bar{g}_1^3)$ make the replacement $g \to \bar{g}_1$ in the factor multiplying the integral. This is important in the sequel. Independence of $\bar{\Gamma}^{(4)}(p^2; g_2, m, \mu_2)$ of μ_2 then requires[4]

$$\mu_2 \frac{d}{d\mu_2} \bar{\Gamma}^{(4)}\left(p^2; \bar{g}_2\left(\bar{g}_1, \frac{\mu_2}{\mu_1}, \frac{\mu_1}{m}\right), m, \mu_2\right) = 0 \ .$$

Performing the differentiation and setting $\mu_2 = \mu_1$, we have

$$\left(\mu_1 \frac{\partial}{\partial \mu_1} + \bar{\beta}\left(\bar{g}_1, \frac{\mu_1}{m}\right) \frac{\partial}{\partial \bar{g}_1}\right) \bar{\Gamma}^{(4)}(p^2; \bar{g}_1, m, \mu_1) = 0 \ ,$$

where

$$\bar{\beta}\left(\bar{g}_1, \frac{\mu_1}{m}\right) = \left[\mu_2 \frac{\partial}{\partial \mu_2} \bar{g}_2\left(\bar{g}_1, \frac{\mu_2}{\mu_1}, \frac{\mu_1}{m}\right)\right]_{\mu_2 = \mu_1} \tag{18.8}$$

is referred to as the "Beta function". Explicitly we have to order $O(\bar{g}^3)$ (we now set $\mu_1 = \mu$, $\bar{g}_1 = \bar{g}$)

$$\bar{\beta}\left(\bar{g}, \frac{\mu}{m}\right) = \frac{3\bar{g}^2/2}{(4\pi)^2} \int_0^1 d\beta \frac{2\beta(1-\beta)\mu^2}{m^2 + \beta(\beta-1)\mu^2} + O(\bar{g}^3) \ . \tag{18.9}$$

Notice that for $\mu \to \infty$, this β-function tends to

$$\bar{\beta}\left(\bar{g}, \frac{\mu}{m}\right) \to \frac{3\bar{g}^2}{(4\pi)^2} \ , \tag{18.10}$$

which is the CS value (17.69). Observe also that the limit $\mu \to \infty$ can also be seen as the limit $m \to 0$.

So far, Eq. (18.6) merely defines the "Beta" function as a function of the parameter μ. We now *define* a "running coupling constant" as the solution to the *differential equation*

$$\mu \frac{d}{d\mu} \bar{g} = \bar{\beta}\left(\bar{g}, \frac{\mu}{m}\right) \ . \tag{18.11}$$

It is interesting to compare the solution for this "running" coupling constant with the "parametrized" coupling constant (18.4). For the latter we have for large μ,

$$\bar{g}(\mu) \to g\left[1 + \frac{3g}{(4\pi)^2} \ln\left(\frac{\mu}{m}\right)\right] \ ,$$

whereas with (18.10), Eq. (18.11) asymptotically solves the differential equation

$$\mu \frac{d\bar{g}}{d\mu} = \frac{3\bar{g}^2}{(4\pi)^2}$$

with the solution analogous to (17.83),

$$\bar{g}(\mu) \simeq \frac{g}{1 - \frac{3g}{(4\pi)^2} \ln\left(\frac{\mu}{M}\right)} \ , \tag{18.12}$$

---

[4]S. Weinberg, *The Quantum Theory of Fields* (Cambridge University Press, 1996).

where M is some large scale at which $\bar{g} = g$ is defined. We have in the present case an IR fix-point at $\bar{g} = 0$. From above, we see that this amounts to significant deviations from g only for μ of the order $\mu \approx M exp(\frac{(4\pi)^2}{3g})$.

Example 2

Recall the calculation of the QED renormalization constant Z_A from (16.38). The Bogoliubov recursive renormalization and the Taylor subtraction procedure of Chapter 16 make no reference to the "normalization conditions" to be imposed on the 2- and 3-point functions. Similarly, a "minimal subtraction" procedure in dimensional regularization consists of simply subtracting the divergent "$1/\epsilon$" term in (16.70) also making no such reference.

By the same token we could have chosen to expand $\pi_{reg}(k^2)$ around $k^2 = -\mu^2$, instead of the photon mass $k^2 = 0$ in (16.36):

$$\pi(k^2)_{reg} = \pi(-\mu^2)_{reg} + \pi'(k^2; \mu) . \tag{18.13}$$

We then have for the (correctly normalized, transversal) photon 2-point function, instead of (16.37),

$$\tilde{D}_F^{\mu\nu}(k) = \frac{-Z_A^{-1}\mathcal{P}^{\mu\nu}(k)}{k^2(1 - \pi(-\mu^2)_{reg}) - k^2\pi'(k^2; \mu)} ,$$

where Z_A is given by (16.38):

$$Z_A = \frac{1}{1 - \pi(0)_{reg}} . \tag{18.14}$$

$\tilde{D}_F^{\mu\nu}(k)$ is independent of μ. Define

$$Z_A(\mu) = \frac{1}{(1 - \pi(-\mu^2)_{reg})} \tag{18.15}$$

and

$$\bar{Z}_A(\mu) = \frac{Z_A(\mu)}{Z_A} = \frac{1 - \pi(0)_{reg}}{1 - \pi(-\mu^2)_{reg}} . \tag{18.16}$$

Then $\bar{Z}_A(\mu)$ is a finite, cutoff independent constant for $\Lambda \to \infty$, and

$$\bar{\alpha}(\mu) = \bar{Z}_A(\mu)\alpha = Z_A(\mu)\alpha_0 . \tag{18.17}$$

Notice that we have used the Ward identity $Z_\alpha = Z_A$. This leaves us with the μ independent expression

$$D_F^{\mu\nu}(k)_{ren} = \frac{-\bar{Z}_A(\mu)\mathcal{P}^{\mu\nu}(k)}{k^2(1 - \pi(k^2; \bar{\alpha}, \mu)_{ren})} \tag{18.18}$$

with

$$\pi(k^2; \bar{\alpha}, \mu)_{ren} = Z_A(\mu)\pi'(k^2, Z_A^{-1}(\mu)\bar{\alpha}(\mu), m - \delta m, \mu) . \tag{18.19}$$

By construction this propagator is correctly normalized at $k^2 = 0$. Hence we also have

$$\bar{Z}_A(\mu) = \frac{\bar{\alpha}(\mu)}{\alpha} = (1 - \pi(0; \bar{\alpha}, \mu)_{ren}) \ .$$

Equation (18.18) is a special case of the statement

$$\frac{\bar{Z}_A(\mu_1) P^{\mu\nu}(k)}{k^2(1 - \pi(k^2; \bar{\alpha}_1, \mu_1)_{ren})} = \frac{\bar{Z}_A(\mu_2) P^{\mu\nu}(k)}{k^2(1 - \pi(k^2; \bar{\alpha}_2, \mu_2)_{ren})} \ . \tag{18.20}$$

The above example shows that a change in the subtraction procedure requires a change in the multiplicative renormalization constant in order to keep the physics unchanged. This multiplicative constant $\bar{Z}(\mu)$ is cutoff independent, but depends on μ.

Since the lhs in (18.18) is independent of μ we have again a differential equation expressing this independence of μ of the corresponding 1PI function. The requirement

$$\frac{d}{d\mu}\left(\frac{1 - \pi(k; \bar{\alpha}, \mu)_{ren}}{\bar{Z}_A(\mu)}\right) = 0 \tag{18.21}$$

implies

$$\left(\mu\frac{\partial}{\partial\mu} + \bar{\beta}\left(\bar{\alpha}, \frac{\mu}{m}\right)\frac{\partial}{\partial\bar{\alpha}} - 2\bar{\gamma}_A\left(\bar{\alpha}, \frac{\mu}{m}\right)\right)(1 - \pi(k; \bar{\alpha}, \mu)_{ren}) = 0 \tag{18.22}$$

with the running coupling constant satisfying the differential equation,

$$\mu\frac{d}{d\mu}\bar{\alpha} = \bar{\beta}\left(\bar{\alpha}, \frac{\mu}{m}\right) \tag{18.23}$$

and $\bar{\gamma}_A(\bar{\alpha}, \frac{\mu}{m})$ defined in an analogous way to (18.8) by

$$\bar{\gamma}_A\left(\bar{\alpha}, \frac{\mu}{m}\right) = \frac{\mu}{2}\frac{d}{d\mu}\ln\bar{Z}_A\left(\bar{\alpha}(\mu), \frac{\mu}{m}\right) \ . \tag{18.24}$$

Notice that (18.24) involves a *total* derivative. This equation may be formally integrated to give

$$\ln\bar{Z}_A^{\frac{1}{2}}(\mu) = \int_0^\mu \frac{d\mu'}{\mu'}\bar{\gamma}_A\left(\bar{\alpha}(\mu'), \frac{\mu'}{m}\right) \ .$$

Equation (18.22) is the *renormalization group equation* for the vacuum polarization in QED.

Explicit expressions to order $O(\bar{\alpha}^2)$

From (18.14), (18.15) and (17.76) we have

$$Z_A(\mu) = Z_A + \frac{2\bar{\alpha}}{\pi}\int_0^1 d\beta\beta(1 - \beta)\ln\left(1 + \beta(1 - \beta)\frac{\mu^2}{m^2}\right) + O(\bar{\alpha}^2) \ .$$

From here, (18.16) and (18.17) follow, to get,

$$\bar{Z}_A(\mu) = 1 + \frac{2\bar{\alpha}}{\pi} \int_0^1 d\beta \beta(1-\beta)\ln\left(1 + \beta(1-\beta)\frac{\mu^2}{m^2}\right) + O(\bar{\alpha}^2) \,,$$

$$\bar{\alpha}(\mu) \simeq \alpha + \frac{2\alpha^2}{\pi} \int_0^1 d\beta \beta(1-\beta)\ln\left(1 + \beta(1-\beta)\frac{\mu^2}{m^2}\right) \,.$$

And from (18.23) and (18.24) we further have,

$$\frac{\bar{\beta}\left(\bar{\alpha}, \frac{\mu}{m}\right)}{\bar{\alpha}} = 2\bar{\gamma}\left(\bar{\alpha}, \frac{\mu}{m}\right) = 4\frac{\bar{\alpha}}{\pi} \int_0^1 d\beta \frac{[\beta(1-\beta)]^2 \frac{\mu^2}{m^2}}{1 + \beta(1-\beta)\frac{\mu^2}{m^2}} + O(\bar{\alpha}^2) \tag{18.25}$$

and from (18.13) and (18.19) finally follows,

$$\pi(k^2; \bar{\alpha}, m, \mu)_{ren} = \frac{2\bar{\alpha}}{\pi} \int_0^1 d\beta \beta(1-\beta)\ln\left(\frac{m^2 - \beta(1-\beta)k^2}{m^2 + \beta(1-\beta)\mu^2}\right) + O(\bar{\alpha}^2) \,.$$

Note that the Ward-identity relation (17.80) is preserved. Notice also that, analogous to (18.10) in the ϕ^3 case, we have in the $\mu \to \infty$ (or zero mass) limit

$$\bar{\beta}\left(\bar{\alpha}, \frac{\mu}{m}\right) \to \frac{2\bar{\alpha}^2}{3\pi} \,, \tag{18.26}$$

which is the CS value (17.78), and

$$\tilde{\alpha}(\mu) \approx \frac{\alpha(M)}{1 - \frac{2\alpha(M)}{3\pi}\ln\left(\frac{\mu}{M}\right)} \,, \tag{18.27}$$

where the singularity is referred to as *Landau singularity*. The *increase* of $\alpha(\mu)$ with μ (*decrease* with distance) has its analogue in electrostatics in a dielectric medium, where the presence of a charge polarizes the medium, such as to screen the charge.

Example 3

Consider the photon propagator (18.18). Using (18.17) we have

$$\alpha D_{ren}^{\mu\nu}(k^2) = \bar{\alpha}(\mu)\frac{-\mathcal{P}^{\mu\nu}(k)}{k^2[1 - \pi(k^2; \bar{\alpha}, m, \mu)_{ren}]} \,.$$

Since the lhs does not depend on μ, the right-hand side does not depend on μ either; nor does it involve a multiplicative wave function renormalization constant in the numerator. We thus have the RG equation without anomalous dimension, or

$$\left(\mu\frac{\partial}{\partial\mu} + \bar{\beta}\left(\bar{\alpha}, \frac{\mu}{m}\right)\frac{\partial}{\partial\bar{\alpha}}\right)\left(\frac{1 - \pi(k^2; \bar{\alpha}, m, \mu)_{ren}}{\bar{\alpha}}\right) = 0 \,. \tag{18.28}$$

Notice that since the Ward identity $Z_\alpha = Z_A$, this is nothing but Eq. (18.21). Equation (18.28) is a convenient starting point for the calculation of the *beta* function (17.82) of de Rafäel and Rosner.

18.2 Asymptotic solution of RG equation

Let $\Gamma^{(n)}(p; g, m)$ be a conventionally renormalized 1PI n-point function in ϕ^4 theory. As we have seen, normalizing the coupling constant at a different point μ, will require for a proper normalization of the vertex function the introduction of a finite renormalization constant $\bar{Z}_\phi(\mu)$, implying the appearance of a term $\bar{\gamma}(\bar{g}, \mu)$ in the RG equation. The generalization of (18.20) to an arbitrary *proper* function reads

$$\bar{Z}_\phi^{-\frac{n}{2}}(\mu_1)\tilde{\bar{\Gamma}}^{(n)}(\{p_i\}; \bar{\alpha}_1, m, \mu_1) = \bar{Z}_\phi^{-\frac{n}{2}}(\mu_2)\tilde{\bar{\Gamma}}^{(n)}(\{p_i\}; \bar{\alpha}_2, m, \mu_2). \tag{18.29}$$

In different terms: For the *proper* functions, Eq. (16.6) reads in compact notation in the present case,

$$Z^{-n}\left(\bar{e}, \mu, \frac{\Lambda}{m}\right)\tilde{\Gamma}^{(n)}\left(...; \bar{e}, \mu, \frac{\Lambda}{m}\right) = \Gamma_{reg}^{(n)}(...; e_0, m_0, \Lambda)$$

or

$$\mu\frac{d}{d\mu}\left\{Z^{-n}\left(\bar{e}, \mu, \frac{\Lambda}{m}\right)\tilde{\Gamma}^{(n)}\left(...; \bar{e}, \mu, \frac{\Lambda}{m}\right)\right\} = 0 . \tag{18.30}$$

From here, we arrive at

$$\left(\mu\frac{\partial}{\partial\mu} + \bar{\beta}\frac{\partial}{\partial\bar{g}} - n\bar{\gamma}\right)\bar{\Gamma}^{(n)}(\{p_i\}; \bar{g}, m, \mu) = 0 , \tag{18.31}$$

with $\bar{\beta}$ and $\bar{\gamma}$ defined as in (18.8) and (18.24), respectively.

Consider now the renormalized 1PI function $\bar{\Gamma}^{(n)}(p_i; \bar{g}, m, \mu)$, but with the momenta scaled by a factor e^t. It satisfies the RG equation (18.31) with p_i replaced by $e^t p_i$.

$$\left(\mu\frac{\partial}{\partial\mu} + \bar{\beta}\frac{\partial}{\partial\bar{g}} - n\bar{\gamma}\right)\Gamma^{(n)}\left(\{e^t p_i\}; \bar{g}, m, \mu\right) = 0 .$$

We now follow arguments as already used in Chapter 17. By dimensional analysis we have

$$\bar{\Gamma}^{(n)}\left(\{e^t p_i\}; \bar{g}, m, \mu\right) = e^{Dt}\bar{\Gamma}^{(n)}\left(\{p_i\}; \bar{g}, e^{-t}m, e^{-t}\mu\right)$$

or

$$\frac{\partial}{\partial t}\bar{\Gamma}(\{e^t p\}, \bar{g}, m, \mu) = \left(-\mu\frac{\partial}{\partial\mu} - m\frac{\partial}{\partial m} + D\right)\bar{\Gamma}(\{e^t p_i\}, \bar{g}, m, \mu) .$$

We use this equation in order to exchange the μ-differentiation for the t-differentiation in the RG-equation, which now reads:

$$\mathcal{D}\Gamma^{(n)}\left(\{e^t p\}; \bar{g}, m, \mu\right) = 0 ,$$

where

$$\mathcal{D} = \left(-\frac{\partial}{\partial t} - m\frac{\partial}{\partial m} + \bar{\beta}\frac{\partial}{\partial\bar{g}} + D - n\bar{\gamma}\right) .$$

Correspondingly we have

$$\left[-\frac{\partial}{\partial t} - m\frac{\partial}{\partial m} + \bar{\beta}\left(\bar{g}, \frac{\mu}{m}\right)\frac{\partial}{\partial\bar{g}} + D - n\bar{\gamma}\left(\bar{g}, \frac{\mu}{m}\right)\right]\bar{\Gamma}(\{e^t p_i\}; \bar{g}, m, \mu) = 0 .$$

Consider now this equation in the limit $\mu \to \infty$, $\bar{g}$ fixed. This limit is equivalent to $m \to 0$. From (18.9) and (18.25) it is interesting to observe that in our examples of ϕ^4-theory and QED, the *second order* $\bar{\beta}$-function approached respectively their CS value in the $\mu \to \infty$ limit. Indeed, from (18.26) we have for QED,

$$\bar{\beta}\left(\bar{\alpha}, \frac{\mu(t)}{m}\right) \to \beta_{CS}(\bar{\alpha}) = \frac{2\bar{\alpha}^2}{3\pi} + O(\bar{\alpha}^3)$$

$$\bar{\gamma}\left(\bar{\alpha}, \frac{\mu(t)}{m}\right) \to \gamma_{CS}(\bar{\alpha}) = \frac{\bar{\alpha}}{3\pi} + O(\bar{\alpha}^3)$$

and from (18.10) for ϕ^4-theory,

$$\beta\left(\bar{g}, \frac{\mu(t)}{m}\right) \to \bar{\beta}_{CS}(\bar{g}) = \frac{3\bar{g}^2}{(4\pi)^4} + O(\bar{g}^3) .$$

This could be expected to be generally true since the $\mu \to \infty$ limit corresponds, by dimensional analysis, to the $m \to 0$ limit. This would reconcile the *generally valid* homogeneous RG equation with the *asymptotically valid* homogeneous CS equation.

From the parallel treatment leading to (17.56) we infer for $t \to \infty$ and $\frac{\mu}{m} \to \infty$,

$$\Gamma(\{e^t p_i\}; g, m) \to e^{Dt} \Phi(\{p_i\}; \bar{g}(g,t), e^{-t}m) e^{-\int_0^t dt' \gamma(\bar{g}(g,t'))} ,$$

where $\bar{g}(g,t)$ is the running coupling constant satisfying (17.54).

Chapter 19

Spontaneous Symmetry Breaking

The so-called *spontaneous* breaking of a *continuous* symmetry plays a central role in systems with an infinite number of degrees of freedom such as condensed matter, where it is responsible for phase transitions, as well as in QFT, where it is known as the Higgs mechanism in the *Weak Interactions*. A fundamental theorem due to Goldstone plays here a central role.

19.1 The basic idea

The perturbative expansion as given by (11.21) implicitly contains the assumption of a vanishing vacuum expectation value of the fields in question. In the case of fermionic fields this property is guaranteed by the Lorenz invariance of the vacuum. In the case of the scalar field, this is not sufficient. There may exist, however, further symmetries which may be invoked to guarantee a vanishing expectation value $\langle 0|\phi(x)|0\rangle = 0$ in this case. Thus consider for instance the Lagrangian of a real scalar field,

$$\mathcal{L} = \frac{1}{2}(\partial_\mu \phi)^2 - V_{eff}(\phi)$$

with the "effective potential"

$$V_{eff}(\phi) = \frac{1}{2}\mu^2\phi^2 + g\phi^4 \tag{19.1}$$

as represented in Fig. 19.1. We see that the potential (as well as the lagrangian) is symmetric under the *global* and *discrete* transformations

$$\phi(x) \to -\phi(x). \tag{19.2}$$

Let $\mathcal{U}$ be the unitary operator inducing this transformation

$$\mathcal{U}\phi(x)\mathcal{U}^{-1} = -\phi(x).$$

$$V_{\text{eff}}(\phi)$$

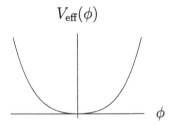

Fig. 19.1. $V_{eff} = \frac{1}{2}\mu^2\phi^2 + g\phi^4$ for $\mu = g = 1$.

If the ground state is invariant under this transformation, then

$$\mathcal{U}|\Omega\rangle = |\Omega\rangle\,,$$

and hence,

$$\langle\Omega|\phi(x)|\Omega\rangle = \langle\Omega|\mathcal{U}^\dagger\phi(x)\mathcal{U}|\Omega\rangle = -\langle\Omega|\phi(x)|\Omega\rangle\,,$$

implying the vanishing of the vacuum expectation value of the field.

$$V_{\text{eff}}(\phi)$$

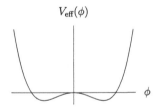

Fig. 19.2. $V_{eff} = -\frac{1}{2}\mu^2\phi^2 + g\phi^4$ for $\mu = g = 1$.

If we however change the sign of the mass term in $\mathcal{L}$, then the $V_{eff}(\phi)$ will become of the form of Fig. 19.2. Though the potential still respects the symmetry under the transformation (19.2), the now two-fold degenerate ground state no longer does, corresponding to the non-vanishing vacuum expectation values

$$\langle\Omega|\phi(x)|\Omega\rangle = \pm a$$

where to zeroth order of $\hbar$, the parameter a represents the location of the minima $\pm\sqrt{\mu^2/2g}$ of V_{eff}. One says that the symmetry (19.2) has been *spontaneously* broken.

A negative mass term in the lagrangian would correspond to a "tachyon". In order to avoid it, we must now expand $\phi(x)$ about one of the potential minima. With respect to the new field, the resulting mass term in the effective lagrangian will have the correct sign to represent a physical particle of mass $\sqrt{2}\mu$.

On quantum level spontaneous symmetry breakdown may occur through quantum fluctuations which may turn the effective potential at tree ($\hbar = 0$) level with a non-degenerate vacuum, into an effective potential with degenerate vacua. Effective potentials to 1-loop order and spontaneous symmetry breakdown will be the subject of the following chapter.

At this point, it should be emphasized that we have considered an example exhibiting SSB of a *discrete* symmetry. A very interesting new phenomenon occurs if we have a SSB of a *continuous* symmetry. A well-known example is provided by the Heisenberg ferromagnet, to be discussed in the next section. If we have the temperature as an additional parameter, then a symmetry preserving effective potential may turn into a symmetry breaking potential as we tune the temperature past some critical value. In this case one speaks of a *phase transition*. In the Heisenberg ferromagnet this transition is accompanied by the appearance of so-called "spin-waves", zero mass excitations, the analogon of "Goldstone bosons", to be discussed in Section 3.

19.2   More about spontaneous symmetry breaking

A symmetry of the Hamiltonian is spontaneously broken, if the ground state of the system does not exhibit this symmetry. There exist many examples of spontaneous symmetry breakdown (SSB) in nature. Our interest will now concentrate on the spontaneous breakdown of *continuous* symmetries. In classical physics this turns out to imply the existence of a continuum of degenerate ground states. The bending of the Euler stick provides an example: Exerting a force on one end of the stick, directed along it, the stick will suddenly bend, once the force exceeds a certain critical value when subject to a small perturbation. The new configuration now represents the state of lowest energy. The direction in which the stick will bend is, however, chosen "spontaneously". All these directions are energetically equivalent and may be rotated into each other without the expenditure of energy.

On the other hand, in quantum mechanics with a *finite* number of degrees of freedom the ground state turns out to be unique due to tunneling, and there exists no SSB. In order to witness SSB we must turn to quantum mechanical systems with an *infinite number of degrees of freedom*. Ferromagnetism provides here an instructive example.

The ground state of a ferromagnet is characterized by the fact, that there exists a non-vanishing magnetization $\vec{M}$ also in the absence of an externally applied magnetic field. This is, however, the case only for temperatures below a critical temperature T_c, called the Curie-temperature, i.e.

$$\vec{M} = 0\,, T > T_c\,;\quad \vec{M} \neq 0\,, T < T_c.$$

The macroscopic state of this ferromagnet is described by the free energy $F[T, V; \mathcal{M}]$, which is a function of the temperature and volume, and a functional of the magnetization $\mathcal{M}(\vec{r})$. It is convenient to define the corresponding free energy density $\mathcal{F}$ by

$$F[T, V, \vec{\mathcal{M}}] = \int_V d^3r \mathcal{F}(\vec{\mathcal{M}}(\vec{r}), T).$$

We now assume that near the critical temperature, the ferromagnet exhibits a *weak* and *nearly constant* magnetization $\vec{\mathcal{M}}(r)$. Assuming *rotational invariance* of the free energy density (though not of the ground *state*) one is led to make the Ansatz

$$\mathcal{F}((\vec{r}), T) = \left(\vec{\nabla} \cdot \vec{\mathcal{M}}(\vec{r})\right)^2 + a(T)|\vec{\mathcal{M}}(\vec{r})|^2 + b(T)\left(\vec{\mathcal{M}}(\vec{r}) \cdot \vec{\mathcal{M}}(\vec{r})\right)^2 \qquad (19.3)$$

where the coefficients a and b are allowed to depend on the temperature. It turns out that this expansion in powers of the magnetization provides a good qualitative description near the Curie temperature, provided we attribute the following properties to the coefficients:

$$b(T) > 0 \ , \ T \simeq T_c$$
$$a(T) > 0 \ , \ T > T_c$$
$$a(T) < 0 \ , \ T < T_c$$

In order to determine the properties of the ground state, we need to minimize the free energy (19.3) with respect to the magnetization. Since $(\vec{\nabla} \cdot \vec{M})^2$ is a semi-positive quantity, this minimum will correspond to a constant magnetization. Under this condition, we have the two possible situations, depending on whether the temperature lies above or below the critical one: For $T < T_c$ we have the free energy depicted below with $\rho = |\vec{\mathcal{M}}|$, representing a cut of a three-dimensional picture obtained by rotating the figure about the vertical axis.

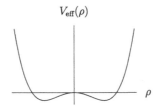

$V_{\text{eff}}(\rho)$

ρ

Fig. 19.3. V_{eff} with minima at $\rho = \pm\sqrt{\frac{-a}{2b}}$.

The figure thus actually represents the bottom of a bottle of champagne (Mexican hat), and hence describes a ground state in which the magnetization can take any one of the orientations related by rotations about the vertical axis. They all correspond to the same energy, and hence we have an infinite set of degenerate ground states associated with this SSB. Are these ground states all unitarily equivalent, i.e. does there exist a unitary operator turning one vacuum state into another?

The spontaneous magnetization of a ferromagnet corresponds to a state in which the spin-components at each of the lattice sites do not add to zero when averaged over a infinitesimal, though macroscopic region of space. Taking the volume V of the ferromagnet for the time being to be finite, we imagine a situation of maximal magnetization, in which all of the spins s_j are oriented along a given axis, which we take to be the z-axis.

$$\boxed{\uparrow_{s_1}\uparrow_{s_2}\cdots\uparrow_{s_i}\cdots\uparrow_{s_j}\cdots\uparrow_{s_N}}$$

For the corresponding Hamiltonian we take the (rotationally invariant) expression

$$H = -\sum_{i,j\epsilon V} J_{ij}\vec{S}_i\cdot\vec{S}_j$$

corresponding to an (isotropic) *Heisenberg* ferromagnet.

The fact that there exists a spontaneous magnetization means in quantum mechanical terms that the ground state expectation value of the spin-operator at the ith site does not vanish:

$$\langle\Omega|\vec{S}_i(t)|\Omega\rangle_V \neq 0\,,$$

where V denotes the volume of the sample. Of course, this can only be the case if the ground state does not respect the rotational symmetry, since otherwise the expectation value would vanish according to the Wigner–Eckart theorem. Now consider the operator

$$\mathcal{U}(\bar{\theta}) = e^{i\bar{\theta}\cdot\vec{S}}\,,\quad \vec{S} = \sum_{i\epsilon V}\vec{S}_i\,.$$

This operator induces a global rotation of the spins by an angle θ about the $\hat{\theta}$-axis:

$$\mathcal{U}(\theta)|\Omega\rangle_V = |\Omega_\theta\rangle_V\,,\quad \vec{S}|\Omega\rangle_V \neq 0\,.$$

Nevertheless

$$[\mathcal{U}(\bar{\theta}), H] = 0\,,$$

that is, no energy needs to be expended in order to arrive at the new state. An alternative view will be helpful: We need not turn all spins at once; thus we could begin by first turning just $N/2$ of the spins through an angle; this will of course require the input of energy. This amount of energy will be recovered, once we subsequently turn the remaining $N/2$ spins.

$$\underbrace{\uparrow\uparrow\cdots\uparrow}_{N/2}\;\underbrace{\nearrow\nearrow\cdots\nearrow}_{N/2}$$

For N finite, the different ground states thus obtained are unitarily equivalent. Thus choose the rotation to be about the z-axis. We then have for the ith spin,

$$\langle\Omega_0|e^{i\bar{\theta}\cdot\vec{S}_i}|\Omega_0\rangle_V = d_{\frac{1}{2}\frac{1}{2}}^{\left(\frac{1}{2}\right)}(\theta)$$

and for the total number of N spins,

$$\langle\Omega_0|\Omega_\theta\rangle_V = \langle\Omega_0|e^{i\bar{\theta}\cdot\vec{S}}|\Omega_0\rangle_V = \left[d_{\frac{1}{2}\frac{1}{2}}^{\left(\frac{1}{2}\right)}(\theta)\right]^N = \left(\cos\frac{\theta}{2}\right)^N\,.$$

That is, for a ferromagnet with a finite number of sites, the old and new ground state have a non-vanishing projection. As this number, however, tends to infinity, we find in the limit $N \to \infty$,

$$\langle \Omega_0 | \Omega_\theta \rangle_{V \to \infty} = 0 \,, \quad 0 < \theta < 2\pi \,. \tag{19.4}$$

Hence, in the limit of an infinite number of degrees of freedom, the different ground states are orthogonal to each other, so that they define unitarily *inequivalent* Hilbert spaces (sectors) of the theory. Physically this reflects the fact that the attempt to flip the spins will require an infinite amount of energy, no matter how we try to realize it.

We conclude that the symmetry of the Hamiltonian under joint rotation of the spins by an angle θ has been broken, as witnessed by the appearance of a continuous set of inequivalent ground states. This is only possible for quantum mechanical systems with infinite degrees of freedom.

19.3   The Goldstone Theorem

The above observation has dramatic consequences, which are the content of the following theorem:[1]

Goldstone Theorem:

Given causality, Poincaré invariance and a translationally invariant ground-state with positive norm, the spontaneous breakdown of a continuous symmetry implies the existence of zero modes with the property $\omega(k) \to 0$ as $k \to 0$, the number of such zero modes being equal to the dimension of the group associated with the broken symmetry.

For the ferromagnet, these zero modes are the well-known *spin-waves*. In the context of QFT, such excitations would correspond to zero mass particles. Since up to date the photon and neutrinos are the only observed zero mass particles in nature, the theorem seems to imply at first sight the absence of SSB in particle physics. This is not quite true, as we shall indicate later on.

Let us prove the above theorem. Starting point is the assumption that there exist operators Q^a such that

$$\mathcal{U}^{-1} H \mathcal{U} = H, \quad \mathcal{U} = e^{i\vec{\theta} \cdot \vec{Q}}$$

and

$$e^{i\vec{\theta} \cdot \vec{Q}} |\Omega\rangle = |\Omega\rangle_\theta \neq |\Omega\rangle \,.$$

In different terms,

$$Q^a |\Omega\rangle \neq 0 \,.$$

Since however

$$[H, Q^a] = 0 \,,$$

this implies a *continuous* degeneracy of the ground states.

---

[1] J. Goldstone, *Nuovo Cimento* **19** (1961) 154.

Now, according to Noether's theorem there exists a conserved current $j^{\mu,a}$ for every generator Q^a of a symmetry of a lagrangian,

$$Q^a(t) = \int d^3r\, j^{0,a}(\vec{r},t) \tag{19.5}$$

where

$$\partial_\mu j^{\mu,a}(x) = 0\,.$$

We now show that[2] for a translationally invariant ground state the generator defined by (19.5) cannot exist, if the symmetry is spontaneously broken. To this end, we consider in particular the norm

$$\|Q^a(t)|\Omega\rangle\| = \langle\Omega|Q^a(t)Q^a(t)|\Omega\rangle = \int d^3r\,\langle\Omega|j_0^a(\vec{r},t)Q^a(t)|\Omega\rangle\,.$$

Because of the translational invariance of the vacuum, we can translate away the space and time-dependence, thus obtaining

$$\langle\Omega|Q^a(t)Q^a(t)|\Omega\rangle = \langle\Omega|j_0^a(\vec{0},0)Q^a(0)|\Omega\rangle \int_{V\to\infty} d^3r\,.$$

If the generator Q^a does not annihilate the vacuum, then $Q^a|\Omega\rangle$ is a state of infinite norm, and hence does not lie in the Hilbert space of the theory. This is a restatement of our observation, that the ground states in (19.4) are orthogonal to each other, and hence inequivalent.

In order to gain control of the situation, we define the operator

$$Q_R^a(t) = \int_{|\vec{r}|\le R} d^3r\, j_0^a(\vec{r},t)\,.$$

Because of the assumed locality (microscopic causality) we nevertheless have that for the commutator $[Q_R^a(t), A(0)]$, with $A(x)$ some (non-singlet) operator local with respect to the Noether currents, that a *non-vanishing* limit $R\to\infty$ exists:

$$\lim_{R\to\infty} [Q_R^a(t), A(t,\vec{r})] = B^a(t,\vec{r})\,. \tag{19.6}$$

From the conservation of the current it further follows that the right-hand side in (19.6) will not depend on the time t in the limit $R\to\infty$. We now take the vacuum expectation value of both sides of this equation with $R\to\infty$:

$$\langle\Omega|Q^a(t)A(t,\vec{r})|\Omega\rangle - \langle\Omega|A(t,\vec{r})Q^a(t)|\Omega\rangle = \langle\Omega|B^a(t,\vec{r})|\Omega\rangle\,. \tag{19.7}$$

Since, by assumption, Q^a does not annihilate the vacuum, the right-hand side of (19.7) need not vanish. We are thus led to suppose that there exists an operator $A(x)$ in the theory such that

$$\exists A\ :\ \langle\Omega|B^a|\Omega\rangle \ne 0\,.$$

---

[2] J. Bernstein, *Rev. Mod. Phys.* **46** (1974) 7.

We now assume as usual that the eigenstates $|n\rangle$ of the generator of translations (4-momentum) are complete

$$P^\mu |n\rangle = p_n^\mu |n\rangle \;,\; \Sigma |n\rangle\langle n| = 1 \;,$$

and in particular, that

$$P^\mu |\Omega\rangle = 0 \,.$$

We make use of this completeness in order to rewrite (19.7) as

$$\sum_n \left[\langle\Omega|Q^a(t)|n\rangle\langle n|A(t,\vec{r})|\Omega\rangle - \langle\Omega|A(t,\vec{r})|n\rangle\langle n|Q^a(t)|\Omega\rangle \right] = \langle\Omega|B^a(t,\vec{r})|\Omega\rangle \,. \quad (19.8)$$

Now, the generator Q^a cannot change the spatial momentum of the states. Hence the sum only extends over states with vanishing 3-momentum: $\vec{p}_n = 0$. Translational invariance in time further allows us to explicitly extract the time dependence as follows:

$$\sum_{\{\vec{p}_n=0\}} \left[\langle\Omega|Q^a(0)|n\rangle\langle n|A(0)|\Omega\rangle e^{-iE_n t} - \langle\Omega|A(0)|n\rangle\langle n|Q^a(0)|\Omega\rangle e^{iE_n t} \right] = \langle\Omega|B^a(0)|\Omega\rangle \,.$$

Since the right-hand side is, however, time-independent, only states of zero *energy* can contribute to the sum, and since the rhs does not vanish, these states must lie in the spectrum of the operator A. They evidently correspond to zero-mass excitations, the so-called *Goldstone bosons!* This concludes the proof.[3]

19.4   Realization of Goldstone Theorem in QFT

Let us consider a simple field-theoretical example realizing the above Goldstone-phenomenon. To this end, consider the Lagrangian

$$\mathcal{L} = \partial_\mu \varphi^\dagger \partial^\mu \varphi + \mu^2 \varphi^\dagger \varphi - g(\varphi^\dagger \varphi)^2 \qquad (19.9)$$

describing the self-interaction of *charged* scalar fields. We take g to be positive, so that the spectrum is bounded from below. The Lagrangian evidently is invariant under the global phase transformation

$$\varphi(x) \to \varphi(x) e^{i\alpha} \,. \qquad (19.10)$$

The Euler–Lagrangian equations read

$$\left(\vec{\nabla}^2 - \frac{\partial^2}{\partial t^2} + \mu^2 \right) \varphi = 2g\varphi |\varphi|^2 \,.$$

For $g \to 0$ this equation states that $-\vec{p}^2 + E^2 + \mu^2 = 0$. Note that this is not the usual Einstein relation between energy and momentum of a particle with mass μ, since the μ^2 term has the wrong sign! Hence, when quantizing the field around

---

[3]If Q^a is a spin zero operator (this excludes supersymmetry), then the Goldstone bosons must also carry spin zero.

$\varphi = 0$ we have a *tachyon*, and we cannot regard the field φ as interpolating a physical particle. Now, including the $\mu^2 \varphi^* \varphi$ term in the potential, we have from (19.9)

$$\mathcal{L} = \partial_\mu \varphi^\dagger \partial^\mu \varphi - V_{eff}(|\varphi|)$$

with the effective potential

$$V_{eff}(|\varphi|) = g \left(|\varphi|^2 - \frac{\mu^2}{2g} \right)^2 - \frac{\mu^4}{4g} . \tag{19.11}$$

This potential shows a continuum of minima parametrized in zeroth order of $\hbar$ as follows:

$$\langle \Omega | \varphi | \Omega \rangle = \sqrt{\frac{\mu^2}{2g}} e^{i\alpha} .$$

Hence we have SSB, and the ground state is no longer invariant under the phase transformation (19.10), so that we have a continuous infinity of classical ground states.

It is convenient to parametrize the field φ as follows,

$$\varphi(x) = \frac{1}{\sqrt{2}} \rho(x) e^{\frac{i}{\lambda}\theta(x)} . \tag{19.12}$$

With respect to ρ the potential has the form of Fig. 19.3.

We expanded ρ around the local minimum of $V_{eff}(\frac{\rho^2}{2})$,

$$\rho(x) = \lambda + \phi(x) , \quad \lambda = \sqrt{\frac{\mu^2}{g}} \tag{19.13}$$

so that we have now,

$$\langle \Omega | \phi(x) | \Omega \rangle = 0 .$$

We make the following remarkable observations:

(i) ϕ describes an excitation with the physical mass $\sqrt{2}\mu$, as seen from the lagrangian

$$\mathcal{L} = \frac{1}{2} \partial_\mu \left(\rho e^{-\frac{i}{\lambda}\theta(x)} \right) \partial^\mu \left(\rho e^{\frac{i}{\lambda}\theta(x)} \right) - V_{eff} \left(\frac{\rho}{\sqrt{2}} \right)$$

$$= \frac{1}{2}(\partial_\mu \phi)^2 - \frac{(\sqrt{2}\mu)^2}{2}\phi^2 + \frac{1}{2}(\partial_\mu \theta)^2$$

$$+ \frac{1}{2\lambda^2}\phi^2(\partial_\mu \theta)^2 + \frac{1}{\lambda}\phi(\partial_\mu \theta)^2 + \frac{\mu^4}{4g} - \lambda g \phi^3 - \frac{g}{4}\phi^4 .$$

(ii) θ describes a zero-mass boson. It is the Goldstone boson associated with the spontaneously broken $U(1)$ symmetry.

Now, as we already remarked, zero-mass, spin-zero particles have not been found in nature. Does this exclude spontaneous symmetry breaking in high-energy physics? The following lagrangian provides an important example for circumventing Goldstone's theorem.

Evasion of the Goldstone Theorem

How can we have a spontaneously broken symmetry without zero mass scalar particles? The massless degree of freedom θ appeared above as a phase of the scalar field φ. Hence let us gauge the Lagrangian (19.9) by minimally coupling the scalar field (tachyon) to a gauge field and adding a Maxwell term. We thus obtain

$$\mathcal{L} = -\frac{1}{4}F_{\mu\nu}F^{\mu\nu} + |\mathcal{D}_\mu\varphi|^2 + \mu^2|\varphi|^2 - g(\varphi^\dagger\varphi)^2 \,, \tag{19.14}$$

where

$$\mathcal{D}_\mu = \partial_\mu + ieA_\mu \,.$$

Make again the change of variable (19.12), the phase $\exp(\frac{i}{\lambda}\theta)$ now is, however, no longer observable, since it may be gauged away by the transformation

$$\varphi(x) \to \varphi(x)e^{-\frac{i}{\lambda}\theta(x)} \equiv \frac{\rho(x)}{\sqrt{2}}$$

$$A_\mu(x) \to A_\mu(x) + \frac{1}{e\lambda}\partial_\mu\theta(x) \equiv B_\mu \,.$$

Only ρ and B_μ thus survive as physical degrees of freedom. We again set $\rho(x) = \lambda + \phi(x)$ in the lagrangian, which now reads

$$\mathcal{L} = -\frac{1}{4}(\partial_\mu B_\nu - \partial_\nu B_\mu)^2 + \frac{1}{2}e^2\lambda^2 B_\mu^2$$
$$+ \frac{1}{2}(\partial_\mu\phi)^2 - \frac{1}{2}(\sqrt{2}\mu)^2\phi^2 + \mathcal{L}_I$$

with

$$\mathcal{L}_I = e^2\lambda B_\mu^2\phi + \frac{e^2}{2}B_\mu^2\phi^2 - \lambda g\phi^3 - \frac{g}{4}\phi^4 + \text{ const.}$$

Note again the change in sign of the ϕ-mass term as compared to (19.14).

We recognize that the above lagrangian now describes the dynamics of a physical scalar particle ϕ of mass $\sqrt{2}\mu$ and a massive vector field B of mass $m_B = e\lambda$. There is no longer any vestige of the "would be Goldstone boson", which has been "eaten" by the gauge field, thereby turning it "fat". This evasion of the Goldstone boson through the "back door" of a gauge field interaction was discovered by *Higgs–Kibble–Guralnik and Brout*, and is generally referred to in high-energy physics as the *Higgs mechanism*.[4] It plays a fundamental role in the *Weak Interactions*. The reason for the breakdown of Goldstone's theorem lies in the existence of *negative metric states* associated with the *longitudinal photons*, which are the ones that became fat in the course of this "dinner".

---

[4] P. Higgs, *Phys. Lett.* **12** (1964) 132; F. Englert and R. Brout, *Phys. Rev. Lett.* **13** (1964) 132; G. Guralnik, C. Hagen and T. Kibble, *Phys. Rev. Lett.* **13** (1964) 1964.

Chapter 20

Effective Potentials

In the past chapters we have concentrated on the vacuum expectation value of the product of field operators, and in particular on *connected* Green functions. The basic constituents of a general Feynman diagram are however the *proper* Green functions. Since they are interlaced by simple propagators within a general Feynman diagram, their associated momentum-space integrals can be carried out independently. The generator of the proper Green functions will provide the definition of the *effective potential* to arbitrary order in $\hbar$, as we shall see.

20.1 Generating functional of proper functions

We have seen in Chapter 13 that the functional defined by (13.1) is the generating functional of the Green functions (connected and disconnected) of the QFT in question. It thus formally has the expansion

$$\mathcal{Z}[j] = 1 + \sum_{n=1}^{\infty} \frac{i^n}{n!} \int d^4x_1 \ldots d^4x_n G^{(n)}(x_1, \ldots, x_n) j(x_1) \ldots j(x_n)$$

with

$$G^{(n)}(x_1, \cdots x_n) = <\Omega|T\phi(x_1)\cdots\phi(x_n)|\Omega>$$

and $|\Omega>$ the normalized physical vacuum. It will be convenient to define a new functional $W[j]$ via

$$\mathcal{Z}[j] = e^{iW[j]} .$$

It is well known that $\ln \mathcal{Z}[j]$ is the generating functional of the *connected* Green functions, i.e.

$$\left(\frac{\delta}{i\delta j(x_1)} \cdots \frac{\delta}{i\delta j(x_n)} W[j]\right)_{j=0} = -i\langle\Omega|T\phi(x_1)\ldots\phi(x_n)|\Omega\rangle_C.$$

We conclude that the generating functional $W[j]$ formally also has the expansion

$$iW[j] = \sum_{n=1}^{\infty} \frac{i^n}{n!} \int d^4x_1 \ldots d^4x_n G_c^{(n)}(x_1 \cdots x_n) j(x_1) \cdots j(x_n).$$

Generating functional of proper Green Functions

The *amputated* Green functions have already been defined in (17.28). They are obtained by stripping off the fully dressed propagators (including the wave function renormalization constants) associated with the external lines of a *connected* Green function. They have already been seen to be the entities from which we obtain the so-called *proper* (1-particle irreducible (1PI), connected) Green functions. We have represented these functions by doubly shaded (hatched) blobs.

As was shown by Jona-Lasinio,[1] the generating functional of *proper* Green functions is the Legendre transform of the generating functional of connected diagrams. We obtain this generating functional as follows: Consider the (classical) field $\varphi_c(x)$ defined by

$$\varphi_c(x) = \frac{\delta W[j]}{\delta j(x)} . \tag{20.1}$$

The right-hand side of this equation can be viewed as the vacuum expectation value of the field operator $\phi(x)$ in the presence of the external source $j(x)$:

$$\varphi_c(x) = \langle \Omega[j] | \phi(x) | \Omega[j] \rangle . \tag{20.2}$$

Hence for a ϕ^4-type potential with a minimum at $\phi = 0$ we expect this expectation value to vanish when turning off the external source j. We shall make use of this in the sequel. Following a procedure familiar from Thermodynamics, we trade the external source j for φ_c by performing the *Legendre transformation*

$$\Gamma[\varphi_c] = \left(W[j] - \int d^4x j(x) \frac{\delta}{\delta j(x)} W[j] \right)_{j=J[\varphi_c]} , \tag{20.3}$$

where $j(x) = J(x|\varphi_c)$ stands for the solution of Eq. (20.1) for j in terms of φ_c.

Note that this procedure is analogous to the transition from the Lagrangian to the Hamiltonian formulations, with

$$-H[\varphi, \pi] = \left(L[\varphi, \dot{\varphi}] - \int d^3x \dot{\varphi}(x) \frac{\delta L}{\delta \dot{\varphi}(x)} \right)_{\dot{\varphi}=F[\pi]} ,$$

where $\dot{\varphi}(x)$ is a solution of

$$\pi(x) = \frac{\delta L}{\delta \dot{\varphi}(x)} .$$

As we now demonstrate, $\Gamma[\varphi_c]$ defined by (20.3) is the generating functional of *proper* Green functions $\Gamma^{(n)}(x_1, \ldots, x_n)$, defined by the expansion (notice the absence of i^n)

$$\Gamma[\varphi_c] = \sum_{n=2}^{\infty} \frac{1}{n!} \int d^4x_1 \ldots d^4x_n \Gamma^{(n)}(x_1, \ldots, x_n) \varphi_c(x_1) \ldots \varphi_c(x_n). \tag{20.4}$$

[1] G. Jona-Lasinio, *Nuovo Cimento* **34** (1964) 1790.

The Legendre transformation (20.3) can be written in the form

$$\Gamma[\varphi_c] = \left[W[j] - \int d^4x\, j(x)\varphi_c(x) \right]_{j=J[\varphi_c]} . \tag{20.5}$$

From here we obtain

$$\frac{\delta}{\delta\varphi_c(x)}\Gamma[\varphi_c] = \int d^4y \left\{ \left[\frac{\delta W[j]}{\delta j(y)} - \varphi_c(y) \right]_{j=J[\varphi_c]} \cdot \frac{\delta J(y|\varphi_c)}{\delta\varphi_c(x)} \right\} - J(x|\varphi_c) .$$

Making use of (20.1), this reduces to

$$\frac{\delta\Gamma[\varphi_c]}{\delta\varphi_c(x)} = -J(x|\varphi_c) . \tag{20.6}$$

From (20.1) and (20.6) we have the *identity*

$$\varphi_c(x) \equiv \frac{\delta W[j]}{\delta j(x)} \bigg|_{j=J[\varphi_c]} \tag{20.7}$$

and

$$\frac{\delta^2\Gamma[\varphi_c]}{\delta\varphi_c(x)\delta\varphi_c(y)} = -\frac{\delta J[x|\varphi_c]}{\delta\varphi_c(y)} \tag{20.8}$$

respectively. Differentiating identity (20.7) with respect to $\varphi_c(y)$, we get

$$\delta^4(x-y) = \frac{\delta}{\delta\varphi_c(y)} \left(\frac{\delta}{\delta j(x)} W[j]\big|_{j=J[\varphi_c]} \right)$$

$$\equiv \int d^4u\, \frac{\delta^2 W[j]}{\delta j(u)\delta j(x)} \bigg|_{j=J[\varphi_c]} \cdot \frac{\delta J(u|\varphi_c)}{\delta\varphi_c(y)} ,$$

or making use of (20.8),

$$\delta^4(x-y) = -\int d^4u\, \frac{\delta^2\Gamma[\varphi_c]}{\delta\varphi_c(u)\delta\varphi_c(y)} \cdot \frac{\delta^2 W[j]}{\delta j(u)\delta j(x)} \bigg|_{j=J[\varphi_c]} . \tag{20.9}$$

Hence the kernel $\frac{\delta^2\Gamma[\varphi_c]}{\delta\varphi_c(z)\delta\varphi_c(y)}$ is the inverse of

$$\frac{\delta^2 W[j]}{i\delta j(z) i\delta j(y)} \bigg|_{j=J[\varphi_c]} .$$

Now, in the absence of spontaneous symmetry breaking we have $\varphi_c = 0$ for $j = 0$. Hence, with the identifications

$$\Gamma^{(2)}(y-x) = \frac{\delta^2\Gamma[\varphi_c]}{\delta\varphi_c(y)\delta\varphi_c(x)} \bigg|_{\varphi_c=0} ,$$

$$\frac{\delta^2 W[j]}{i\delta j(z) i\delta j(y)} \bigg|_{j=0} = -i < \Omega|\phi(z)\phi(y)|\Omega >= \Delta_F(z-y)$$

we have in this case

$$\int d^4 z \Gamma^{(2)}(y-z)\Delta_F(z-x) = \delta^4(y-x) , \qquad (20.10)$$

or going to momentum-space,

$$\tilde{\Gamma}^{(2)}(p) = p^2 - m^2 - \Sigma(p) . \qquad (20.11)$$

The interpretation of a general $\Gamma^{(n)}$ as a proper Green function may be obtained by successive differentiation at $\varphi_c = 0$ of the "master identity" (20.9). Thus, for instance, one further differentiation of this identity with respect to $\varphi_c(z)$ and setting $\varphi_c = 0$ leads to

$$\int d^4 u \Gamma^{(3)}(u,y,z)G_c^{(2)}(u,z) = i \int d^4 u \int d^4 v \Gamma^{(2)}(u,y)G_c^{(3)}(x,u,v)\Gamma^{(2)}(v,z)$$

i.e.

$$i\Gamma^{(3)}(x,y,z) = \int d^4 x' \int d^4 y' \int d^4 z' G_c^{(3)}(x',y',z')\Gamma^{(2)}(x',x)\Gamma^{(2)}(y',y)\Gamma^{(2)}(z',z)$$

or using (20.10),

$$i\Gamma^{(3)}(x,y,z) = \langle \Omega | T\varphi(x)\varphi(y)\varphi(z)|\Omega\rangle_{amp} .$$

Thus $i\Gamma^{(3)}$ coincides with the *amputated* 3-point function, which *happens* to be also 1-particle-irreducible. In general one can show that it holds for 1PI Green functions.

$$i\Gamma^{(n)}(x_1,\cdots,x_n) = \langle \Omega | T\varphi(x_1)\cdots\varphi(x_n)|\Omega\rangle_{1PI} . \qquad (20.12)$$

20.2   The effective potential

Consider the Lagrange density

$$\mathcal{L} = \frac{1}{2}(\partial_\mu \phi)^2 - \frac{1}{2}m_0^2 \phi^2 - V(\phi) \qquad (20.13)$$

with

$$V(\phi) = \frac{g_0}{4!}\phi^4 .$$

In the *tree graph* approximation we have from (20.11),

$$\Gamma^{(2)}(x-y) = \int \frac{d^4 p}{(2\pi)^4} \tilde{\Gamma}^{(2)}(p)e^{ip(x-y)} = \int \frac{d^4 p}{(2\pi)^4}(p^2 - m_0^2)e^{ip(x-y)}$$
$$= -(\Box + m_0^2)\delta^4(x-y) ,$$

and from (20.12),

$$i\Gamma^{(4)}_{tree}(x_1 \ldots x_4) = -ig \int d^4z \prod_{i=1}^{4} \delta^4(z - x_i) \tag{20.14}$$

as the only contributions to $\Gamma[\varphi_c]_{tree}$. Hence, recalling (20.4),

$$\Gamma[\varphi]_{tree} = S_{cl}[\varphi] \tag{20.15}$$

we conclude that the generating functional $\Gamma[\varphi]$ coincides to lowest order in $\hbar$ with the *classical action*. It can thus be regarded as providing a generalization of the classical action to all orders of $\hbar$. One refers to it as the *effective action*:

$$\Gamma[\varphi] = S_{eff}[\varphi] .$$

Expanding the integrand of the effective action in powers of derivative of the fields, one has in general, making use of translational invariance,

$$\Gamma[\varphi] = \sum_{n=2} \frac{1}{n!} \int d^4x \left[\int \prod_{i=1}^{n-1} d^4z_i \Gamma^{(n)}(z_1 \cdots z_{n-1}, 0) \varphi(z_1 + x) \cdots \varphi(z_{n-1} + x)\varphi(x) \right] ,$$

which we can write in the form

$$\Gamma[\varphi] = \int d^4x \left\{ -V_{eff}(\varphi) + \frac{1}{2}Z(\varphi)(\partial_\mu\varphi)^2 + \cdots \right\} . \tag{20.16}$$

Comparing with (20.15), we see that $V_{eff}(\varphi)$ generalizes the classical (tree graph) potential, *including the mass term*, to all orders in $\hbar$. It is thus referred to as the *effective potential*. It is evidently obtained from the integrand of the effective action, by considering *constant* field configurations σ. In that case we have from (20.4)

$$V_{eff}(\sigma) = -\sum_{n=2}^{\infty} \frac{1}{n!} \sigma^n \tilde{\Gamma}^{(n)}(0, \cdots, 0) = i \sum_{n=2}^{\infty} \frac{1}{n!} \sigma^n \tilde{G}^{(n)}(0, \cdots, 0)_{amp} , \tag{20.17}$$

where $\tilde{\Gamma}^{(n)}(0, \cdots, 0)$ is the momentum-space 1PI n-point function defined via the Fourier transform

$$\int d^4x_1 \cdots d^4x_n e^{i(p_1 \cdot x_1 + \cdots p_n \cdot x_n)} \Gamma^{(n)}(x_1 \cdots x_n) = \tilde{\Gamma}^{(n)}(p_1 \cdots p_n)(2\pi)^4\delta^4(p_1 + \cdots + p_n) .$$

The effective potential plays a central role in the discussion of *spontaneous symmetry breaking*.

20.3   The 1-loop effective potential of ϕ^4-theory

1-loop approximation

In this approximation, the generating functional $\Gamma[\phi_c]$ of 1PI-irreducible Green's functions involves an infinite number of terms, as expressed diagrammatically below:

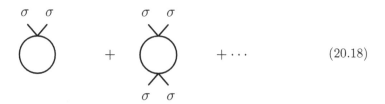

Fig. 20.1. Sum of 1-loop diagrams for V_{eff}.

Although this series is in general not summable, it may be summed up for vanishing external momenta (constant field configurations σ), where the vertices represent simply zero-momentum insertions. In this case the above series *formally* reduces in momentum space to the sum,

$$V_{eff}^{(1-loop)}(\sigma;\Lambda) = \frac{1}{2}m_0^2\sigma^2 + V(\sigma) - \int_0^\Lambda \frac{d^4K}{(2\pi)^4} \sum_{n=1}^\infty \frac{(-1)^n}{2n}\left(\frac{V''(\sigma)}{K^2+m_0^2}\right)^n , \quad (20.19)$$

where a *Wick rotation* has been performed, a UV cutoff Λ for the momentum integration has been introduced, and use has been made of (20.17). In arriving at this expression the following combinatoric factors have been taken into account:

1. A factor $\frac{1}{n}$: There exist $(n-1)!$ permutations of the n pairs of external lines leading to topologically inequivalent graphs (the n cyclic permutations are topologically equivalent).

2. The factor $\frac{1}{2}$ in $\frac{g_0}{2}\sigma^2$ arises from the $\frac{4!}{2!}$ possible contractions of the Bose fields with the remaining fields. This cancels the factor $\frac{1}{4!}$ in $V(\varphi)$ and leaves $\frac{1}{2}$.

3. An overall factor of $\frac{1}{2}$. It is a consequence of an incomplete cancellation of the factor $1/n!$ in the perturbative expansion, since diagrams with labeled vertices differing by a flow of "virtual" charge in the loops are identical since the field $\phi(x)$ is neutral (unlike the case of electrons).

Notice that for $n \geq 2$ the terms in the sum above exhibit for $m_0 = 0$ an ever increasing logarithmic divergence at $k^2 = 0$. Formal summation however reduces this divergence to a soft one of the logarithmic type:

$$V_{eff}^{(1-loop)}(\sigma;\Lambda) = \frac{1}{2}m_0^2\sigma^2 + V(\sigma) + \frac{1}{2}\int_0^{\Lambda^2} \frac{K^2 dK^2}{16\pi^2} \ln\left(1 + \frac{\frac{g_0}{2}\sigma^2}{K^2+m_0^2}\right) , \quad (20.20)$$

where use has been made of (15.1) and (15.2) with $D = 4$ and a cutoff has been introduced. The quadratic UV divergence already present in (20.19) continues however to exist, and requires renormalization. Since, as seen from (20.19) the divergence occurs only for the first two terms in the series, it can be cured on tree-graph level by the addition of suitable counter-terms to the Lagrangian (20.13).

20.4   WKB approach to the effective potential

For our forthcoming discussion it will be useful to develop a slightly different view of the effective potential. To this end we begin with a consideration which appears to be unrelated to our discussion in the preceding section, leaving to the end of this section the demonstration that we obtain in this way two equivalent approaches to the calculation of the effective potential. We shall consider again the Lagrange density (20.13). The corresponding action in the presence of an external source $j(x)$ is given by

$$S[\phi, j] = \int d^4x \left(\frac{1}{2}(\partial_\mu \phi)^2 - \frac{1}{2}m_0^2\phi^2 - V(\phi) + j\phi \right). \tag{20.21}$$

We search for the minimum of this action. It corresponds to a field configuration satisfying the *classical* equation of motion

$$(\Box + m_0^2)\phi_{c\ell}(x) + V'(\phi_{c\ell}(x)) = j(x). \tag{20.22}$$

Notice that $\phi_{c\ell}(x)$ is a functional of j. Expanding a general field configuration about this classical field,

$$\phi(x) = \phi_{c\ell}(x) + \xi(x),$$

substituting this into (20.21), and expanding in powers of ξ, we obtain using (20.22),

$$S[\phi; j] = S_{c\ell}[j] - \int d^4x \frac{1}{2}\xi(\Box + m_0^2 + V''(\phi_{c\ell}))\xi - \int d^4x \sum_{p \geq 3} V^{(p)}(\phi_{c\ell})\frac{\xi^p}{p!}, \tag{20.23}$$

with

$$S_{c\ell}[j] = \int d^4x \left\{ -\frac{1}{2}\phi_{c\ell}(\Box + m_0^2)\phi_{c\ell} - V(\phi_{c\ell}) + j\phi_{c\ell} \right\}.$$

The expansion (20.23) corresponds to an expansion in powers of the Planck constant $\hbar$. This is easily seen by absorbing the $\frac{1}{\hbar}$ in $\ln Z = iS/\hbar$ into the ξ-integration by making the change of variable

$$\frac{1}{\hbar^{1/2}}\xi \to \xi$$

in the corresponding Feynman path integral, which leads to

$$Z[j] = e^{\frac{i}{\hbar}S_{c\ell}[j]} \int \mathcal{D}\xi \, e^{-i\int d^4x\left[\xi\left(\Box + m_0^2 + V''(\phi_{c\ell})\right)\xi - \sum_{p\geq 3}^{\infty} \hbar^{\frac{p}{2}}\frac{\xi^p}{p!}V^{(p)}(\phi_{c\ell})\right]}.$$

The WKB approximation in quantum mechanics corresponds to dropping the contribution from the last term describing the non-Gaussian fluctuations. Making use of [2] the fundamental formula

$$\int \mathcal{D}\xi \, e^{-i\int d^4\xi(x)\mathcal{D}(x)\xi(x)} = \left(Det\,\mathcal{D}(x)\right)^{-1/2} = e^{-\frac{1}{2}Tr\ln\mathcal{D}}, \tag{20.24}$$

---

[2]This formula generalizes the N-dimensional integral

$$\int \prod_0^N da_n e^{i\sum_n \lambda_n a_n^2} = \left(\prod_n \lambda_n\right)^{-\frac{1}{2}}$$

to the continuum.

we have in this approximation (we assume that for $j \to 0$, $\phi_{c\ell} \to 0$)

$$\mathcal{Z}[j] = \frac{Z[j]}{Z[0]} \simeq e^{\frac{i}{\hbar}S_{c\ell}[j]} Det \left(\frac{(\Box + m_0^2 + V''(\phi_{c\ell}))}{(\Box + m_0^2)} \right)^{-\frac{1}{2}}. \tag{20.25}$$

We thus have for the generating functional of connected Green functions,

$$W[j] \simeq S_{c\ell}[j] + i\frac{\hbar}{2}Tr\ln \left(1 + \frac{1}{\Box + m_0^2}V''(\phi_{c\ell}) \right) + \theta(\hbar^2). \tag{20.26}$$

In order to obtain the connection with the effective potential discussed in Section 5, we now proceed to obtain from here the generating functional of the proper functions via the Legendre transform (20.3). We need to express $W[j]$ as a function of $\varphi_c(x)$ defined in (20.1). From (20.26) we have

$$\varphi_c(x) = \frac{\delta S_{c\ell}[j]}{\delta j(x)} + O(\hbar).$$

Now,

$$\frac{\delta S_{c\ell}[j]}{\delta j(x)} = \int d^4z [-\frac{\delta \phi_{c\ell}(z)}{\delta j(x)}(\Box + m_0^2)\phi_{c\ell}(z) - V'(\phi_{c\ell})\frac{\delta \phi_{c\ell}(z)}{\delta j(x)} + j(x)\frac{\delta \phi_{c\ell}(z)}{\delta j(x)}$$
$$+ \phi_{c\ell}(x)\delta(z-x)] = \phi_{c\ell}(x),$$

where use has been made of the equation of motion (20.22). Hence, with (20.26) and (20.1) we have,

$$\varphi_c(x) = \phi_{c\ell}(x) + \theta(\hbar). \tag{20.27}$$

Since $\phi_{c\ell}$ represents a stationary configuration of $S[\phi_{c\ell}, j]$, we have, setting $\varphi_c(x) = \phi_{c\ell}(x) + \hbar\xi(x)$,

$$S[\varphi_c, j] = S[\phi_{c\ell} + \hbar\xi, j] = S[\phi_{c\ell}, j] + O(\hbar^2),$$

so that we also have

$$W[j] = S[\varphi_c, j] + \frac{i\hbar}{2}Tr\ln \left(1 + \frac{V''(\varphi_c)}{\Box + m_0^2} \right) + \theta(\hbar^2). \tag{20.28}$$

We thus have achieved our first goal, to order $\hbar$. Performing the subtraction of the integral in (20.3), the $j\varphi_c$ term $\int \phi j$ contained in $S[\varphi, j]$ in (20.28) cancels so that to this order

$$\Gamma[\varphi_c] = S[\varphi_c, 0] + i\frac{\hbar}{2}Tr\ln (\Box + m_0^2 + V''(\varphi_c)) + O(\hbar^2) \tag{20.29}$$

$$= \int d^4x \left\{ \frac{1}{2}(\partial_\mu \varphi_c)^2 - \frac{1}{2}m_0^2\varphi_c^2 - V(\varphi_c) + \frac{i\hbar}{2}\langle x|\ln \left(1 + \frac{V''(\varphi_c)}{\Box + m_0^2} \right)|x\rangle \right\}.$$

Hence $\Gamma[\phi]$ is only a function of φ_c as guaranteed by the Legendre transformation.

As we now show, this is identical with the effective potential (20.20). Indeed, expanding the logarithm, we have

$$\text{Tr}\langle x|\ln\left(1+\frac{1}{\Box+m_0^2}V''(\varphi_c)\right)|x\rangle = \sum_{n=1}\frac{(-1)^{n+1}}{n}\int d^4x\langle x|\frac{1}{\Box+m_0^2}V''(\varphi_c)|x\rangle$$

$$= \sum_{n=1}\frac{(-1)^{n+1}}{n}\int d^4x\int\frac{d^4k'}{(2\pi)^4}\int\frac{d^4k}{(2\pi)^4}\langle x|k'\rangle\langle k'|\frac{1}{\Box+m_0^2}V''(\varphi_c)|k\rangle\langle k|x\rangle$$

$$= \sum_{n=1}\frac{(-1)^{n+1}}{n}\int\frac{d^4k}{(2\pi)^4}\left(\frac{V''(\varphi_c)}{-k^2+m_0^2}\right)^n$$

$$= i\sum_{n=1}\frac{(-1)^{n+1}}{n}\int\frac{d^4K}{(2\pi)^4}\left(\frac{V''(\varphi_c)}{K^2+m_0^2}\right)^n$$

$$= i\int\frac{d^4K}{(2\pi)^4}\ln\left(1+\frac{V''(\varphi_c)}{K^2+m_0^2}\right),$$

or with (20.16), we have, introducing a cutoff,

$$V_{eff}(\sigma) = \frac{m_0^2}{2}\sigma^2 + V(\sigma) + \frac{1}{2}\int_0^\Lambda\frac{K^2dK^2}{32\pi^2}\ln\left(1+\frac{V''(\varphi_c)}{K^2+m_0^2}\right) \qquad (20.30)$$

in agreement with (20.20).

Notice that the overall factor of $\frac{1}{2}$ arose here from the determinant in (20.24) appearing with the power $-\frac{1}{2}$ (unlike $+\frac{1}{2}$ in the fermionic case), which in turn is linked to the fact, that $\phi(x)$ is a real (charge neutral) bosonic field.

Our alternative derivation has shown that the effective potential is nothing but the *effective action* one obtains by expanding the classical action about the *constant* field configuration minimizing the action, dropping all terms of higher order than quadratic in the fluctuations, and then performing the remaining gaussian functional integration.

We now give an example of SSB for the case of ϕ^4 theory.

20.5 The effective potential and SSB

Let us examine the effective potential (20.30) with regard to spontaneous symmetry breaking. This will give us also the opportunity to gain insight into the renormalization program, as well as the renormalization group aspects.

The $m \neq 0$ effective potential

We rewrite (20.20) in the form

$$V_{eff}^{(1-loop)}(\sigma;\Lambda) = \frac{1}{2}m_0^2\sigma^2 + \frac{g_0}{4!}\sigma^4 + \int_0^{\Lambda^2}\frac{xdx}{32\pi^2}\left[\ln\left(x+m_0^2+\frac{g_0\sigma^2}{2}\right) - \ln(x+m_0^2)\right].$$

$$(20.31)$$

The integrals are elementary. We have for $\Lambda \to \infty$,

$$\int_0^{\Lambda^2} x dx \ln (x+a) = \left[\frac{y^2}{2} \ln y - \frac{y^2}{4} - a(y \ln y - y) \right]_a^{\Lambda^2 + a} \tag{20.32}$$

$$\approx \frac{1}{2} \Lambda^4 \ln \Lambda^2 - \frac{1}{2} a^2 \ln \frac{\Lambda^2}{a} - \frac{1}{4} \Lambda^4 + a\Lambda^2 - \frac{1}{4} a^2 .$$

We thus find for the effective potential,

$$V_{eff}(\sigma) = \frac{1}{2} m_0^2 \sigma^2 + V(\sigma) \tag{20.33}$$

$$+ \frac{1}{32\pi^2} \left[-\frac{a^2}{2} \ln \frac{\Lambda^2}{a} + \frac{m_0^4}{2} \ln \frac{\Lambda^2}{m_0^2} + \frac{g_0 \sigma^2}{2} \Lambda^2 - \frac{m_0^2}{2} \left(\frac{g_0 \sigma^2}{2} \right) - \frac{1}{4} \left(\frac{g_0 \sigma^2}{2} \right)^2 \right]$$

where

$$a = m_0^2 + \frac{g_0 \sigma^2}{2} .$$

To simplify the notation set

$$z \equiv \frac{g_0 \sigma^2}{2}, \qquad a = m_0^2 + z .$$

To renormalize this potential we take the point of view of Section 16.2 of Chapter 16 by taking g_0 and m_0 to be physical parameters g and m, respectively, and adding the counter terms

$$\delta V(\sigma; \Lambda) = \frac{1}{32\pi^2} \left[B(\Lambda)z + C(\Lambda)z^2 \right] . \tag{20.34}$$

We can then write the (pre-) renormalized (see (16.6)) effective potential in the form

$$\tilde{V}_{eff}(\sigma; \Lambda) = \frac{1}{2} m^2 \sigma^2 + V(\sigma) + \frac{1}{32\pi^2} \left\{ \frac{1}{2} (m^2 + z)^2 \left[\ln \left(1 + \frac{z}{m^2} \right) + \ln \left(\frac{m^2}{\Lambda^2} \right) \right] \right.$$

$$- \frac{1}{2} m^4 \ln \left(\frac{m^2}{\Lambda^2} \right) + z\Lambda^2 - \frac{1}{2} m^2 z - \frac{1}{4} z^2 \right\}$$

$$+ \frac{1}{32\pi^2} \left[B(\Lambda)z + C(\Lambda)z^2 \right] . \tag{20.35}$$

The coefficients $B(\Lambda)$ and $C(\Lambda)$ are associated with the mass and coupling constant renormalizations. We fix the arbitrary constants by imposing the mass-shell condition

$$\tilde{\Gamma}^{(2)}(p)|_{p^2=0} = -m^2 ,$$

or equivalently (see (20.17))

$$\left. \frac{d^2 \tilde{V}_{eff}}{d\sigma^2} \right|_{\sigma=0} = m^2 . \tag{20.36}$$

This gives,

$$B(\Lambda) = -\frac{1}{32\pi^2}\left(m^2\ln\left(\frac{m^2}{\Lambda^2}\right) + \Lambda^2\right) \tag{20.37}$$

and leaves only the mass term quadratic in σ.

The usual normalization condition for the 4-point function at *vanishing* external 4-momenta requires (see (20.14) for sign)

$$\tilde{\Gamma}^{(4)}(0,0,0,0) = -g \tag{20.38}$$

or (see (20.17) for sign)

$$\frac{d^4\tilde{V}_{eff}}{d\sigma^4}\Big|_{\sigma=0} = g . \tag{20.39}$$

The easiest way to evaluate the left-hand side is to expand the logarithm in powers of z. A simple calculation yields,

$$C(\Lambda) = -\frac{1}{32\pi^2}\left(\frac{1}{2}\ln\left(\frac{m^2}{\Lambda^2}\right) + \frac{3}{4}\right). \tag{20.40}$$

Putting things together, we are left with the renormalized effective potential

$$V_{eff}(\sigma) = \frac{1}{2}m^2\sigma^2 + \frac{g\sigma^4}{4!} \tag{20.41}$$

$$+ \frac{1}{32\pi^2}\left[\frac{m^4}{2}\left(1 + \frac{g\sigma^2}{2m^2}\right)^2\ln\left(1 + \frac{g\sigma^2}{m^2}\right) - \frac{g\sigma^2}{2}\left(\frac{3}{2}\frac{g\sigma^2}{2} + m^2\right)\right].$$

The immediate question arises: is there spontaneous symmetry breakdown? A plot of V_{eff} shows that one has to go to extremely high values of g to see a vestige of a spontaneous symmetry breakdown.

The $m = 0$ *effective potential*

As the result (20.40) shows, the implementation of normalization condition (20.38) for $m = 0$ is not possible. As before, we take m_0 and g_0 to be *physical* parameters g and m in (20.33), respectively. Setting $m_0 = m$ in (20.33), we then obtain,

$$V_{eff}(\sigma, \Lambda) = \frac{g}{4!}\sigma^4 + \frac{1}{32\pi^2}\left[\frac{g\Lambda^2}{2}\sigma^2 + \frac{g^2}{8}\left(\ln\left(\frac{g\sigma^2}{2\Lambda^2}\right) - \frac{1}{2}\right)\sigma^4\right]. \tag{20.42}$$

The UV divergence is again cured by the addition of suitable counterterms

$$\delta V_{eff}(\Lambda) = \frac{1}{32\pi^2}\left[\frac{1}{2}B_0(\Lambda)\sigma^2 + \frac{1}{4!}C_0(\Lambda)\sigma^4\right]$$

which in turn are fixed by imposing conditions on V_{eff},

$$\tilde{V}_{eff}(\sigma;\Lambda) = \frac{g}{4!}\sigma^4 + \frac{1}{32\pi^2}\left[B_0(\Lambda) + g\Lambda^2\right]\frac{\sigma^2}{2!} \tag{20.43}$$

$$+ \frac{1}{32\pi^2}\left[C_0(\Lambda) + 3g^2\left(\ln\left(\frac{g\sigma^2}{2\Lambda^2}\right) - \frac{1}{2}\right)\right]\frac{\sigma^4}{4!}.$$

The coefficients $B_0(\Lambda)$ and $C_0(\Lambda)$ are associated with the mass and coupling constant renormalization. We *try* to fix the arbitrary constants by requiring the mass-shell condition

$$\tilde{\Gamma}^{(2)}(p)|_{p^2=0} = 0 ,$$

or equivalently (see (20.17))

$$\frac{d^2\tilde{V}_{eff}}{d\sigma^2}\bigg|_{\sigma=0} = 0 \qquad (m \overset{!}{=} 0) . \tag{20.44}$$

This gives,

$$B_0(\Lambda) = -g\Lambda^2 .$$

As we already observed, the usual normalization condition for the 4-point function at *vanishing* external 4-momenta cannot be implemented due to the logarithmic dependence of (20.42) on σ. We thus need to choose a different normalization point $\sigma = \mu$. A straightforward calculation yields

$$\bar{g}(\mu) \equiv \left(\frac{d^4\tilde{V}_{eff}}{d\sigma^4}\right)_{\sigma=\mu} = g + \frac{1}{32\pi^2}\left[C_0(\Lambda) + 3g^2\left(\ln\frac{g\mu^2}{2\Lambda^2} - \frac{11}{3}\right)\right] . \tag{20.45}$$

The identification of g with $g(\mu)$ now requires

$$C_0(\Lambda) = -3g^2\left(\ln\frac{g\mu^2}{2\Lambda^2} - \frac{11}{3}\right) ,$$

which is in fact is the choice of counterterm made by S. Coleman and E. Weinberg.

Replacing g by $g(\mu)$ in (20.43) one finally obtains for the renormalized effective potential

$$V_{eff}(\sigma, \bar{g}, \mu) = \bar{g}(\mu)\frac{\sigma^4}{4!} + \frac{\bar{g}(\mu)^2}{32\pi^2}3\left(\ln\frac{\sigma^2}{\mu^2} - \frac{25}{6}\right)\frac{\sigma^4}{4!} \tag{20.46}$$

which is the result obtained by Coleman and Weinberg.[3]

Is there SSB?

The question of immediate interest is, whether spontaneous symmetry breakdown (SSB) occurs. The location σ_{ex} of the extrema is obtained by solving the equation

$$V'_{eff}(\sigma_{ex}, \mu) = \frac{g}{3!}\sigma_{ex}^3 + \frac{g^2\sigma_{ex}^3}{64\pi^2}\left(\ln\frac{\sigma_{ex}^2}{\mu^2} - \frac{25}{6}\right) + \frac{g^2\sigma_{ex}^3}{264\pi^2} = 0 .$$

$\sigma_M = 0$ corresponds to a maximum, while σ_m, given by

$$\ln\frac{\sigma_m^2}{\mu^2} = -\frac{32\pi^2}{3g} + O(1) ,$$

that is

$$\sigma_m^2 \sim \mu^2 e^{-\frac{32\pi^2}{3g}}$$

---

[3]S. Coleman and E. Weinberg, *Phys. Rev. D* **7** (1973) 1888.

corresponds to two minima with

$$V_{eff}(\pm\sigma_m) = -e^{-\frac{64\pi^2}{3g}} \frac{g^2}{256\pi^2} \frac{25}{6} < 0 \, .$$

$V_{eff}(\sigma_m)$ thus corresponds to a double well potential with minima at $\pm\sigma_m$ very close to the maximum at $\sigma_M = 0$ with $V_{eff}(0) = 0$. Contrary to the case of the ferromagnet discussed in Section 4 of Chapter 19, the double-well potential arises here from the logarithmic term, giving rise to a very soft SSB. However, the strictly non-perturbative nature of this result makes this conclusion questionable, and can be expected to be an artifact of the 1-loop approximation.

RG equation for V_{eff}

Using the notation introduced in connection with (18.7), we have from (20.45)

$$\bar{g}_2 = \bar{g}_1 + \frac{3\bar{g}_1^2}{32\pi^2} \ln\left(\frac{\mu_2^2}{\mu_1^2}\right). \tag{20.47}$$

For the $\bar{\beta}$-function defined in (18.8) we have,

$$\bar{\beta}(\bar{g}_1) = \left[\mu_2 \frac{d\bar{g}(\mu_2)}{d\mu_2}\right]_{\mu_2=\mu_1} = \frac{6\bar{g}_1^2}{32\pi^2} \, .$$

Hence, to order $O(\bar{g}_1^3)$, $V_{eff}(\sigma; \mu_1)$ is seen to satisfy the renormalization group equation

$$\left(\mu_1 \frac{\partial}{\partial\mu_1} + \bar{\beta}(\bar{g}_1)\frac{\partial}{\partial\bar{g}_1}\right)\bar{V}_{eff}(\sigma, \mu_1) = O(\bar{g}_1^3) \, .$$

Notice that this beta function differs from (18.10). It is however positive, so $\bar{g}_1 = 0$ continues to be an IR atractor. Also notice the absence of an anomalous dimension. As already in the case of our Example 1 in Chapter 18, the renormalization constant can be set $Z_\phi = 1$ up to order $\bar{g}^2$.

Index

Printed in the United States
by Baker & Taylor Publisher Services